LECTURES ON
DEFORMATION
QUANTISATION

From Moyal Product to
Kontsevich's Formality Theorem

PEKING UNIVERSITY SERIES IN MATHEMATICS

ISSN: 2010-2240

Series Editors: Huijun Fan, Pingwen Zhang, Bin Liu,
and Jiping Zhang *(Peking University, China)*

Published:

Vol. 8: *Lectures on Deformation Quantisation:*
From Moyal Product to Kontsevich's Formality Theorem
by Georgy Igorevich Sharygin
(Lomonosov Moscow State University, Russia)

Vol. 7: *Selected Papers of Weiyue Ding*
edited by You-De Wang (Chinese Academy of Sciences, China)

Vol. 6: *Lecture Notes on Calculus of Variations*
by Kung-Ching Chang (Peking University, China)

Vol. 5: *Arbitrage, Credit and Informational Risks*
edited by Caroline Hillairet (Ecole Polytechnique, France),
Monique Jeanblanc (Université d'Evry, France) and
Ying Jiao (Université Lyon I, France)

Vol. 4: *Elliptic, Hyperbolic and Mixed Complex Equations with*
Parabolic Degeneracy: Including Tricomi–Bers and
Tricomi–Frankl–Rassias Problems
by Guo Chun Wen (Peking University, China)

Vol. 3: *Approaches to the Qualitative Theory of Ordinary Differential*
Equations: Dynamical Systems and Nonlinear Oscillations
by Tongren Ding (Peking University, China)

Vol. 2: *Numerical Methods for Exterior Problems*
by Lung-An Ying (Peking University & Xiamen University, China)

Vol. 1: *An Introduction to Finsler Geometry*
by Xiaohuan Mo (Peking University, China)

Peking University Series in Mathematics — Vol. 8

LECTURES ON
DEFORMATION
QUANTISATION

From Moyal Product to
Kontsevich's Formality Theorem

Georgy Igorevich Sharygin
Lomonosov Moscow State University, Russia

NEW JERSEY · LONDON · SINGAPORE · BEIJING · SHANGHAI · TAIPEI · CHENNAI

Published by

World Scientific Publishing Co. Pte. Ltd.
5 Toh Tuck Link, Singapore 596224
USA office: 27 Warren Street, Suite 401-402, Hackensack, NJ 07601
UK office: 57 Shelton Street, Covent Garden, London WC2H 9HE

Library of Congress Control Number: 2024047443

British Library Cataloguing-in-Publication Data
A catalogue record for this book is available from the British Library.

Peking University Series in Mathematics — Vol. 8
LECTURES ON DEFORMATION QUANTISATION
From Moyal Product to Kontsevich's Formality Theorem

Copyright © 2025 by World Scientific Publishing Co. Pte. Ltd.

All rights reserved. This book, or parts thereof, may not be reproduced in any form or by any means, electronic or mechanical, including photocopying, recording or any information storage and retrieval system now known or to be invented, without written permission from the publisher.

For photocopying of material in this volume, please pay a copying fee through the Copyright Clearance Center, Inc., 222 Rosewood Drive, Danvers, MA 01923, USA. In this case permission to photocopy is not required from the publisher.

ISBN 978-981-12-9780-9 (hardcover)
ISBN 978-981-12-9781-6 (ebook for institutions)
ISBN 978-981-12-9782-3 (ebook for individuals)

For any available supplementary material, please visit
https://www.worldscientific.com/worldscibooks/10.1142/13973#t=suppl

Typeset by Stallion Press
Email: enquiries@stallionpress.com

Preface

These lecture notes are based on the course, that I taught in PKU, Beijing in the Fall semester 2019. So I think it's proper that I should start this preface with expressing my deepest gratitude to this institution for giving me an opportunity to teach the subject that I like and providing me with wonderful living conditions and perfect work environment.[1]

The main purpose of my course was to acquaint the students with the purposes and the main results of the quantisation theory and demonstrate them various approaches and technics that are used there. I must admit that I definitely underestimated the difficulty of this task (or rather overestimated my powers), the amount of ideas and results in the field being far beyond the capacity of any one semester-long series of lectures. However I did my best to cope with it and now I hope that my readers will find in my notes material worth perusing.

The notes begin with the discussion of classical concepts of quantisation theory, such as Weyl's quantisation formula and then through the Moyal quantisation we are brought to the central question of deformation quantisation program (successfully solved in 1997 by Kontsevich): *Is it possible to find an $\hbar$-linear associative $\star$-product on the space $C^\infty(M)[[\hbar]]$, where M is a Poisson manifold and $\hbar$ is a formal variable, such that in low degrees it is given by*

$$f \star g = fg + \frac{\hbar}{2}\{f,g\} + \cdots?$$

We proceed via classical deformation theory first to the proof of this conjecture for symplectic manifolds due to Cahen, de Wilde, Gutt, Lecomte and others. Although ideologically quite simple and technically rather awkward (or even intimidating), this construction can be quite handy, as it gives certain insight into the structure of the deformation series lacked by other constructions, such as the renowned Fedosov deformation formula, which we describe further.

---

[1] Later I taught a Russian version of this course in the Independent University in Moscow; some of the ideas presented in these notes were introduced at this time. Thus I feel obliged to add words of gratitude to IUM too.

The last few lectures of the course are dedicated to the proof of Kontsevich deformation theorem and its corollaries. This theorem is based on the concept of L_∞-algebras and morphisms, so we give a brief introduction into this subject, which is closely related with Algebraic Topology and Homological Algebra. We also briefly discuss the relation of this construction with Physical theories, such as Poisson sigma-model etc.

Although the covered material is very vast, I tried to keep the course as self contained as possible, so I used few lectures to give brief introductions to symplectic and Poisson geometry, Hochschild homology and other necessary theories. Exercise sessions also were a part of the course and the problems discussed at these sessions are included into the course notes. The sections of the present notes roughly correspond to the lectures that I gave twice a week in PKU and there was one exercise session every two weeks, so there are only seven sets of exercises, approximately three times less than the number of lectures.

I also decided that it is worth to move part of the material, originally covered in my lectures, namely, the renowned Tamarkin's proof of Kontsevich's theorem to an appendix: this material stands off quite a bit, as it asks for a considerably higher degree of knowledge of the Homological algebra and related topics, on one hand, and on the other hand, there is not much impact of this sections on the remaining part of the course. This is also justified by the fact that while compiling these lecture notes I enjoyed my time and tried to fill in many details that were only mentioned in the course that I taught which made this text considerably larger than I expected.

All the material used in these notes come from various original papers and sometimes standard textbooks (especially, when we deal with the standard facts from Geometry, Algebra etc.). I tried to give reference to the papers, from which one or another fact was taken, and I hope the readers will forgive me, if I missed some of sources of my text. I also would like hereby to declare that all the original ideas and results covered by this lecture notes belong to the authors of the research papers even if I forget to mention them, while all the misprints and misinterpretations of these results are mine.

Introduction

From Quantum Physics to Algebra

It has long been known that the most productive and exciting mathematical theories are results of attempts by mathematicians to understand and improve the methods and constructions, used by more applied sciences, such as Physics, Biology or Social studies. One can draw this observation so far as to say, that Mathematics owes to these attempts the very existence of such important branches as Calculus, Differential equations, or Statistical analysis; some people even argue that the creation of Geometry itself was due to the immediate and urgent necessity to measure the land lots in a most fair and reasonable way.

Another universally recognised characteristic feature of a truly important mathematical theory is its central position with respect to the existing branches of Mathematics: it should appeal to the tastes of a vast range of mathematicians coming from various backgrounds, from Analysis to Algebra and from Probability Theory to Topology: everybody should be able to see the beauty of this theory and find inside its bounds a place for the application of his or her efforts.

From this point of view, the subject that we are going to discuss in these lecture notes has all what it takes to become pure classic with time, if it hasn't yet become one: the mathematical study of various quantisation constructions can be traced back to the pioneering works of Erwin Schrödinger, Werner Heisenberg, Paul Dirac, Hermann Weyl and other great physicists and mathematicians of the first half of the XXth century, the creators of modern Quantum Physics in all its variegated incarnations. From the very start of this theory it has been closely related with the Representation theory, Analysis and Differential equations; it took over 100 years and the efforts of numerous scientists to bring the subject to its modern state, when it has more to do with Algebra, Topology and even to Category theory than one would probably safely assume (from a naïve point of view, of course). In order to uncover this mystery, let us begin with recalling basic principles of the first quantum physical theory ever, Quantum Mechanics.

If one were to put the basic idea of Quantum Physics in four words, they would probably be: "observable functions turn noncommutative". More accurately, *quantisation*, i.e. the passage from a classical to the corresponding quantum system consists of *finding for every observable function, involved in the system, a linear operator on a certain Hilbert space, so that the commutator of the operators P_1, P_2, P_3 corresponding to momenta p_1, p_2, p_3 of a quantum particle and the operators Q_1, Q_2, Q_3, corresponding to its coordinates q_1, q_2, q_3, satisfy the canonical commutation relation*

$$[P_i, Q_j] = -\hbar\sqrt{-1}\delta_{ij}.$$

Here $\hbar = 1.054571817 \cdot 10^{-34} m^2 \mathrm{kg}/s$ is the (reduced) Planck's constant; in this book however, the letter $\hbar$ will almost always stand for an abstract "infinitesimal parameter": after all, in Mathematics we can dispense of the mere physical reality when aspiring to the theoretical heights. For the purposes of this discussion, one can even assume that $\hbar^2 = 0$, something, that can only happen in pure Mathematics.

It follows from the commutation relation for P_i and Q_j that for every two (noncommutative) polynomials of the quantum coordinates and momenta, $f(P,Q)$, $g(P,Q)$ (here and elsewhere $P = (P_1, P_2, P_3)$, $Q = (Q_1, Q_2, Q_3)$ for short and similarly $p = (p_1, p_2, p_3)$, $q = (q_1, q_2, q_3)$) their commutator up to degree 1 in $\hbar$ is given by the formula

$$[f(P,Q), G(P,Q)] = -\hbar\sqrt{-1}h(P,Q) + \cdots,$$

where $\cdots$ denote the terms of degree 2 and higher in $\hbar$ and

$$h(p,q) = \{f(p,q), g(p,q)\}.$$

The expression $\{f(p,q), g(p,q)\}$ denotes the Poisson bracket, well-known from the classical mechanics. The simplest way to define it is by the formula

$$h(p,q) = \sum_{i=1}^{3}\left\{\frac{\partial f}{\partial p_i}\frac{\partial g}{\partial q_i} - \frac{\partial f}{\partial q_i}\frac{\partial g}{\partial p_i}\right\}.$$

In other words, if we consider the usual polynomials $f(p,q)$ and $g(p,q)$ of the (usual) momenta and the coordinates of the particle, corresponding to the noncommutative polynomials $f(P,Q)$, $g(P,Q)$ (e.g. one can send $\hbar$ to 0 to this end), and take the Poisson bracket of $f(p,q)$ and $g(p,q)$, then the first term of the commutator will be equal to $\{f,g\}$, in which we replace p_i by P_i and q_j by Q_j. Observe, that the result of this reverse procedure will depend on the way, we write down $h(p,q)$, i.e. on the order of p_i and q_j in the monomials; but the difference will always be of degree 2 and higher in $\hbar$.

As we see, finding the quantum analogs of classical observable functions involves the condition that the leading term of their commutator should be determined by the Poisson bracket. On the other hand, it is also clear that finding the operators that represent quantum momenta and coordinates is not sufficient to determine the quantum counterparts of all other observable functions. And it is not just due to the analytic problems that arise from considering non-polynomial functions: in fact, the order of momenta and coordinates in a monomial, which is totally arbitrary in the commutative (classical) case, acquires tremendous importance in quantum world. Therefore any recipe of quantisation can not be reduced to just choosing the quantum momenta and quantum coordinates; any such recipe should give "at once" the quantisation of all observable functions from a reasonably large class.

The first quantisation methods that were used at the dawn of the last century dealt with the functions on Euclidean spaces, in which situation one can use the analytical methods (Fourier analysis and the theory of pseudo-differential operators) to solve the problem in one or another class of functions. Namely, let $\mathcal{F}$ be the Fourier transform on $\mathbb{R}^n$. Then for any function $f(p,q)$ on $\mathbb{R}^{2n}$ (in Physics one usually assumes that $n = 3$ or $n = 4$) from the chosen class and any test function φ, usually from the Schwarz space on $\mathbb{R}^n$, we put

$$Op(f)(\varphi) = \mathcal{F}^{-1}(f\mathcal{F}(\varphi)).$$

This method and its numerous updates and modifications are powerful enough to allow one deal with the Quantum Mechanics on flat (Euclidean) spaces. This has been sufficient for the physicists for many years. However, by the middle of the century physical theories started to move more and more from flat (pseudo) Euclidean spaces to manifolds, locally Euclidean spaces, which could be topologically and geometrically nontrivial: this was caused by the development of cosmology, which by that time became free from the assumption that our spacetime is flat.

Fourier analysis on general manifolds although not impossible, is much harder and its results are not as powerful as one needs to implement successfully the same trick in full generality. In fact, one can fairly simply generalise the construction we described above to the situation, when momenta are linear functions on the fibres of the cotangent bundle of a manifold; one can say that momenta in this case run through a Euclidean space, that depends on the position of the particle in the curved space-time. In this case the quantisation is done with the help of the theory of pseudo-differential operators on manifolds. And when we say that this is "simple", we by no means intend to downplay the analytical difficulties one has to overcome in the process. We only mean that the main ideas used in the process are direct generalisations of the classical Fourier analysis in $\mathbb{R}^n$. However, these methods are hardly applicable in the case, when both the coordinates and the momenta of a particle are from similar curved space, which is a fairly natural assumption if we want to have a theory in which coordinates and momenta would be interchangeable.

This question started to haunt physicists in the second half of the last century, when the abstract Differential Geometry finally made its appearance in Quantum Physics. Abstract manifolds of fairly general nature became the playground for modern physical theories, so that one could not any more remain satisfied with the restrictions imposed upon the Quantum theories by poorly implemented quantisation principles. So from about mid 1960s there started to appear more general constructions, aiming to overcome the defects and imperfections of the traditional quantisation methods. The purpose of these efforts was to extend the domain of the Quantum Physics, which by that time have proved to be a powerful set of theories, to more abstract realms. The important condition to satisfy was that observable functions in this new quantisation methods should be all smooth functions on the manifold M of dimension n, in which the quantum phenomena were studied. The condition on commutator of the quantised functions then implies that there should be a Poisson bracket defined for any pair of such functions, so M should be a symplectic or, still more generally, a Poisson manifold.

Historically the first attempts to move in this direction were the *geometric quantisation* constructions, due to Jean-Marie Souriau (see Chapter V of his book [Sou70]), Alexandre Kirillov (the creator of the "orbit method", directly connected with the geometric quantisation program, see his survey [Ki85]) and Bertram Kostant (see his paper [Kos70] in the Lecture notes in Mathematics series). The geometric quantisation constructions are still quite popular, their advantage is that *they provide not just the quantisation of the product of functions, but they give a representation of the quantised algebra of functions as operators on a Hilbert space*. In effect, such representation is an important part of the quantisation, since it is the spectral data of the corresponding operators that have important physical applications. On the other hand, the corresponding constructions of geometric quantisation are rather delicate and ask for a considerable number of additional conditions to work perfectly, so that many theoretically important situations fall beyond the confines of the domain of their application.

Luckily, Mathematics has an old established traditional ways to overcome such limitations by discarding parts of the conditions that are perceived as redundant to the main task (or just look too hard to implement). In the case of quantisation, one of such (arguably) redundant conditions is the necessity to provide the representation of the quantised algebra of observable functions: on one hand, algebras of operators on Hilbert spaces are quite delicate and impose lots of additional restrictions on the algebraic side of the story. On the other hand, one can divide the quantisation problem (i.e. the task of finding operators on Hilbert space that satisfy certain commutation relations) into two parts:

(i) construct the associative algebra $\mathcal{A}$ so that all the observable functions f, g that we may need, can be thought of as the elements of $\mathcal{A}$ and their commutators are given by $\hbar$ times the Poisson brackets $\{f, g\}$ up to an infinitesimally small correction term.

(ii) Construct a representation of this algebra in a Hilbert space.

The second topic on this list (i.e. the representation of a noncommutative algebra) is the subject, extensively studied in Representation

Theory, so one can expect that this renowned theory will give a solution of the part (*ii*) automatically, as soon as the algebra $\mathcal{A}$ is given. So it is only natural that we shall concentrate at the part (*i*) of the list. And, as it is often done in Mathematics, we are only interested in finding the algebra $\mathcal{A}$ up to an isomorphism of algebras (here and further in this section we only consider the isomorphisms of algebras that induce the identity map on M).

Cutting some angles and speaking loosely, one can say that it is this part (*i*) (i.e. the construction of the algebra $\mathcal{A}$) that is the main subject of the *deformation quantisation*. More accurately in order to fit into the deformation quantisation program we need to impose some additional conditions on the structure of $\mathcal{A}$: *we will assume that as a linear space $\mathcal{A}$ is isomorphic to the algebra of formal power series (possibly, Laurent series) in the variable $\hbar$ with coefficients in* $C^\infty(M)$.

The main advantage of introducing formal power series into consideration is that the question of convergence in this case turns into purely algebraic property, so that now all analytic difficulties are put aside and one can concentrate on the algebraic essence of the quantisation program. From now on we will be working with formal power series in $\hbar$, i.e. we fix a linear isomorphism $\mathcal{A} \cong C^\infty(M)[[\hbar]]$; in algebraic language this means that $\mathcal{A}$ is a flat $\mathbb{C}[[\hbar]]$-module (and this now can be used in more general context of sheaves). On the other hand it is natural to assume that the multiplication in $\mathcal{A}$ should be $\hbar$-linear, i.e.

$$(\varphi(\hbar)a) \star b = \varphi(\hbar)(a \star b) = a \star (\varphi(\hbar)b),$$

for all $a, b \in \mathcal{A}$ and any formal power series $\varphi(\hbar) \in \mathbb{C}[[\hbar]]$; this should be so since the original meaning of $\hbar$ was just some physical constant. Thus the product in $\mathcal{A}$ is uniquely determined by the product of the functions from $C^\infty(M) \subset \mathcal{A}$: for any two smooth functions f, g on M we have a formal power series

$$f \star g = \sum_{k=0}^{\infty} \hbar^k B_k.$$

Here $B_k \in C^\infty(M)$ are some coefficients; it is clear, that B_k should depend on f and g. Moreover, since we assume that the $\star$-product is

Introduction xiii

$\hbar$-linear, $B_k = B_k(f, g)$ should be given by a bilinear function of its arguments. The condition on the commutator of f and g in $\mathcal{A}$ now takes the form of the following equation

$$f \star g - g \star f = \hbar\{f, g\} + o(\hbar),$$

where $o(\hbar)$ denotes the sum of all terms in which $\hbar$ appears in degree 2 and higher; this allows one fix (up to an equivalence relation) the second term in the formula for $f \star g$: it is clear that if we put $B_1(f, g) = \frac{1}{2}\{f, g\}$, then the condition on the commutator will automatically hold. One can show that this choice is in fact unique up to an isomorphism $\mathcal{A} \cong \mathcal{A}'$.

Next let us consider the first term in this equation, i.e. the term independent of $\hbar$ can be guessed from the following consideration: if we have the operators P_i, Q_j, quantising the momenta and the coordinates of a particle, then it is natural to use them instead of the variables p_i, q_j in order to define the quantisation of an observable polynomial $f(p, q)$; in this case, of course, we will be obtaining different results, depending on the order, in which we write these operators inside monomials, but the difference will always depend on $\hbar$, so that the element independent of this formal variable is well-defined. If we identify this term with f, then clearly the same should go for the product $f \star g$, i.e. the term independent of $\hbar$ in the corresponding formula, should be just $B_0(f, g) = fg$.

Finally, the remaining part of the series for $f \star g$ cannot be fixed by any of the simple methods; the only natural condition that one can impose on the maps B_k, $k \geq 2$ is that they should be given by some sort of local expressions of f and g; this is usually done by saying that B_k are *(bi)linear bidifferential operators* on M, i.e. in a local coordinate system $(x_1, \ldots, x_n)$ on M, the map B_k should be given by expressions like:

$$B_k(f, g) = \sum_{\alpha, \beta} A_{\alpha\beta}(x) \frac{\partial^{|\alpha|} f}{\partial x_1^{\alpha_1} \cdots \partial x_n^{\alpha_n}} \frac{\partial^{|\beta|} g}{\partial x_1^{\beta_1} \cdots \partial x_n^{\beta_n}},$$

where the summation is done over all multi-indices $\alpha = (\alpha_1, \ldots, \alpha_n)$, $\beta = (\beta_1, \ldots, \beta_n)$ for $\alpha_i, \beta_j \geq 0$ and $|\alpha| = \sum \alpha_k$. If we sum up these conditions, we will get the exact wording of the problem we study in these lecture notes:

Let M be a smooth manifold and let $\{f,g\}$ be a Poisson bracket on the space of smooth functions on M. Find an associative $\hbar$-linear product $\star$ on the space of formal power series $C^\infty(M)[[\hbar]]$ such that for any $f,g \in C^\infty(M)$ we have

$$f \star g = fg + \frac{\hbar}{2}\{f,g\} + \sum_{k=2}^{\infty} \hbar^k B_k(f,g),$$

where B_k are suitable bidifferential operators on M. Classify the $\star$-products of this sort up to an isomorphism (inducing the identical map on M).

This problem is now formulated in terms of Differential Geometry and pure Algebra, and is known under the name of *the deformation quantisation problem*. In these lecture notes we give a brief introduction to the methods and survey the major results that have lead to the solution of this problem.

The history of deformation quantisation

The deformation quantisation problem in the form, close to what we have just seen, was first formulated in the papers [BFFLS78₁] and [BFFLS78₂]. By that time, the geometric quantisation had already proved an interesting and important construction both from the point of view of pure Mathematics and from the point of view of its relations with Physics. On the other hand, its limitations and difficulties related with implementation of its results were also well known, and many mathematicians and theoretical physicists were looking for a simpler theory, presumably related with the deformation theory of algebras, developed by Gerstenhaber about 10 years earlier, see [Ge64], [Ge66], [Ge68] and [Ge74].

It turned out that the first instances of this algebraic approach had already appeared by that time in the papers of Jose Moyal [Mo49] in 1949 and even earlier in the thesis of Hilbrand Groenewald [Gr46] in 1946; we mean the construction that is now known by the name of *Moyal product*. This construction is based on the observation that the asymptotic series for the symbol of the product of two pseudo-differential operators on $C^\infty(\mathbb{R}^n)$ can be interpreted as the

formal power series depending on the symbols f and g of the factors. Namely if
$$\omega = dp_1 \wedge dq_1 + \cdots + dp_n \wedge dq_n$$
is the standard symplectic structure on $\mathbb{R}^{2n}$, then for any two symbols $f, g \in C^\infty(\mathbb{R}^{2n})$ (the doubling of the dimension is due to the fact that symbols depend not only on coordinates, but also on momenta), this asymptotic series looks as follows
$$f \star g = m\left(\exp\left(\frac{\hbar}{2}\omega_{ij}\partial_i \otimes \partial_j\right)(f \otimes g)\right).$$

Here m is the product of functions and $\hbar$ plays the role of infinitesimal parameter, while $\partial_i = \frac{\partial}{\partial x_i}$, $i = 1, \ldots, 2n$ are the partial derivatives with respect to a given linear coordinate system $(x_1, \ldots, x_{2n})$ on $\mathbb{R}^{2n}$. Now it is easy to derive the explicit formulas for the coefficients $B_k(f, g)$ of this product and one can prove independently of the analytical interpretation that it is associative and satisfies the condition on its commutator. Thus, this formula is an example of the $\star$-product in the sense of the deformation quantisation problem as we have earlier defined it; in effect Moyal product was one of the sources of inspiration for the definition of the deformation quantisation.

Another classical construction, that can be interpreted as the deformation quantisation of some Poisson algebra, is the construction of universal enveloping algebra $U\mathfrak{g}$ of a Lie algebra $\mathfrak{g}$: by Poincaré–Brikhoff–Witt (PBW) theorem $U\mathfrak{g}$ is linearly isomorphic to the polynomial algebra $S(\mathfrak{g})$ so that the commutator of the elements in $U\mathfrak{g}$ transforms into the Poisson structure on $S(\mathfrak{g})$. Now if we slightly modify the construction of $U\mathfrak{g}$ by introducing the formal parameter $\hbar$ and passing to a suitable completion with respect to it, we can use the same PBW theorem to pull the product on $U\mathfrak{g}$ to a $\star$-product on $S(\mathfrak{g})[[\hbar]]$, thus providing yet another example of the deformation quantisation.

For some time these two constructions remained the only completely understood constructions of the deformation quantisation. However, in a rather short time after the publication of the papers [BFFLS78$_1$] and [BFFLS78$_2$] many new examples and constructions of deformation quantisation were proposed. They include the constructions of ⋆-product on cotangent bundles of a smooth manifold [DL83$_1$], in particular of the parallelisable manifolds and Lie groups [CG82], [Gu83], as well as the deformation quantisation of symplectic manifolds with trivial third de Rham cohomology group, see [NV81]. Then in 1983 the first significant step towards the complete solution of the problem was made: in papers [DL83$_2$] and [DL85] Marc de Wilde and Pierre Lecomte proposed a construction of ⋆-products on the smooth functions of an arbitrary symplectic manifold.

As anyone familiar with the methods used in the cited papers, can observe, the study of the deformation quantisation has from the very beginning been closely related with the development of the cohomological methods; this is not to be wondered at, since it has been closely related with the deformation theory, where these methods have always been very popular. One can say, that these two theories have been borrowing the methods and ideas from each other. Thus the first construction of ⋆-product on symplectic manifolds of de Wilde and Lecomte was based on an accurate analysis of the differentiable Hochschild cohomology of $C^\infty(M)$ as well as on the previously established results of Jacques Vey [Vey75], who described the cohomological obstructions for the deformation of Poisson structures. It lay completely within the paradigm of the traditional perturbative deformation theory, when the deformation is constructed by induction on degrees of the deformation parameter, so that at each stage one has to solve the equation, that determines the next stage of the deformation.

This approach dominated the deformation quantisation for quite some time and still is quite popular, as it is very intuitively natural and close to the standard methods of Physics. However, the construction of de Wilde and Lecomte is quite complicated, the coefficients B_k that it gives are hard to grasp, and which is even worse, it is of a very little help, when one works with a generic Poisson manifold rather than with a symplectic manifold. Therefore other constructions were actively sought by researchers throughout the world.

The first alternative construction to be found was the well-known *Fedosov quantisation*. First announced in 1986 in a short paper written in Russian [Fed86], it remained almost unnoticed abroad until about ten years later, when the English variant of the paper and Fedosov's book were finally published, [Fed94], [Fed96]. The main idea of Fedosov was fairly simple and brilliant: when we work with symplectic manifolds, there is no problem with local quantisation: in effect, locally we always can choose Darboux' coordinates, in which symplectic form has constant coefficients, and hence we can apply Moyal quantisation. So the only remaining problem is to define a globalisation procedure in the case when there exist no global Darboux' coordinates. In de Wilde and Lecomte's approach this had been done by comparing different local formulas and proving that the cohomological obstructions for their glueing together in a single $\star$-product can be neglected: this followed from the fact that these obstructions could take only a limited number of values. Fedosov's idea on the contrary was to go further down from local to what one can call micro-local level, observing that Moyal formula can be applied to functions on the cotangent space at every point: just regard the value of symplectic form at this point as the constant symplectic structure on this fibre. Now one can say that the globalisation amounts to finding a global section of this $\star$-product bundle; Fedosov did this by means of choosing a flat (or *Abelian*) connection in it.

Fedosov's construction showed the intimate relation of deformation quantisation with the differential geometry of the tangent bundle on the manifold; more generally: intuitively, one can say that functions on cotangent space at a point $x \in M$ are functions on an infinitesimal neighbourhood of x in M. This brings forth the following idea: one can divide the problem of deformation on a manifold into subproblems of local quantisation, i.e. finding the $\star$-product on an infinitesimal neighbourhood of a point, and the subproblem of globalisation: if for every point $x \in M$ and for a suitable infinitesimal coordinate system at x (e.g. Darboux system in symplectic case) we have a $\star$-product (depending on the choices we made), then we need to find a global section of this new bundle, so that the $\star$-product is pulled down to the base. The globalisation problem can now be attacked by means similar to Fedosov's construction of flat connections.

As for the other subproblem, i.e. the construction of local $\star$-products for a generic Poisson structure on $\mathbb{R}^d$, its difficulty arises from the fact that the rank of a Poisson bivector in a generic case depends on the point, so the local geometric structure can be quite complicated. The solution of this problem was eventually found by Kontsevich in 1997. In physical literature, however, one can encounter the opinion that Kontsevich suggested a mathematically-rigorous solution, while, as argue some experts, the physical approach to this question, based on Feynmann's continual integral formalism had been around in the studies on Poisson σ-model even before that. However I think there is no reason to doubt Kontsevich's priority in finding the quantisation of a generic Poisson structure, since the distance between the physical ideas and the actual Kontsevich's construction is very large and it is not easy to trace the latter from the point of view of the former. In effect, the relation between the physical approach and Kontsevich construction had not been accurately explained until few years after the publication of Kontsevich's paper, when this was done by Alberto Cattaneo and Giovanni Felder.

Whatever be one's opinion about the origin of Kontsevich's construction, one can hardly overestimate the impact its publication made on Mathematics as a whole. In effect, its importance is far beyond merely bringing mathematical lustre to a physical theory. The changes it has inspired by introducing into the deformation theory and other related subjects the methods from Algebraic Topology, Category theory and related topics, such as the higher homotopy algebra, the operad theory etc. can hardly be described in these introductory notes. In effect, it was after this that the importance of these constructions became acknowledged by mainstream mathematical community, which fuelled the development (and even sometimes the creation) of such popular fields as the derived algebraic geometry, higher category theory and other fields, it has inspired many scientists to further investigate the existing physical constructions and encouraged further incursions from Physics into the domain of Mathematics. One of the modest purposes of our lecture notes is to acquaint the reader with this circle of ideas, if only by providing him or her with a brief outlook of these wonderful theories: of course, one cannot squeeze all the necessary material into this book, but we still hope that the reader will get a taste of this constructions and will be inspired to pursue the study of the subject further on his or her own.

Composition of the text

As we have explained earlier, even if we had restricted our attention just to the ideas and methods directly related to or inspired by Kontsevich's quantisation formula, we would have to write a much longer book than the present one, which is already in my opinion a pretty long text. So, while preparing my lectures and later, when I was writing this manuscript, I had to choose, which theories I was going to discuss, and which theories I was about to omit completely or by major part. In many cases my choices were dictated not by my tastes but rather by what one can call "pedagogical logic", i.e. by my impression of what the reader of this book can be familiar with and what theory needs explanation, what he or she would find more attractive or more important etc. In particular, I skipped the traditional formalism of Artin algebras and related topics, while speaking about the deformation theory: of course, these constructions are already an indispensable part of modern approach to the deformations, it stems from the original papers by Gerstenhaber, see [Ge64]–[Ge74] and is used in a much wider context, which includes the deformation of virtually any algebraic structure. However, such abstractions make the book barely accessible to unprepared readers, especially the readers with physical background; and keeping the exposition as close to the roots of the deformation quantisation theory, which lie in Physics rather than in Mathematics, was one of the goals I pursued while writing the text.

The idea that while we talk about the existing theory of deformation quantisation, we should keep in mind its origins, can be regarded as one of the ultimate guidelines I tried to follow when writing this book. To put it in other words, and in a little bit fancier form, let me say that I believe in the basic law of biogenetics formulated by Ernst Haeckel, which says that "ontogenesis recapitulates phylogenesis in abbreviated form", i.e. that the development of an individual organism reproduces in an abbreviated form the evolution of the whole species, and I am sure that this law can and should be applied to the spheres of human intellectual activities. In particular I am sure that knowing the history of one or another theory including the initial constructions and approaches that were used at the first stages of its study are important for anybody who wants to learn this theory at a level, necessary for conducting independent research. Therefore a good textbook in my opinion should be written in a way reflecting the process by which the subjects, covered by it, have been developing; of course this condition is sometimes hard to achieve, especially in the case of such old renowned branches of Mathematics as Algebra, Geometry or Calculus, but in case of more recent theories such as the Deformation Quantisation this principle can be applied with much more ease. This said, it must be clear now that I tried my best to follow these principles while working on the text.

This phylogenetic approach is reflected in the structure of the present book: the order of exposition that we follow here reflects the history of the problem that we study to the extent which in my opinion is sufficient to allow the reader to grasp the basic principles and ideas that lie in the foundations of this theory. One may say that the structure of the text is pretty linear, if not just straightforward: there is no complicated interconnections between distant sections of the text, it is supposed that the book should be read in the order it is written. If you feel that you know the material covered in a section well enough, then skipping the section and continuing reading from the next one is possible: we try to stick to the established notation and definition, which you may find elsewhere. In order to make the text self-sufficient and accessible for a young reader, I have included several sections, containing more or less standard material, covered in many textbooks. This concerns first of all Sections 4–7, where we recall the definitions and basic properties of symplectic and Poisson manifolds, Lie algebras and also give a crash-course in Homological algebra. Of course this material is rather deficient and one cannot regard it as an introduction into any of these subjects, however I hope, it is sufficient to provide the reader with enough information to follow our further exposition.

There are twenty sections, two appendices and seven problem sections in this book. Each section and both appendices are divided into subsections, and each of them roughly corresponds to a single lecture. Problem sections mark logical subdivision of the text into seven informal chapters; I preferred to keep them informal, because too much subdivision would in my opinion break down the composition too much, so I withstood the temptation to introduce chapters into the play. Contents of these quasi-chapters are roughly denominated by the titles of the corresponding problem sections:

I General quantisation principles;

II Poisson structures, quantisation and cohomology;

III Hochschild homology and cohomology: Obstructions theory;

IV Obstructions and deformation quantisation of symplectic manifolds;

V Fedosov quantisation;

VI Higher homotopy algebras, Kontsevich's theorem;

VII Kontsevich's quantisation: Properties and applications.

The appendices contain a short introduction into operads theory necessary to give Tamarkin's proof of Kontsevich's theorem, as well as the proof itself. Let us now describe the contents of the text in more details.

Contents of the book

In accordance with the historical principle, we begin the book by recalling in the first two sections the basic ideas from Classical and Quantum Mechanics, such as the Hamilton equations, Poisson brackets, fundamental commutation relations, uncertainty principle, canonical quantisation etc. Of course, these constructions are widely known and thoroughly explained in details in numerous comprehensive textbooks and monographs, some of which we have added into the list of references at the end of our book. Therefore one cannot and should not regard our text as an introduction into these subjects: the purpose for introducing these sections into the book is to put the deformation quantisation into the context of wider theory that includes Quantum Physics and Classical Mechanics.

After we have established the relation of the quantisation principles with Poisson brackets and Hamilton equations, we move to the first example of deformation quantisation, that is the Moyal product: in Section 3 we give a detailed exposition of this construction, starting from the asymptotic decomposition formula for the symbol of the composition of pseudodifferential operators. We omit the analytical theory that is the basis for this formula, since explaining it in sufficient detail would take too much efforts and time. Instead in Section 3.3 we give a formal derivation of the Moyal formula from the "Fourier transform of Dirac's δ-function" (formula (3.12)):

$$\delta(x) = \int_{\mathbb{R}^n} e^{2\pi i x \cdot \xi} d\xi.$$

From the point of view of Analysis the right-hand side of this formula is ill-defined, since the integral does not converge for all values of $x \in \mathbb{R}^n$; however informally speaking, one can observe that unless $x = 0$, the integrated function is oscillating hence its average value over the space can be taken to be 0, and if $x = 0$, then the integral can be taken equal to plus infinity. This point of view can be made more rigorous with the help of special functions and Fourier transform, however, the formal point of view is quite sufficient for our purposes.

In order to discuss the deformation quantisation in full generality, we need to deal with generic symplectic and Poisson structures on manifolds, so we give the necessary definitions in Section 4. Just as in the first two sections, we do not try to give a thorough exposition of this theory: it is well explained in many good textbooks. Our purpose was just to recall the necessary definitions and results, which will be used further in the book, such as the definition of Hamilton vector fields and Poisson brackets, associated with a symplectic structure, or the definition Schouten brackets on polyvector fields, and the equivalence of the Jacobi identity for the Poisson structure and the equation

$$[\pi, \pi] = 0$$

for the corresponding bivector (here [,] denote the Schouten brackets).

Sections 6, 7 and in some degree Section 8 serve the same purpose as Section 4: most constructions of deformation quantisation use various methods of homological algebra, one can even say that the ideas of homological algebra play central role in most research on this subject, even if some constructions were not originally explicitly formulated in the terms of (co)homology theories (this is the case for instance of the original Fedosov's construction). So we cannot move further without using these methods, and recalling them is the main purpose of these three sections (more accurately, in Section 6 we give the most general definitions that we need, and in Sections 7 and 8 we discuss more specific questions, concerned with Hochschild homology and cohomology: in Section 7 we give the general definitions of various structures related with Hochschild homology and cohomology, while Section 8 is dedicated to the proof of the Hochschild–Kostant–Rosenberg theorem about the structure of differentiable Hochschild cohomology of the smooth functions on a manifold, including the additional structures on it such as the Gerstenhaber brackets and the ∪-product).

Of course, one cannot and should not expect that we managed to give in a smallest degree complete account of the methods and results of Homological algebra in just two sections; however we hope that with a rare exception we have given definitions of all terms and have discussed all the major constructions, used in the text. The biggest omission in our account of these results is the theory of spectral sequences: this is the major tool for calculating various cohomology, but being very technical, this theory demands (and deserves) much more attention than what we can pay it in this book. Therefore we refer the reader to suitable textbooks (see for instance [MCl01]), and we suggest that our readers would skip the details of reasonings based on the use of spectral sequences, whenever we refer to this method. Luckily, most part of our material can be understood pretty easily without it, except the appendices, which ask for a considerably greater level of knowledge in homological algebra.

In some sense the purpose of Section 5 is also to remind the reader some standard notions and definitions, this time from the theory of Lie algebras. However, this section deals with one more important construction: in it we show that the universal enveloping algebra can be regarded as an example of deformation quantisation. This statement is based on the well-known fact, the Poincaré–Birkhoff–Witt (PBW) theorem, which identifies the universal enveloping algebra $U\mathfrak{g}$ of a Lie algebra $\mathfrak{g}$ with the symmetric algebra $S(\mathfrak{g})$ of $\mathfrak{g}$ (as vector spaces). This identification can be regarded as a deformation of the multiplication in $S(\mathfrak{g})$: if $\sigma : S(\mathfrak{g}) \to U\mathfrak{g}$ is the linear isomorphism, induced by the PBW theorem, then we put

$$f \star g = \sigma^{-1}(\sigma(f) \cdot \sigma(g)).$$

Modifying the definition of $U\mathfrak{g}$ a little bit, we can rephrase this construction completely in the framework of the deformation quantisation, thus obtaining the so-called "Campbell–Baker–Hausdorff quantisation" formula, named after Campbell–Baker–Hausdorff theorem, that allows one to express $Z \in \mathfrak{g}$ in the formula

$$\exp(X)\exp(Y) = \exp(Z)$$

as a function of X and Y. We do not explain this formula here as we do not actually need it; instead we use an analogous formula that belongs to Berezin: we use it to show, why $f \star g$ defined above is in effect given by a series of bidifferential operators. It is remarkable, that the construction we use here, can in fact be reversed, in the sense that one can first construct a $\star$-product, and then use it to prove the PBW theorem: we do this twice in the course of the lectures, first in Section 9.5, where we show the existence of $\star$-product in $S(\mathfrak{g})$ by somewhat straightforward computations of the obstructions (see below), and later we return to this idea again in Section 20 where the existence of $\star$-product follows from the general Kontsevich's construction.

Moving on from the discussion of Hochschild cohomology, in Section 9 we discuss the traditional obstructions calculus for the deformation theory: working step by step on the degrees of $\hbar$ we show that the $\star$-product exists if certain elements in the third Hochschild cohomology of $C^\infty(M)$ vanish, and in this case the equivalence classes of $\star$-product are described by the second Hochschild cohomology of this algebra. We further illustrate this construction by applying it to $M = \mathfrak{g}^*$ (see Section 9.5), and to $M = T^*X$ (Section 10), where one can use a grading on the Hochschild cohomology to prove vanishing of the obstructions. The deformation obtained in the latter case can be regarded as the algebra of (symbols of) pseudodifferential operators on X: this is a direct analogy with the interpretation of Moyal algebra on $\mathbb{R}^{2n}$ equipped with canonical symplectic form as the algebra of (pseudo) differential operators on $\mathbb{R}^n$. We also describe some further generalisation of the obstruction calculus, leading to a variant of Neroslavski-Valssov construction in Section 9.6 and pay attention to some special cases of the deformation of cotangent bundles (in particular, Lie groups) in Section 10.2.

After this we turn to the first construction of the $\star$-product on a generic symplectic manifold, suggested by Marc de Wilde and Pierre Lecomte in 1983 (see [DL83$_2$]). Their construction is also derived from the study of obstruction classes, however this time one begins with the obstructions for the deformation of the Poisson bracket: it turns out that the existence of $\star$-product can be comparatively easily derived from the existence of a nontrivial deformation of the Poisson structure. The existence and classification of the latter deformations is also related to the study of certain cohomology theories, this time the Chevalley–Eilenberg cohomology of the Lie algebra of functions on the symplectic manifold which we regard as the Lie algebra with respect to the Poisson bracket; one can further use the Gelfand and Fuks cohomology of formal vector fields to describe this cohomology, as was explained by Vey in [Vey75].

We outline these constructions in Section 11, where we also give the description of the corresponding cohomology spaces up to degree 3: this has been first done by Vey in the cited paper; the proof involves some computations with spectral sequences, so we cannot reproduce it in full detail, instead we describe two explicit constructions of the well-known Vey's class S^3_Γ. In particular, this gives us an opportunity to speak about symplectic connections: we prove their existence in Section 11.3.1. Later in the book we use a symplectic connection in Fedosov's construction, where it plays one of the central roles. After this preparatory work, the de Wilde and Lecomte's construction is discussed in Section 12: first we give a more detailed description of the low degree Chevalley-Eilenberg cohomology of the Poisson algebra of a symplectic manifold M, paying attention to the classes that can appear when the first Pontryagin class of TM vanishes. After this, one can prove the existence of nontrivial deformations of the Poisson structure on M and conclude the existence of an associative $\star$-product in $C^\infty(M)[[\hbar]]$: this is the $\star$-product that underlies the deformed Poisson bracket, obtained earlier. Remark, that the principal step towards the proof of this theorem consists of the observation that locally every symplectic manifold is in fact exact: the corresponding symplectic form ω is locally equal to the de Rham differential of a 1-form: $\omega = d\alpha$. This allows one to get a local deformation of the Poisson structure in a universal (coordinate-free) way, and hence the corresponding deformations can be "glued" together to a global one. This approach, in particular, shows that there's a description of the equivalence classes of $\star$-products in terms of the degree 2 de Rham cohomology of M.

The next three sections (i.e. Sections 13, 14 and 15) are devoted to the discussion of Fedosov quantisation construction for a symplectic manifold M; in some sense its main idea is very similar to the PBW-quantisation: if we find a linear embedding Q of $C^\infty(M)[[\hbar]]$ into a noncommutative algebra W in such a way that the image is in effect a subalgebra of W, then the $\star$-product can be written as

$$f \star g = Q^{-1}(Q(f) \cdot Q(g)).$$

The trick, used by Fedosov is to look for the subalgebra of the form $\ker(D)$ for a derivation D of W. In his constructions D is a flat (or as he says *Abelian*) connection on the algebra of (global) sections of the Weyl algebra bundle over M: one can regard the value ω_x of symplectic structure ω at a point $x \in M$ as the coefficients of the constant symplectic structure on $T_xM \cong \mathbb{R}^{2n}$ and consider the corresponding Weyl algebra of $\mathbb{R}^{2n}$

$$W(T_xM, \omega_x) = T^\otimes(T_xM)/([v,w] = \hbar\omega_x(v,w)),$$

where $v, w \in T_xM$ are arbitrary elements (we omit here various technical details one needs to make sure that all the formulas we encounter in the process of the proof are well-defined). Remark, that the flatness of D immediately suggests that there should be a homological structure underlying this construction, even though Fedosov himself makes no references to it.

However simple, even naïve the idea of Fedosov quantisation might seem, it is very powerful and plays a major role in research, when one needs to study global properties of deformation quantisation on a manifold (in Poisson case as well), so we explain it in detail: in Section 13 we construct the Weyl algebra bundle and recapitulate Fedosov's construction of Abelian (flat) connections on it;

in Section 14 we complete the proof of the existence of deformation quantisation by constructing a linear isomorphism Q of the space of flat sections of Weyl algebra bundle and $C^\infty(M)[[\hbar]]$. After this we show that the $\star$-product induced by this isomorphism does in effect satisfy all the conditions we impose onto the deformation quantisation $\star$-products. It is also in this section that we explain Fedosov's classification theorem: every Abelian connection D is uniquely characterised by its "Weyl curvature" $F \in \Omega^2(M)[[\hbar]]$, $dF = 0$, such that

$$D^2(a) = [F, a] = 0,$$

for all sections a of Weyl algebra bundle (the last equality follows from the fact that $\Omega^2(M)[[\hbar]]$ consists of central sections of Weyl algebra bundle); uniqueness means that for every F one can find a connection with this Weyl curvature, and that Abelian connections D and D' are gauge equivalent iff their Weyl curvatures F and F' are cohomologous. At the same time Fedosov showed that the $\star$-products, induced by D and D', are equivalent iff these connections are gauge equivalent. Thus we see that Fedosov's $\star$-products are classified by the space $H^2_{dR}(M)[[\hbar]]$. It is also worth mentioning, that Fedosov's $\star$-products are characterised by the fact that it coincides with Moyal product in local Darboux coordinates at any point $x \in M$.

In Section 15 we describe some applications of Fedosov quantisation and variations of Fedosov's construction. We begin with the index theory: judging by Fedosov's original papers, his ultimate purpose for considering deformation quantisation was to make the theory of pseudodifferential operators (and hence the index theory) more algebraic and more accessible for direct computations. In fact as we have mentioned earlier, deformation quantisation of cotangent bundle T^*X of a smooth manifold X is naturally identifiable with the algebra of symbols of pseudo-differential operators on X, so one can ask if invariants of such operators can be expressed in terms of the deformed algebra. By his pioneering work Fedosov initiated research in this direction: in his papers he showed that one can define trace on the algebra of compactly supported functions on a symplectic manifold endowed with the $\star$-product, he had defined earlier. In terms of this trace he gave the definition of the index of an "elliptic element" in the algebra and proved a version of the (local) index theorem for such element. We outline this constructions in the Section 15.1 after a brief recollection of the classical index theory, including Atiyah–Singer theorem.

Next we speak about the variants of Fedosov quantisation for manifolds with additional structures, such as the action of a Lie group G on the symplectic manifold, or the complex structure. In the former case (see Section 15.2 and the paper [RW16]) we describe the classification of G-invariant $\star$-products and the "quantum momentum maps" in the terms of equivariant cohomology of M and G. And Section 15.3 deals with the deformation quantisation of holomorphic functions: in holomorphic case one cannot work with global functions and sections of vector bundles, since in most cases the only globally defined holomorphic functions on a complex manifold are constants (and there are similar restrictions for the sections). Hence one has to develop a suitable sheaf-theoretic version of the deformation theory. This was done by Ryszard Nest and Boris Tsygan ([NT01]) who modified Fedosov's construction so that one can take the bidifferential operators in the deformation series from a suitable subalgebra (determined by a *Lie algebroid* in a generic case), in particular from the space of purely holomorphic operators, and showed that under some mild condition there always exists a deformation quantisation of the sheaf of holomorphic functions. A remarkable property of their construction is that the $\star$-products in this case are classified by the classes from a direct summand inside the degree 2 de Rham cohomology, and not by the whole $H^2_{dR}(M)$: the "missing" part of the Fedosov's classes is restored from these data with the help of a series of nonlinear maps τ_k. These rather mysterious maps are (up to my knowledge) still not completely understood: in Nest and Tsygan's paper they appeared from an existence theorem, and as much as I could find, there is still no explicit formula available in the literature that would somehow clarify the construction of the maps τ_k and relate them with some other theory.

The remaining five sections of the book as well as the appendices deal with various aspects of Kontsevich's quantisation theorem. Unlike the traditional approach, based on the standard deformation procedure, when the corresponding series is obtained by solving the consequent obstruction problem, Kontsevich's idea was to regard the deformation series as a single object of special kind. Namely, one can regard the formal sum

$$B = \sum_{k=1}^{\infty} \hbar^k B_k,$$

where $B_1(f,g) = \frac{1}{2}\{f,g\}$, as a degree 2 element in the (formally completed) Hochschild differentiable cochain complex of the smooth functions algebra of a manifold M; then the associativity condition for the corresponding $\star$-product is equivalent to the equation (the bracket in this formula denotes the Gerstenhaber bracket on Hochschild complex)

$$\delta B + \frac{1}{2}[B, B] = 0.$$

This is a canonical example of the *Maurer–Cartan equation*, ubiquitous in the deformation theory. Thus the deformation quantisation problem is equivalent to the construction and classification of the *Maurer–Cartan elements*, i.e. solutions of this equation. On the other hand, the equation $[\pi, \pi] = 0$ (for Schouten brackets $[,]$) satisfied by the Poisson bivector π can also be interpreted as the Maurer–Cartan equation, if we introduce trivial differential on polyvector fields. As one knows, the algebra of polyvector fields is equal to the differentiable Hochschild cohomology of $C^\infty(M)$. Thus the question is, whether one can find a method to relate the set of Maurer–Cartan elements in polyvector fields and in differentiable Hochschild complex?

It turns out that there is a context, which allows one to transfer the Maurer–Cartan elements from one differential graded Lie algebra to another.[2] We mean the theory of *strong homotopy Lie algebras* or the L_∞-algebras, see Section 16, especially 16.3.3. It turns out that any homomorphism of DGLA, that induces an isomorphism in cohomology, can be inverted as a morphism of L_∞-algebras and moreover, any homomorphism of L_∞-algebras allows one to transfer Maurer–Cartan elements from the DGL algebra in its domain to

---

[2] Recall, that *differential graded Lie algebra* or *DGLA* for short is a Lie algebra, equipped with a grading and a differential, so that the bracket preserves the grading and satisfies the Leibniz rule with respect to the differential.

the DGL algebra in its range. However, although the Hochschild–Kostant–Rosenberg theorem does in fact show that Schouten brackets are induced from Gerstenhaber brackets, the map that induces this isomorphism (the standard Hochschild–Kostant–Rosenberg map or HKR map for short) is not a homomorphism of Lie algebras. To put shortly, one can say that Kontsevich found a way to extend the HKR map to an L_∞-morphism, and hence to construct a $\star$-product for any Poisson bivector π.

We explain the major ingredients of this construction in Sections 16, 17 and 18: first, we describe the topological origins of the strong homotopy algebras theory and give the necessary definitions. Of course we cannot explain all the topological details in this book, however, we included this material as we believe that it will help the interested reader to see the topological connotations of this theory, which are indispensable for anyone who would like to study it further in future. In Section 17 we discuss the properties of L_∞-algebras and L_∞-morphisms; in particular we explain that *quasi-isomorphisms* of DG Lie algebras (more generally, of any L_∞-algebras) induce bijections on the sets of equivalence classes of Maurer-Cartan elements in them. After this we give the definition of formal L_∞ algebras and eventually formulate the *Kontsevich's formality theorem*. Proof of this theorem is sketched in Section 18: we closely follow Kontsevich's original paper and describe the construction of compactified configuration spaces, admissible graphs and weight integrals. However we give only few details of the original Kontsevich's proof; instead we refer the reader to Appendix B, where we give Tamarkin's proof of this theorem.

Sections 19 and 20 deal with some further corollaries and modifications of Kontsevich's theorem. First we discuss its relation with *Poisson σ-model* and Feynmann path integrals: as was shown by Cattaneo and Felder in [CF01], these constructions can be used rather formally to define a $\star$-product on $C^\infty(M)$; in this case the associativity of the product will follow as a simple consequence of the properties of path integrals, such as their invariance with respect to some reparametrisation of the domain of the maps. We further explain the way Kontsevich's formality theorem is extended from local (i.e. $M = \mathbb{R}^d$) case, which is given by the original Kontsevich formality construction, to the case of an arbitrary Poisson manifold M:

Kontsevich sketched a globalisation procedure in his original paper, deriving it from the properties of infinite jet spaces, in particular from the fact that the bundle of formal affine coordinate systems over a manifold has a global section. After a short discussion of these constructions, we reproduce Kontsevich's reasoning in Section 19.2.2. We also describe an alternative construction, due to Cattaneo, Felder and Tomassini [CFT02], in which one uses an analog of Fedosov connection to globalise the $\star$-product.

In the last section of the book, Section 20, we discuss a corollary of Kontsevich's construction: if we apply this construction to the case $M = \mathfrak{g}^*$ with the Kirillov-Kostant-Souriau Poisson structure, then it is not difficult to see that the resulting associative algebra will be isomorphic to the universal enveloping algebra $U\mathfrak{g}$ (we omit the formal variable $\hbar$ here). One of the important constructions associated with $U\mathfrak{g}$ is the well-known Duflo's formula, which determines an algebra isomorphism between the center of $U\mathfrak{g}$ and the Poisson center of $S(\mathfrak{g})$. It turns out that this formula can be derived from Kontsevich's construction! This is due to the remarkable properties of Kontsevich's L_∞-map: in addition to being a quasi-isomorphism of L_∞-algebras, it induces an isomorphism of the associative algebras, that is, it intertwines up to a chain homotopy the $\wedge$-product in the space of polyvector fields and the $\cup$-product in Hochschild complex. This isomorphism is well known on the level of cohomology, but the fact that the given L_∞-map provides the necessary result on the level of complexes is pretty unexpected; it signifies that the chosen map is indeed more than just an L_∞-morphism, and in fact captures some properties of the whole algebraic structure on both sides, which includes the associative product. This allows one show that the *tangent morphism* of Kontsevich's morphism induces a homomorphism of centers and it only remains to identify this morphism with Duflo's map. We sketch the proof of this fact in Section 20.6.2: it is not difficult to see that the resulting formula should look like an exponent of a formal power series with certain coefficients, just like the formula for the Duflo map; then as Kontsevich remarked, a direct computation shows that there can be only one such map. Observe that one can make this proof more explicit, by calculating some of the coefficients in the exponent, which was done by other researchers, see, e.g. [CvdB10].

Finally in the appendices we discuss another celebrated result, *Tamarkin's proof of the formality theorem*. Unlike Kontsevich's approach that relies heavily on explicit formulas for the components of formality morphism, Tamarkin derives the formality of the polydifferential operators on $\mathbb{R}^d$ from the general properties of Gerstenhaber algebras.[3] In effect, he shows that *any homotopy Gerstenhaber algebra whose cohomology is equal to the algebra of polyvector fields on $\mathbb{R}^d$ is formal*. This follows from the general theory that we develop in appendix, the core of this theory lies in the operad theory, which we explain.

Namely, first of all in Appendix A we speak about the general operad theory. Initially considered as a rather exotic construction from Topology, operads saw a revival of general interest in 1990s, known as *the Renaissance of Operads*, due to the fact that during this period many new applications of operads in Mathematical Physics and related mathematical theories were discovered. Sticking to the aforementioned principle that our exposition should follow the historic pattern, we begin with a brief discussion of the topological origins of operads: we start with the problem of characterisation of the iterated loop spaces, which lead J.-P. May to the introduction of the little cube operads; so we follow this pattern and give the definition of an abstract (topological) operad and algebra over operad, modelled upon the little cubes operad and the iterated loop spaces.

After this we proceed with the discussion of algebraic operads and constructions with them. In particular, we give the definitions of quadratic operads and the Koszul property, suggested by Ginzburg and Kapranov in [GK95]. Speaking a little loosely, this property means that certain natural construction applied to the quadratic operad does in fact produce a resolution of this operad. We also discuss three canonical examples of the algebraic operads: the operad of associative algebras **ASS**, the operad **COM** of commutative algebras and **LIE**, the operad of Lie algebras, as well as the corresponding constructions of their Koszul resolutions.

After this preliminary work (which is of course important in its own right, independently of the applications we give), we proceed

---

[3]Recall that Gerstenhaber algebra is a graded space with two operations: wedge-product and a bracket, satisfying certain conditions, see Definition B.4.

with Tamarkin's proof in Appendix B. We first discuss the notion of *intrinsic formality* of an algebra over a Koszul operad and prove a nice cohomological criterion for this property. After this, we can try to prove the intrinsic formality for the algebra of polyvector fields on $\mathbb{R}^d$. However, if we regard this algebra as a differential Lie algebra, the criterion cannot be applied! In order to overcome this difficulty Tamarkin suggested to regard the full algebraic structure that exists on polyvector fields: it turns out that if we take into consideration the $\wedge$-product of polyvectors, i.e. regard this space as a Gerstenhaber algebra and not just as Lie algebra, then the intrinsic formality criterion will work!

Thus all what remains to be done it is to find a structure of the homotopy Gerstenhaber algebra on the differentiable Hochschild complex, which would correspond to the analogous structure on polyvector fields. This is the hardest part of the reasoning, related with the well-known *Deligne's conjecture*, which states that the Hochschild complex bears a structure of algebra over a "suitable version of" the operad of cochains on the topological operad of little squares. Tamarkin proved this fact with the help of Kazhdan and Etingof's theorem, that relates (if one is to formulate it in elementary terms) the categories of Lie bialgebras over $\Bbbk[[\hbar]]$ and cocommutative modulo $\hbar$ Hopf algebras with same coefficients.

In addition to the regular sections, there are seven exercise sections in the book. These exercises are aimed at allowing the reader to take a look at various applications of the theory developed earlier, as well as to get further insight into the theory. In some situations we relegate the proofs of some auxiliary results or important generalisations of the theorems we discuss at the lectures to these sections, so looking into the exercise lists (and doing the exercises) is an important part of learning the material.

Further reading

It would be next to impossible to give a whatsoever thorough list of papers and books that deal with the deformation quantisation and related topics, especially those that were written after the publication of Kontsevich's result. Luckily the access to Yandex, Bing, Google and other Internet search engines facilitates the process of looking for the literature to a great extent nowadays.

Therefore we are not going to give a comprehensive list of papers and textbooks for the follow-up reading and confine our efforts to giving just a couple of titles of papers and textbooks dealing with the several major topics of research connected with the subjects covered by this book. The interested reader is encouraged to "google" the topics and authors that we mention in this section, this is much more efficient than limiting oneself to the papers we can include in our list: first, the number of papers available in various electronic archives is much greater than anything one can put on a paper, and it is growing every month at a rapid rate; second, the list that we can give will inevitably be strongly influenced by the author's taste, which is not what I regard as a desirable result.

This book begins with a short excursion into Quantum Mechanics and a brief discussion of the various methods for constructing the quantisation of the algebra of functions, usually on Euclidean space. All the methods that we described are covered in the books, which we refer to in the first three sections: [Ar89], [Ku18], [EMS04], [Fey05]. There are plenty of other books on Classical Mechanics (see for instance [G80], or [Mor07]) as well as on Quantum Mechanics, see for instance the classical books by Paul Dirac [Di30] or John von Neumann [vN32]; there are also more modern textbooks like [Sh11]. And of course I cannot fail to mention the Quantum Mechanics part from the multi-volume treatise by Landau and Lifschitz [LL77]. One might also pay attention to the books on Quantum field theory, which brings the ideas and methods of Quantum Mechanics into more modern and mathematically rigorous context: [BS83], [K93], [Ra01] (these books are also the source of constructions, related with the Kontsevich quantisation, see Section 19.1); see also the two volumes introduction into the Quantum Fields Theory by Manoukian [Man16]. For further reading we can recommend the book [Del99], which explains the Mathematical foundations of string theory, that has been serving as a source of inspiration for many mathematicians for the last thirty years; another important book is [BGM03], where the branes theory is explained as well as its relations with the deformations and Mirror Symmetry. Finally, some applications of the theory that we discuss in this text are described in the books [CGP11], [CMOPRl17].

In the book there are few important sections, which are not directly related with the main topic of our interest, i.e. with the deformation quantisation, but in which we explain various important mathematical theories, used further to solve the deformation quantisations problem and illustrate the results obtained from this solution. These are the sections on Lie groups and Lie algebras theory, on Symplectic and Poisson Geometry, and especially on Homological algebra. All these sections are provided with rather comprehensive lists of references for an interested reader and we can hardly add anything new here without wondering too far away from the purposes of this book, which is concerned with what one can call physically-motivated applications of these subjects. Just remark that the number of textbooks and monographs on such comparatively old and renown areas as the theory of Lie groups and algebras, or on Homological algebra is exceedingly large, second probably only to the Mathematical classics, like Calculus, Geometry or abstract Algebra. The interested reader can read the books [Se65], [He84] on Lie algebras and Lie groups and their applications; also, since these subjects are closely related to the Representation theory, and Algebraic Groups Theory, we would also recommend the books [Hu78] and [OV90]. As for the Homological algebra, one can find a more or less comprehensive introduction into this subject in any book on Algebraic Topology, including [Sp66] or [BT82]; there are also more specialised books, like [Wei94], [GM03] and [MCl01], dealing specifically with one or another aspect of Homological algebra and its applications; Hocschild homology and cohomology are discussed at length in the books [Lo98] and [Wi19]. As an additional reading I would probably recommend the classical ground-breaking article by Grothendieck [Gro57] and the fundamental treatise by Lurie [Lu09], which very much summarises the results of the modern developments in direction of Homological Algebra and Category Theory.

The principal purpose of this book is to discuss three constructions of the deformation quantisation and their properties. Each of these constructions have been studied extensively in the literature, and there exist many survey articles, dedicated to various aspects of this theory. Of these articles and surveys, we can recommend [Gu11], [Bo08], [CKTB06]; the reader should also pay attention to the references therein. Let us mention separately the construction of Gelfand and Fuks cohomology that we used in the proof of the De Wilde and Lecomte's quantisation; a good survey of this theory can be found in Fuks' book [Fu86]. This cohomology theory has many applications in Topology, Index Theory, etc., an interested reader can look it up for instance in the original papers by Gelfand, Gabrièlov and Losik [GGL75], where it was used to construct characteristic classes of triangulated manifolds, or in the paper by Alain Connes and Henri Moscovici [CM98], where Gelfand and Fuks' computations were applied to the transverse Index theory. Higher structures, and in particular operad theory play a crucial role in Kontsevich's theorem (and in Tamarkin's proof thereof); they also deserve a separate mention here. There are plenty of books that give a thorough introduction into these theories, some of which we have mentioned in the corresponding sections below, for instance [Sm01], [MSS02], [Lu09] and [CGP11]. One more important construction that was used in the text, and which has many other important application, is the *Formal Geometry* developed by Gelfand and his disciples, including Kontsevich, see [Kon93]. It is closely related to the Gelfand and Fuks cohomology of infinite-dimensional Lie algebras on one hand, and on the other, to the study of infinite jet spaces and their application to partial differential equations.

Let us conclude this section with a list of books and papers dealing with various aspects and applications of deformation quantisation and related topics. First of all, one can ask, whether the deformation quantisation exists in the case, when there is one or another restriction on the type of bidifferential operators that can be used in the formula; we discuss some of such restrictions in the book, in Section 15, when we talk about invariant deformation and deformation quantisation of holomorphic functions. These restrictions can be further extended: on one hand, one can consider an (affine or projective) algebraic variety X instead of a smooth or complex manifold. In this case, it would be natural to speak about deformations of the sheaf $\mathcal{O}_X$ of regular functions on X. This question has been studied by many authors in the last twenty years, see for example [Kon01], [BK04], [vdB06] and a survey by Yekutieli [Ye08]. On the other hand, one can consider a subalgebra of (bi)differential operators on the Poisson manifold, for instance by asking that all the components of these operators are generated by vector fields, tangent to a given integrable distribution inside the Poisson manifold (the natural condition here is that the Poisson bivector itself is from this subalgebra). This can be formalised in the terms of Lie algebroids over the manifold and this brings forward the problem of deformation quantisation of Poisson algebroids (i.e. of manifolds equipped with a Poisson bivector field, whose components are in the linear span of the anchor map of a Lie algebroid). The study of this problem was initiated by Nest and Tsygan, see [NT01] and continued by other authors, for instance see [Ca05], [CDH07] (observe that in the latter case the authors in fact work with the sheaves of sections on holomorphic Lie algebroids).

Another example of restrictions, imposed on the deformation series is concerned with a group action: if G acts on a Poisson manifold so that the Poisson bracket is preserved, one can ask, whether there exists a $\star$-product, that commutes with this action. This is usually called *equivariant* deformation quantisation; this subject in spite of extensive studies (see, e.g. the papers [Do05], [RW16] and [Sha17]) is still wide open in many important cases. In particular, if there is a Poisson-commutative family of functions (an integrable system in an informal language), one can regard it as a special case of the commutative Lie group action on the manifold. Then finding an equivariant $\star$-product with respect to this group action is a step towards finding a "quantisation" of this integrable system. The study of such "quantum integrable systems" is an interesting branch of the general discussion of deformation quantisation, see for instance [GvSt10], [ST17], [KS18].

Appendix of this book deals with the Operad theory, which ever since the *Renaissance of operads* in 1990s has been an important tool in the Mathematical theories, related with Physics, Topology and Geometry. We have cited some of books on operads and related research earlier; to this list we can add the works [LoVa12], [LaVo14], [Wi15], [ŠW], [DW15] and many others. Observe that much of research are related with Kontsevich's graph-complexes: one can say that the graph calculus appears in Kontsevich's quantisation formula (see Section 18) and operadic constructions are simple examples of this powerful tool, which plays central role in many computations and applications of the Operad theory.

As I explained earlier, the number of papers and books concerned with various aspects of deformation quantisation and related theories is exorbitant (I hope the reader will forgive me this strong adjective), and we could have continued this section for many more pages. Let me now just give a brief list of the topics in Mathematics and Mathematical Physics, which we couldn't cover here, that are actively studied with connection to deformation quantisation: Quantum Groups Theory, Cyclic (co)Homology, Noncommutative Geometry (both algebraic and topological), Index Theory, theory of Quantum Integrable Systems, Higher Spins Theory, various versions of Quantum Field Theory, Mirror Theory and many others. We hope that our book can help the reader to make his or her first steps in the direction of the topic of his or her choice.

Acknowledgements

It is my pleasure to conclude this introduction with the list of people who helped me in one or another way during my work on this book. First of all, this is professor Fan Huijun from PKU, who invited me to teach this course: if not for this, the book would never have been written; Ms Chen Pingping, the secretary of Mathematics department at PKU, who helped me during my stay in Beijing in 2019, and Miss Lin Meng, the secretary of Sino-Russian Math Center, who helped me during my visit in 2022, while I was still working on this text. I would also like to thank Mrs Dobrushina from IUM, who helped me a lot, while I was teaching the Russian version of this course; Prof. Zheglov from Lomonosov Moscow State University, who urged me to write down my notes in the form of a book. And last, but not the least, I am infinitely grateful to my wife Liza and daughters, Klava and Frosya, who bore the sight of grumpy dad, absorbed in typing something in his laptop all day long for two whole years.

Moscow
February 2023

Contents

Preface		v
Introduction		vii

1 General principles of classical and quantum mechanics — 1
- 1.1 Classical mechanics — 1
  - 1.1.1 Hamilton functions in Newton mechanics — 1
  - 1.1.2 Lagrangians and Hamiltonians — 3
- 1.2 Poisson bracket — 4
- 1.3 Quantum mechanics — 6
  - 1.3.1 Example: Quantum harmonic oscillator — 8

2 Weyl quantisation — 10
- 2.1 The general quantisation problem — 10
  - 2.1.1 Wintner's theorem — 11
  - 2.1.2 Uncertainty principle — 13
  - 2.1.3 Canonical quantisation — 14
- 2.2 Weyl quantisation formula — 16
  - 2.2.1 Further generalisations — 17

3 Moyal quantisation — 20
- 3.1 Asymptotic expansions — 20
- 3.2 Moyal deformation series — 22
  - 3.2.1 Moyal products — 24
  - 3.2.2 Properties of the Moyal product — 26
  - 3.2.3 Equivalence of Moyal $\star$-products — 27
  - 3.2.4 The complex case — 29
- 3.3 The integral formula — 30

Exercises 1: General quantisation principles — 32

4 Symplectic and Poisson structures — 35
- 4.1 The deformation quantisation problem in $\mathbb{R}^n$ — 35
- 4.2 Symplectic structures on manifolds — 38

		4.2.1 Definition and examples 39
		4.2.2 Poisson brackets on symplectic manifolds . . . 40
		4.2.3 Darboux' theorem 42
	4.3	Poisson structures on manifolds 44
		4.3.1 Polyvector fields 45
		4.3.2 Schouten bracket 47
		4.3.3 Poisson bivectors 48
		4.3.4 Weinstein's theorem 49
	4.4	Deformation quantisation problem 50

5 Deformations and Lie algebras: A nontrivial example 52
- 5.1 Lie algebras and Poisson structures 52
- 5.2 Universal enveloping algebras and PBW theorem . . . 54
- 5.3 PBW quantisation 58
  - 5.3.1 Proof of Proposition 5.5 60

6 Crash course in homological algebra, I. Examples 65
- 6.1 General nonsense 65
- 6.2 Examples 67
  - 6.2.1 De Rham cohomology 67
  - 6.2.2 Dolbeault cohomology 67
  - 6.2.3 Lichnerowicz–Poisson cohomology 69
  - 6.2.4 Lie algebra homology and cohomology 71
- 6.3 Algebraic structures on cohomology groups 74

Exercises 2: Poisson structures, quantisation and cohomology 76

7 Crash course in homological algebra, II. Hochschild homology 79
- 7.1 Definitions 79
- 7.2 Derived tensor product and derived Hom_R functors 82
- 7.3 Hochschild homology and cohomology as derived functors 83
- 7.4 Polynomial algebras 85
  - 7.4.1 Computation in low dimensions 85
  - 7.4.2 Small resolution 87

	7.4.3	The algebraic Hochschild–Kostant–Rosenberg theorem	88

8 Hochschild cohomology of C^∞-functions: Hochschild–Kostant–Rosenberg theorem — 92

- 8.1 Local and differentiable Hochschild complexes 92
- 8.2 Smooth Hochschild–Kostant–Rosenberg theorem . . . 95
  - 8.2.1 The main theorem 95
  - 8.2.2 Symbol calculus 98
- 8.3 Algebraic structures on Hochschild cohomology 101
  - 8.3.1 The $\cup$-product 101
  - 8.3.2 Gerstenhaber bracket 102

9 Obstructions and deformation theory: Examples — 109

- 9.1 Deformations of associative algebras 109
- 9.2 Obstructions 110
  - 9.2.1 $n = 1$ 110
  - 9.2.2 $n = 2$ 111
  - 9.2.3 $n > 2$ 112
- 9.3 Uniqueness 113
- 9.4 Résumé: Obstructions theory 115
- 9.5 Example 1: Poincaré–Birkhoff–Witt theorem revisited 116
  - 9.5.1 Graded Hochschild cohomology 117
  - 9.5.2 Obstructions calculus 118
  - 9.5.3 The Poincaré–Birkhoff–Witt theorem 120
- 9.6 Example 2: Symplectic manifolds with vanishing $H^3_{dR}(M)$ 122
  - 9.6.1 Gerstenhaber brackets and obstructions 122
  - 9.6.2 Generalised Poisson cohomology 123
  - 9.6.3 Deformation quantisation: A Neroslavski–Vlassov type result 124

Exercises 3: Hochschild homology and cohomology: Obstructions theory — 126

10 Deformation quantisation of cotangent bundles — 129

- 10.1 Graded Hochschild cohomology and obstructions . . . 129

    10.2 Cotangent bundles of parallelisable manifolds 132
        10.2.1 The general situation 133
        10.2.2 Lie groups 136

11 Lie algebra cohomology: Vey class   140
    11.1 Deformation of Poisson structures 140
    11.2 Lie algebras related with a manifold and their
        cohomology 142
        11.2.1 Algebras of formal vector fields on $\mathbb{R}^{2n}$ 143
        11.2.2 Cohomology of formal vector fields 144
        11.2.3 Cohomology of formal symplectic space 146
        11.2.4 Cohomology of $\mathfrak{P}(M)$ 148
    11.3 Symplectic connections and cohomology 149
        11.3.1 Existence of symplectic connections 149
        11.3.2 Vey's class S_Γ^3 152
        11.3.3 Alternative description of S_Γ^3 154

12 Lecomte and de Wilde's theorem: Quantisation of symplectic manifolds   156
    12.1 Cohomology of Poisson Lie algebra $\mathfrak{P}(M)$ in low
        dimensions 158
        12.1.1 Differential forms and differentiable
            cohomology $H^*(\mathfrak{P}(M), C^\infty(M))$ 158
        12.1.2 Chern–Weil forms and cohomology of vector
            fields on manifolds 159
        12.1.3 De Wilde, Gutt and Lecomte's theorem 162
    12.2 Deformations of the Poisson bracket 163
        12.2.1 Exact symplectic structures and
            Lie derivatives 163
        12.2.2 Cohomological properties of
            Lie derivatives 164
        12.2.3 Deformation of the Poisson algebras for exact
            symplectic forms 165
        12.2.4 Deformation of Poisson structures on general
            symplectic manifolds 167
    12.3 Deformations of the Poisson structure and
        $\star$-products: Proof of Theorem 12.1 170

Exercises 4: Obstructions and deformation quantisation of symplectic manifolds — 174

13 Fedosov quantisation: Abelian connections — 178
  13.1 The main idea: An overview 178
  13.2 The Weyl algebras bundle 180
    13.2.1 Fibrewise Weyl algebras 180
    13.2.2 Algebras $W_\hbar(M)$, $\widehat{W}_\hbar(M)$ and $\widehat{W}_\hbar^*(M)$ 182
    13.2.3 Symmetrisation 184
    13.2.4 Alternative constructions 185
  13.3 Fedosov connections 186
    13.3.1 Symplectic connections and the bundle $\widehat{W}_\hbar^{\mathbb{C}}(M,\omega)$ 186
    13.3.2 Fedosov's theorem 188
    13.3.3 Proof of Fedosov's theorem: Change of variables 190
    13.3.4 Proof of Fedosov's theorem: Operators δ and δ^* 191
    13.3.5 Proof of Fedosov's theorem: Iterative process 193

14 Fedosov quantisation and its properties — 196
  14.1 Fedosov's construction: The isomorphism 196
  14.2 Properties of Fedosov's deformation quantisation . . . 200
    14.2.1 Fedosov product is a $\star$-product 200
    14.2.2 Classification of Fedosov products 202

15 Properties, generalisations and applications of Fedosov quantisation (a survey) — 208
  15.1 Trace and algebraic index theorem 208
    15.1.1 Outline of the index theory 208
    15.1.2 Quantum symbol calculus and the algebraic index 212
    15.1.3 The trace on $\widehat{W}_D^{comp}(M)$ 214
    15.1.4 Fedosov's algebraic index formula 215
  15.2 Group action, quantisation and quantum momentum maps 217

15.2.1 Invariant connections and equivariant
⋆-products 218
15.2.2 Quantum momentum maps 220
15.2.3 Classification of quantum momentum maps
and equivariant cohomology 222
15.3 Deformation of holomorphic symplectic
structures 225
15.3.1 Few words about complex geometry: Dolbeault
cohomology, Atiyah class 226
15.3.2 Nest and Tsygan's construction 230

Exercises 5: Fedosov quantisation   **235**

**16 Higher homotopy algebras: Topological background
and definitions**   **238**
16.1 Topological preliminaries: Loop spaces and loop
products 238
16.2 H-spaces and Stasheff's theorem 242
16.3 From topology to algebra: Strong homotopy
algebras 247
16.3.1 A_∞-algebras 248
16.3.2 Cofree coalgebras cogenerated by a graded
space 250
16.3.3 L_∞-algebras 253

**17 Maurer–Cartan equations and Kontsevich's
theorem**   **255**
17.1 Maurer–Cartan equations in differential
Lie algebras 255
17.2 L_∞-morphisms and Maurer–Cartan elements 257
17.2.1 L_∞-morphisms: definitions 257
17.2.2 Properties of L_∞- and A_∞-morphisms 259
17.2.3 L_∞-morphisms and Maurer–Cartan
elements 263
17.2.4 Equivalence of Maurer–Cartan elements 264
17.3 Formality of differential Lie algebras and
Kontsevich's theorem 266

18 Kontsevich's construction — 270
- 18.1 Configuration spaces and their compactifications . . . 271
  - 18.1.1 Definitions 271
  - 18.1.2 Examples 273
  - 18.1.3 The stratification of $\partial \bar{C}_{n,m}$, $\partial \bar{C}_n$, "magnifying glass" 276
- 18.2 Admissible graphs and integrals 279
  - 18.2.1 The graphs $G_{n,m}$ 279
  - 18.2.2 Differential forms and weights 281
- 18.3 The formula 282
- 18.4 Proof of Theorem 18.2 (sketch) 284
  - 18.4.1 The term $\mathcal{U}_1$ is equal to the Hochschild–Kostant–Rosenberg map 284
  - 18.4.2 The L_∞-morphism equation 286

Exercises 6: Higher homotopy algebras, Kontsevich's theorem — 288

19 Kontsevich's quantisation: Modifications and related questions — 293
- 19.1 "Physical" approach to the $\star$-product 294
  - 19.1.1 The path integral formalism 294
  - 19.1.2 Poisson σ-model and the $\star$-product 296
  - 19.1.3 Associativity of the $\star$-product 297
- 19.2 Globalisation of the $\star$-product 299
  - 19.2.1 Few words about jets and stuff 300
  - 19.2.2 Kontsevich's globalisation method (via the formal geometry) 304
  - 19.2.3 Fedosov-type constructions and the globalisation 306

20 Applications of Kontsevich's quantisation: Duflo's isomorphism — 311
- 20.1 The tangent map 311
- 20.2 Centre of a universal enveloping algebra 315
- 20.3 Duflo's isomorphism: The original construction 317
- 20.4 Kontsevich's quantisation of $S(\mathfrak{g})$ 318
- 20.5 Properties of the map $\mathcal{U}_1^\pi$ 320

20.6 Comparison of Kontsevich's and Duflo's
  isomorphisms 324
  20.6.1 The graphs and the "wheels" 324
  20.6.2 The exponent 326

Exercises 7: Kontsevich's quantisation: Properties and applications   **329**

A Operads: History and definitions   **332**
  A.1 Topological preliminaries (a bit of history) 332
  A.2 Algebraic operads 338
    A.2.1 The general operads theory: Definitions and
      examples 339
    A.2.2 Homological algebra of operads: Co-operads
      and resolutions 346
    A.2.3 Quadratic operads 349
    A.2.4 Quadratic duals and Koszul property 352

B Tamarkin's proof of formality theorem   **357**
  B.1 Intrinsic formality: Definition and criterion 357
    B.1.1 Proof of the criterion for intrinsic
      formality 358
  B.2 Gerstenhaber algebras and the corresponding
    operad 362
  B.3 Proof of the formality theorem 364
    B.3.1 Koszulity of the operad $\mathcal{G}$ 365
    B.3.2 Intrinsic formality of polynomial polyvectors
      in $\mathbb{R}^d$ 366
    B.3.3 The $\mathcal{B}_\infty$-operad and its action on Hochschild
      complex 368
    B.3.4 The operad $\widetilde{\mathcal{B}}$ 371
    B.3.5 The theorem of Kazhdan and Etingof 373
    B.3.6 The isomorphism $\widetilde{\mathcal{B}} \cong \mathcal{B}_\infty$ 376

List of References   **378**
  Textbooks, surveys and monographs 378
  Original papers 386

Index   **397**

1 General principles of classical and quantum mechanics

This and the next sections are introductory and deal with the standard facts from classical and quantum mechanics. Both these subjects are traditional pieces of the general knowledge in Mathematics and Physics nowadays and are thoroughly treated in many nice textbooks, some of which we mention below. The reader, familiar with these ideas can skip the first two sections altogether and jump straight to the discussion of Moyal quantisation in Section 3.

1.1 Classical mechanics

The quantisation program, which includes the Deformation Quantisation, began when the physicists started to study the properties of elementary particles and discovered that the laws of classical mechanics fail at the quantum level. The changes that were to be implemented to the principles of Classical Mechanics constitute the basis of Quantum Theory. Thus it is not surprising that Quantum Mechanics resemble the classical theory in many aspects. Moreover proper understanding of many ideas in quantum theory derives from analogies with classical systems. So we think it is right to begin with the brief outline of the principles of Newtonian mechanics. There are a lot of books dedicated to this subject; for instance, one can find a good mathematical treatment in [Ar89].

1.1.1 Hamilton functions in Newton mechanics

Consider a point particle on a straight line with coordinate q, moving with time under a force $\vec{F} = \vec{F}(q)$. Its trajectory is determined by the second Newton's law:

$$m\vec{a} = \vec{F}, \tag{1.1}$$

where $\vec{a} = \ddot{q}$ is the acceleration and m is the mass of the particle. Introducing the momentum vector $p = m\vec{v} = m\dot{q}$ where $\vec{v} = \dot{q} = \frac{d}{dt}q$ is velocity, we can rewrite it as $\dot{p} = F$. In what follows we shall usually drop the vector signs.

Let us transform the equation a little: if we multiply both sides of Equation (1.1) by $\dot{q} = \frac{p}{m}$, we get:

$$\frac{p\dot{p}}{m} = F(q)\dot{q},$$

which readily integrates to

$$\frac{d}{dt}\left(\frac{p^2}{2m} - U(q)\right) = 0, \tag{1.2}$$

where $U(q)$ is the function, determined by equality $U'(q) = \frac{\partial}{\partial q}(U(q)) = F(q)$.

The function $H = H(p,q) = \frac{p^2}{2m} - U(q)$ is called *the Hamilton function* or *the Hamiltonian* of the dynamical system. Its physical meaning is that it depicts the total energy of the point (i.e. both the kinematic energy $\frac{p^2}{2m}$ and the potential energy $-U(q)$), so that the equality (1.2) is just the energy conservation law.

Knowing $H(p,q)$ it is easy to restore the original dynamics: since $\dot{p} = F$ and $\dot{q} = \frac{p}{m}$, we have

$$\begin{cases} \dot{p} = -\frac{\partial H}{\partial q} \\ \dot{q} = \frac{\partial H}{\partial p}. \end{cases} \tag{1.3}$$

These equations are usually called *the Hamilton equations*, and the study of mechanical systems that can be reduced to the study of the differential equations of the form (1.3), is called *Hamiltonian mechanics*.

Of course everything we discuss here can be done almost identically in the n-dimensional situation, i.e. for any point moving in an n-dimensional Euclidean space in accordance with the Newton's law (1.1), it only takes taking care of the coordinate indices in proper places. In this case, one considers the n-dimensional momentum space with coordinates $p_1, \ldots, p_n$ and eventually ends up with the system

$$\begin{cases} \dot{p}_i = -\frac{\partial H}{\partial q^i} \\ \dot{q}^i = \frac{\partial H}{\partial p_i} \end{cases} \tag{1.4}$$

for all $i = 1, \ldots, n$. The only condition, we need is that the vector-valued function force $\vec{F}$ allows *potential*, i.e. there exists V such that

$\vec{F} = \operatorname{grad} V$ (all is done with respect to the Euclidean structure). In fact, this construction allows much further generalisations, which we are going to discuss shortly.

1.1.2 Lagrangians and Hamiltonians

It turns out that Hamilton equations appear in many important cases, and not just in the framework of classical Newton mechanics. One of the more common situations in Physics, which also allows formulation via a Hamilton function, is the following: suppose our system is governed by the *minimal action principle*, i.e. the trajectory of a system in coordinates q when time varies from t_1 to t_2 is such that the *action S*

$$S = \int_{t_1}^{t_2} \mathcal{L}(q, \dot{q}, t) dt$$

takes minimal value; the function $\mathcal{L}(q, \dot{q}, t)$ is called the *Lagrangian function* or just *Lagrangian* of the system. The space, spanned by coordinates q^i (index i can be replaced by a continuous parameter in some cases) as well as the Lagrangian function can be of arbitrary nature so that the integration is possible. This formulation of physical problems is quite general: study of many physical models begins with considering one or another variational principle although in many situations it is necessary to deal with infinite-dimensional space. We here shall assume that $q = (q^1, \ldots, q^n)$ is just a point in a Euclidean space.

As one knows, the solution of the variational problem is given by *Euler–Lagrange equations*:

$$\frac{d}{dt} \frac{\partial \mathcal{L}}{\partial \dot{q}^i} - \frac{\partial \mathcal{L}}{\partial q^i} = 0. \qquad (1.5)$$

Let us now introduce a supplementary set of coordinates $p = (p_1, \ldots, p_n)$ by the formula $p_i = \frac{\partial \mathcal{L}}{\partial \dot{q}^i}$ (in other words, we use these equalities to replace $\dot{q}^i$ by suitable functions of other variables, including $p^1, \ldots, p^n$) and consider the function $H = H(p, q, t)$:

$$H(p, q, t) = \sum_i p_i \dot{q}^i - \mathcal{L}(q, \dot{q}, t).$$

Then $H(p, q, t)$ is called *the Legendre transform* of $\mathcal{L}$ and the following is true.

Proposition 1.1. *Equations* (1.5) *are equivalent to the Hamilton system* (1.4).

Proof. To avoid dealing with redundant indices we shall work with $n = 1$; the general situation is quite similar. Let us consider the complete differential of H (regarding $\dot{q}$ as parameter), using the standard properties thereof

$$dH = \frac{\partial H}{\partial p} dp + \frac{\partial H}{\partial q} dq + \frac{\partial H}{\partial t} dt$$

$$= \dot{q} dp - \frac{\partial \mathcal{L}}{\partial q} dq - \frac{\partial \mathcal{L}}{\partial t} dt.$$

Since this form is invariant, we conclude that $\dot{q} = \frac{\partial H}{\partial p}$ and

$$\frac{\partial H}{\partial q} = -\frac{\partial \mathcal{L}}{\partial q} = [\text{iff Euler–Lagrange Equation (1.5) holds}]$$

$$= -\frac{d}{dt}\frac{\partial \mathcal{L}}{\partial \dot{q}} = -\dot{p}.$$ $\square$

1.2 Poisson bracket

The Hamilton equations give rise to an important algebraic structure on the observable functions (i.e. on smooth functions of physical parameters q, p and time). We mean the *Poisson bracket*. Namely, consider the formula

$$\{f, g\} = \sum_{i=1}^{n} \left\{ \frac{\partial f}{\partial p_i} \frac{\partial g}{\partial q^i} - \frac{\partial f}{\partial q^i} \frac{\partial g}{\partial p_i} \right\}. \tag{1.6}$$

The relation of this construction with the Hamilton equations follows from the next observation:

$$\{H, p_k\} = \sum_{i=1}^{n} \left\{ \frac{\partial H}{\partial p^i} \frac{\partial p_k}{\partial q^i} - \frac{\partial H}{\partial q^i} \frac{\partial p_k}{\partial p_i} \right\} = -\sum_{i=1}^{n} \frac{\partial H}{\partial q^i} \delta_i^k = -\frac{\partial H}{\partial q^k},$$

$$\{H, q^k\} = \sum_{i=1}^{n} \left\{ \frac{\partial H}{\partial p^i} \frac{\partial q^k}{\partial q^i} - \frac{\partial H}{\partial q^i} \frac{\partial q^k}{\partial p_i} \right\} = \sum_{i=1}^{n} \frac{\partial H}{\partial p_i} \delta_i^k = \frac{\partial H}{\partial p_k}.$$

In other words,
$$\begin{cases} \dot{p}_i = \{H, p_i\} \\ \dot{q}^i = \{H, q^i\}. \end{cases} \quad (1.7)$$

More generally, it is easy to see that for any observable function $f = f(p,q)$ its derivation with respect to time along the trajectory is given by
$$\frac{df(p(t), q(t))}{dt} = \{H, f\}. \quad (1.8)$$

Let us summarise the properties of the bracket $\{,\}$:

Proposition 1.2. *The operation $\{,\}$, introduced by the formula (1.6) satisfies the following conditions:*

(i) *Skew symmetry:* $\{f, g\} = -\{g, f\}$;

(ii) *Linearity over scalars:* $\{af + bg, h\} = a\{f, h\} + b\{g, h\}$ *for all scalars a, b;*

(iii) *Leibniz rule:*
$$\{fg, h\} = f\{g, h\} + g\{f, h\};$$

(iv) *Jacobi identity:*
$$\{\{f, g\}, h\} + \{\{g, h\}, f\} + \{\{h, f\}, g\} = 0, \quad (1.9)$$

for any smooth functions f, g and h of p, q.

All these properties are evident, except for the last one, which can be either proved by direct computations, or by the following observation: denote by $J(f, g, h)$ the expression on the left of the Jacobi identity; then it follows from the properties (i)–(iii) that it is totally antisymmetric in f, g and h and satisfies the equality
$$J(f_1 f_2, g, h) = f_1 J(f_2, g, h) + f_2 J(f_1, g, h),$$

so it is enough to prove that $J \equiv 0$ for f, g, h coordinate functions; this can be done now by direct inspection.

A good way to understand Jacobi identity is the following. Fixing H, one can regard the map $f \mapsto \{H, f\}$ as a differentiation of the algebra of observable functions, i.e. as a vector field X_H, given by

$$X_H = \sum_i \left\{ \frac{\partial H}{\partial p_i} \frac{\partial}{\partial q^i} - \frac{\partial H}{\partial q^i} \frac{\partial}{\partial p_i} \right\}.$$

Then (1.9) is equivalent to the equation:

$$[X_F, X_G] = X_{\{F,G\}}, \tag{1.10}$$

where on the left we have the usual commutator of vector fields.

We conclude this section by observing, that the bracket $\{,\}$ that we have just introduced, is uniquely determined (at least on rational functions of p and q) by the properties (i)–(iv) (see Proposition 1.2) and the equation

$$\{p_i, q^j\} = \delta_i^j. \tag{1.11}$$

1.3 Quantum mechanics

It turns out that on quantum level the theory that we have just described cannot be implemented directly and needs to be reinterpreted in an "allegorical way". The traditional explanation of this phenomenon is based for instance on the famous experiment when electrons are dispersed over a wall with one or two narrow gaps, so that the dispersion picture demonstrates the wave nature of the particle. This is very well explained in the celebrated Feynmann's Lecture notes [Fey05].

The conclusions that are derived from these experiments constitute the basic principles of Quantum Mechanics and more generally of the Quantum Physics, that include Quantum Field theory, Quantum Electrodynamics and many other branches. There are plenty of wonderful books that explain these principles in details, for instance the cited above Feynmann's lectures [Fey05] or the more modern exposition in [Ku18]; here we will only formulate the most basic consequences and conclusions derived from these experiments, that constitute the basis of the Quantum Mechanics and many other branches of modern Physics.

The main principle of Quantum Physics is that the behaviour of every quantum particle should be described by a complex-valued *wave function* ψ on the physical space Ω, so that the *probability that the particle is inside an interval* $I \subset \Omega$ *is equal to the integral of* $|\psi|^2$ *over this interval* (a measure μ on Ω is assumed to be given). The space of these functions is endowed with *Hilbert space* structure via the integral:

$$\langle \phi \mid \psi \rangle = \int_\Omega \bar{\phi}\psi d\mu.$$

In particular, since the probability of a particle being in Ω is equal to 1, we shall have

$$\|\psi\|^2 = \langle \psi \mid \psi \rangle = \int_\Omega |\psi|^2 d\mu = 1.$$

The functions ψ with this property are called *quantum states* of the system.

In this context all the observable functions shall be reinterpreted as operators on wave functions (speaking in physical terms, our measurements influence the state of a quantum particle) and the *expectation value* of this function on the quantum state is:

$$\langle F \rangle_\psi = \langle \psi \mid F \mid \psi \rangle = \int_\Omega \bar{\psi} F(\psi) d\mu.$$

In particular, if ψ is an eigenfunction of F with eigenvalue λ, then $\langle \psi \mid F \mid \psi \rangle = \lambda$. The Hamilton function H then corresponds to the energy operator $\hat{H}$, and its eigenvalues are treated as possible *energy levels* of the system; the corresponding eigenfunctions are called *eigenstates* of the system.

We shall also assume that constant functions are represented by scalar operators and that the *in all formulas, involving Poisson brackets, these brackets should be replaced with commutators of the corresponding operators*. In particular, if P_i, Q^j are the operators, corresponding to the momentum and coordinate functions, then the formula (1.11) gives

$$[P_i, Q^j] = \sqrt{-1}\hbar \delta_i^j. \tag{1.12}$$

The factor $\sqrt{-1}\hbar$ before δ_i^j on the right appears for physical reasons. The *Planck constant* $\hbar$ plays the role of deformation parameter, and should appear in front of all commutators that we substitute for the Poisson brackets as it will appear later, but here and in the next lecture we shall omit it from our formulas.

Let us now consider the way our quantum particles evolve with time. According to the general principle, Equation (1.8) turns into

$$\frac{dF}{dt} = [H, F]. \tag{1.13}$$

An alternative form of this equation can be obtained as follows: consider the wave function $F(\psi)$ then

$$\frac{d}{dt}(F(\psi)) = \frac{dF}{dt}(\psi) + F\left(\frac{d\psi}{dt}\right) = [H, F](\psi) + F\left(\frac{d}{dt}(\psi)\right),$$

or

$$[H, F] = \left[\frac{d}{dt}, F\right].$$

In other words, for any operator F, its commutator with H is equal to its commutator with $\frac{d}{dt}$. Hence (restoring the Planck constant and imaginary unit at their lawful place) we conclude that the following equation on wave function is equivalent to (1.13):

$$\sqrt{-1}\hbar \frac{d}{dt}(\psi) = H(\psi). \tag{1.14}$$

Equation (1.14) is called *Schrödinger equation*, while (1.13) is called *Schrödinger equation in Heisenberg's form*.

1.3.1 Example: Quantum harmonic oscillator

The simplest example of quantum mechanical system is the following: consider the 1-dimensional case. Then there is a canonical way to define the operators P and Q on the space of complex-valued functions on the straight line with coordinate x:

$$P(\psi) = \sqrt{-1}\hbar \frac{d\psi}{dx}, \quad Q(\psi) = x\psi.$$

Then, clearly: $[P, Q] = \sqrt{-1}\hbar$, so the condition (1.12) holds.

Now the Hamilton function of the classical *harmonic oscillator* is

$$H(p,q) = \frac{p^2}{2m} + \frac{1}{2}m\omega^2 q^2,$$

where p and q are the momentum and the coordinate of the oscillating particle, ω is the oscillation frequency and m is the mass. Then in full accordance with our principles it is natural to take the following operator as the quantum Hamilton function of oscillator:

$$\widehat{H} = \frac{P^2}{2m} + \frac{1}{2}m\omega^2 Q^2.$$

Thus the Schrödinger equation in this case will look as follows (up to the change of parameters):

$$\frac{\partial \psi}{\partial t} = \frac{\hbar\sqrt{-1}}{2m}\frac{\partial^2 \psi}{\partial x^2} + \frac{1}{2}m\omega^2 x^2 \psi.$$

If we introduce the annihilation and creation operators $\hat{a}^-$, $\hat{a}^+$

$$\hat{a}^+ = \sqrt{\frac{m\omega}{2\hbar}}Q - \frac{1}{\sqrt{2m\omega\hbar}}P, \ \hat{a}^- = \sqrt{\frac{m\omega}{2\hbar}}Q + \frac{1}{\sqrt{2m\omega\hbar}}P,$$

then it is possible to express the eigenstates of the quantum oscillator as the results of iterated action of $\hat{a}^+$ on the *ground state* ψ_0 of the system:

$$\psi_n = (\hat{a}^+)^n \psi_0, \ \widehat{H}(\psi_n) = E_n \psi_n,$$

where ψ_0 is the solution of the equation

$$\hat{a}^- \psi_0 = 0.$$

We leave it as an exercise to the reader to find the precise expressions for F_n (and for ψ_n).

2 Weyl quantisation

The passage from Classical to Quantum Mechanics gives rise to the notion of *quantisation*; in this section we describe the major difficulties which arise in the process of implementing the principles of Quantum Mechanics and constructions used to overcome them; in particular we describe the Weyl quantisation procedure.

2.1 The general quantisation problem

As it is explained in the first section (also see the books [Fey05], [Ku18]), in Quantum Mechanics one deals with the following correspondence, informally called **quantisation**: *every observable function f on the phase space of the system (here we shall assume that it is (an open neighbourhood in) $\mathbb{R}^{2n}$ with coordinates $(p_1, \ldots, p_n, q^1, \ldots, q^n)$) is replaced with an operator F on a Hilbert space $\mathcal{H}$ of $\mathbb{C}$-valued **wave functions** ψ so that constant functions correspond to scalar operators and every Poisson bracket is replaced with the commutator of the corresponding operators (times a scalar multiple)*. Recall that the commutator of two elements in an associative algebra is

$$[a, b] = a \cdot b - b \cdot a$$

where $\cdot$ is the product. The operator F corresponding to an observable function f will be called *the quantisation of f*. In this situation the dynamics of a particle corresponding to the wave function ψ is determined by the Schrödinger Equation (1.14), and the mean value of f on the particle, represented by the wave function ψ is given by the expectation value $\langle F \rangle_\psi = \int \bar{\psi} F(\psi) d\mu$. In what follows we shall always assume integrability of all expression that we meet: although usually there's a lot of analytic work around the quantisation procedure, we shall stick with algebraic side of the story for almost all the time.

Of course this principle is too hard to be implemented in full generality, so from the very beginning one makes few assumptions to relax these conditions. First of all, *we shall assume that the only*

commutation equality which must hold exactly is (1.12):
$$[P_i, Q^j] = \delta_i^j.{}^4 \tag{2.1}$$

This relation is called the *canonical commutation relation*. In classical situation if all the observable functions f are polynomials in p_i-s and q^j-s, it is enough to know the Poisson bracket of p_i and q^j to determine the bracket for all f and g. The same goes for the operators and the commutator, since the Leibniz rule holds for the commutators in an algebra as well, although in a noncommutative form:
$$[a, bc] = [a, b]c + b[a, c]. \tag{2.2}$$

Thus we come up with the following question, which we shall call **the classical quantisation problem**: *suppose, we are given the operators $P_1, \ldots, P_n, Q^1, \ldots, Q^n$, satisfying Equation (2.1), then for every function f of $(p_1, \ldots, p_n, q^1, \ldots, q^n)$ choose an expression in terms of these operators, that will look "similar" to f.*

Observe, that even if $f(p, q)$ is a polynomial, the quantisation problem is totally nontrivial; for instance for $n = 1$, the monomial $p^2 q$ can be represented by three different expressions in terms of P and Q:
$$p^2 q \rightsquigarrow P^2 Q, \text{ or } PQP, \text{ or } QP^2.$$

All these operators are different, since
$$PQP = P^2 Q - P[P, Q] = P^2 Q - P,$$
$$QP^2 = PQP - [P, Q]P = PQP - P = P^2 Q - 2P.$$

Which of the operators on the right should get preference, or should we consider some linear combination of all three of them, is one of the main questions of the quantisation theory. One can say, that *quantisation is a consistent rule that will allow one to choose the expression in terms of P_i and Q^j for any observable function.*

2.1.1 Wintner's theorem

Before we proceed with the quantisation problem, let us discuss few other ideas, related with quantisation, which make the universal solution of this problem quite difficult.

---
[4] Note that we threw the factor $\hbar\sqrt{-1}$ away from the formula for a while.

The standard way to introduce operators $P_1, \ldots, P_n, Q^1, \ldots, Q^n$ that would satisfy Equation (2.1) is to set:

$$P_i(\psi) = \frac{\partial \psi}{\partial x^i}, \ Q^j(\psi) = x^j \psi.$$

Here we suppose that the Hilbert space $\mathcal{H}$ consists of $\mathbb{C}$-valued functions on $\mathbb{R}^n$. As one sees, this definition is not perfect, since the partial derivative operators are not everywhere defined in the Hilbert space of square integrable functions, so we need to restrict them to certain subspaces, which makes the whole construction quite involved and brings forward many analytic subtleties. One of them is that however hard we try, we cannot make these operators bounded. It turns out, that this is not just the property of smooth functions and differentiations, which is the root of this problem. Namely

Proposition 2.1 (Wintner's theorem). *There cannot exist two bounded self-adjoint operators P and Q on a Hilbert space $\mathcal{H}$, which would satisfy the equality*

$$PQ - QP = 1.$$

Proof. Suppose $\|P\| = x$, $\|Q\| = y$; we compute:

$$\|[P^n, Q]\| = \|P^n Q - Q P^n\| \leq \|P^n Q\| + \|Q P^n\| \leq x^n y + y x^n \leq 2 x^n y.$$

On the other hand, by induction we see:

$$[P^n, Q] = n P^{n-1}.$$

Indeed, this formula holds for $n = 1$ and if $[P^{n-1}, Q] = (n-1) P^{n-2}$, then by (2.2)

$$[P^n, Q] = [P \cdot P^{n-1}, Q] = P[P^{n-1}, Q] + [P, Q] P^{n-1}$$
$$= (n-1) P^{n-1} + P^{n-1} = n P^{n-1}.$$

So, we have

$$n x^{n-1} \leq \|n P^{n-1}\| = \|[P^n, Q]\| \leq 2 x^n y,$$

hence $n \leq 2xy$ for all $n \in \mathbb{N}$, a contradiction. □

This result means that one needs to work with unbounded operators, which adds a lot to the difficulty of the analytic side of quantisation.

2.1.2 Uncertainty principle

Another well-known statement that follows from the quantisation principles and has important physical meaning is the *uncertainty principle*. In simple terms, it says that *it is impossible to know with absolute precision the coordinate and the momentum of a quantum particle*.

To show, how this statement follows from the quantisation principles recall that the value of any observable function on quantum particle is given by the expectation value of the corresponding operator on the wave function of this particle, so the coordinate and momentum of a quantum particle are

$$\langle Q \rangle_\psi = \int \bar{\psi} Q(\psi) d\mu, \quad \langle P \rangle_\psi = \int \bar{\psi} P(\psi) d\mu.$$

Regarding the expectation values as scalar operators, let $\Delta_\psi Q = Q - \langle Q \rangle_\psi$, $\Delta_\psi P = P - \langle P \rangle_\psi$ be the fluctuations of coordinate and momentum observables, so that $\langle |\Delta_\psi Q|^2 \rangle_\psi \geq 0$, $\langle |\Delta_\psi P|^2 \rangle_\psi \geq 0$ are the squares of mean quadratic deviations of coordinate and momentum of the particle; here for an operator A on a Hilbert space we put $|A|^2 = A^\dagger A$, where $A^\dagger$ is the adjoint operator. In particular, for self-adjoint operators $|A|^2 = A^2$. Then

$$[\Delta_\psi P, \Delta_\psi Q] = [P, Q] = \hbar \sqrt{-1},$$

since scalar operators commute with everything. We can assume that the operators P and Q are self-adjoint (the standard choice of quantisation with $P = \sqrt{-1}\hbar \frac{\partial}{\partial x}$ shows that this is the case), and so are $\Delta_\psi P$ and $\Delta_\psi Q$. Consider the linear combination $X(t) = \Delta_\psi P + t\sqrt{-1}\Delta_\psi Q$, then we compute

$$0 \leq \langle |X(t)| \rangle_\psi = \int \bar{\psi} X^\dagger(t) X(t)(\psi) d\mu$$
$$= \int \bar{\psi}(\Delta_\psi P)^2(\psi) d\mu + t^2 \int \bar{\psi}(\Delta_\psi Q)^2(\psi) d\mu$$
$$+ t\sqrt{-1} \int \bar{\psi}(\Delta_\psi P \Delta_\psi Q - \Delta_\psi Q \Delta_\psi P)(\psi) d\mu$$

$$\begin{aligned}
&= \langle |\Delta_\psi P|^2 \rangle_\psi + t^2 \langle |\Delta_\psi Q|^2 \rangle_\psi \\
&\quad + t\sqrt{-1} \int \bar\psi [\Delta_\psi P, \Delta_\psi Q](\psi) d\mu \\
&= \langle |\Delta_\psi P|^2 \rangle_\psi - t\langle \hbar \rangle_\psi + t^2 \langle |\Delta_\psi Q|^2 \rangle_\psi \\
&= \langle |\Delta_\psi P|^2 \rangle_\psi - t\hbar + t^2 \langle |\Delta_\psi Q|^2 \rangle_\psi.
\end{aligned}$$

Now since the quadratic expression on the right takes only nonnegative values its discriminant is nonpositive:

$$D = \hbar^2 - 4 \langle |\Delta_\psi P|^2 \rangle_\psi \langle |\Delta_\psi Q|^2 \rangle_\psi \le 0.$$

Hence the conclusion: $\langle |\Delta_\psi P|^2 \rangle_\psi \langle |\Delta_\psi Q|^2 \rangle_\psi \ge \frac{\hbar^2}{4}$; if we take into consideration, that the values on the left are squares of the mean deviations, we obtain the well-known inequality:

$$\Delta p \Delta q \ge \frac{\hbar}{2}. \qquad (2.3)$$

As a matter of fact, the proof we just presented applies in a word for word manner to any pair of non commuting self-adjoint operators.

2.1.3 Canonical quantisation

Let us consider the quantisation problem again: find a consistent rule to associate with each observable function f of (p,q) an operator F, "made up from" the operators (P,Q), satisfying the canonical commutation relation (2.1). In some cases it is not difficult to choose the operator without ambiguity. For instance this is the case, when the function f does not depend on one of the sets of variables, either p, or q, or is a linear combination of such functions. In some important situations this is true, and in that case one can find pretty straightforward quantisation of the mechanical problem. We shall put off such examples to the exercises.

A more general approach to answer this question comes from Harmonic Analysis and the theory of pseudo-differential operators, see e.g. the book [Sh01]. So we assume that the physical phase space of our system is $\mathbb{R}^{2n}$ with coordinates $(\bar p, \bar q)$. Recall that the space of *Schwarz functions* on a Euclidean space $\mathbb{R}^n$ (with coordinates x) is

$$\mathcal{S}(\mathbb{R}^n) = \{ \varphi \in C^\infty(\mathbb{R}^n) \mid \lim_{x \to \infty} \left| \frac{\partial^{|I|} \varphi}{\partial x^I} p(x) \right| = 0, \forall I, \forall p(x) \},$$

where $I = (i_1, \ldots, i_n)$, $i_k \in \mathbb{Z}$, $i_k \geq 0, k = 1, \ldots, n$ is a multi-index, $|I| = i_1 + \cdots + i_n$,

$$\frac{\partial^{|I|}}{\partial x^I} = \frac{\partial^{i_1}}{\partial x_1^{i_1}} \cdots \cdots \frac{\partial^{i_n}}{\partial x_n^{i_n}}$$

and $p(x)$ is a polynomial in coordinates x.

The purpose of introducing the Schwarz space is to consider the *Fourier transform* $\mathcal{F}$, which is a linear automorphism on $\mathcal{S}(\mathbb{R}^n)$:

$$\mathcal{F}(\varphi)(x) = \frac{1}{\sqrt{2^n \pi^n}} \int_{\mathbb{R}^n} \varphi(\xi) e^{-ix\cdot\xi} d\xi, \tag{2.4}$$

we use the standard notation $\sqrt{-1} = i$ and $x \cdot \xi$ is the scalar product in $\mathbb{R}^n$; its inverse is given by

$$\mathcal{F}^{-1}(\psi)(x) = \frac{1}{\sqrt{2^n \pi^n}} \int_{\mathbb{R}^n} \psi(\xi) e^{ix\cdot\xi} d\xi. \tag{2.5}$$

It is known, that $\mathcal{F}$ intertwines the partial derivatives with the operations of multiplication by coordinates. This brings forward the following construction, called *canonical quantisation*.

For any observable function $a(p,q)$ we consider the function $a(x,\xi)$ of auxiliary variables (observe the order of variables) and set for any $u \in \mathcal{S}(\mathbb{R}^n)$:

$$(Op(a)(u))(x) = \mathcal{F}^{-1}(a(x,\xi)\mathcal{F}(u))$$
$$= \frac{1}{\sqrt{2^n \pi^n}} \int_{\mathbb{R}^n} a(x,\xi)\mathcal{F}(\varphi(\xi)) e^{-ix\cdot\xi} d\xi. \tag{2.6}$$

In many books this expression is modified so as to get rid of the factors in front of the integral on the expense of their appearance inside it; then the formula (2.6) is rewritten as

$$(Op(a)(u))(x) = \iint_{\mathbb{R}^{2n}} e^{2\pi i (x-y)\cdot\xi} a(x,\xi) u(y) dy d\xi. \tag{2.7}$$

It is easy to compute $Op(q^j) = x^j \cdot$, $Op(p_k) = \frac{1}{2\pi i}\frac{\partial}{\partial x_k}$ and $Op(q^j p_k) = x^j \frac{1}{2\pi i}\frac{\partial}{\partial x_k}$ etc. (see exercises). The operators $Op(a)$ are studied in Mathematics under the name of *pseudo-differential operators* on $\mathbb{R}^n$

and a in this case is the *(full) symbol of the operator $Op(a)$*; usually one imposes certain analytic conditions on the symbols a, which guarantee convergence of all integrals. Theory of pseudodifferential operators is an important part of Analysis and plays crucial role in the index theory.

2.2 Weyl quantisation formula

The method we described above works quite nicely from the algebraic point of view, but there are few analytic issues which need to be resolved. First, the space of Schwarz functions is not a Hilbert space, nor even a Banach space. This observation eventually brings one to the notion of Sobolev spaces, unbounded operators and other subtle analytic devices, see the book [Sh01].

Another flaw of this method is of a more algebraic nature: the operator associated with a real-valued function by this method is not necessarily self-adjoint (which is a physically meaningful condition, since eigenvalues of the operator play the role of mean values of the observable function). For instance, this is the case for the function $q_k p^k$.

A suitable solution of this latter difficulty is to change the formula (2.7) (and (2.6)) as follows:

$$(Op^W(a)(u))(x) = \iint_{\mathbb{R}^{2n}} e^{2\pi i(x-y)\cdot\xi} a\left(\frac{x+y}{2},\xi\right) u(y) dy d\xi. \quad (2.8)$$

This formula was proposed by Hermann Weyl (see [Wey27], [Wey31]) and is therefore called *Weyl quantisation* formula.

Let us sketch, why Weyl quantisation formula sends real-valued functions into self-adjoint operators. To this end, we consider the following integral operator on $C^\infty(\mathbb{R}^{2n})$ (or rather family of operators, depending on real parameter t, $|t| > 0$):

$$(J^t a)(x,\xi) = |t|^{-n} \iint_{\mathbb{R}^{2n}} e^{-2\pi i t^{-1} y \cdot \eta} a(x+y, \xi+\eta) dy d\eta,$$

then $(J^t)^\dagger = J^{-t}$ and $J^p J^q = J^{p+q}$ if everything is defined.[5] Using (2.7), one can compute that

---
[5] In fact this operator is just the integral expression for $\exp(2\pi i t(\frac{\partial}{\partial x}; \frac{\partial}{\partial \xi}))$, the shift operator.

$$(Op(J^t a)(u)) = \iint_{\mathbb{R}^{2n}} e^{2\pi i(x-y)\cdot\xi} a((1-t)x + ty, \xi)u(y)dyd\xi,$$

and so $Op^W(a) = Op(J^{\frac{1}{2}}a)$, $Op(a)^\dagger = Op(J^1\bar{a})$. Hence

$$\begin{aligned}(Op^W(a))^\dagger &= (Op(J^{\frac{1}{2}}a))^\dagger = Op(J^1(\overline{J^{\frac{1}{2}}a}) \\ &= Op(J^1 J^{-\frac{1}{2}}\bar{a}) = Op(J^{\frac{1}{2}}\bar{a}) = Op^W(\bar{a}),\end{aligned}$$

in particular this operator is self-adjoint if $\bar{a} = a$.

2.2.1 Further generalisations

There's a way to generalise Weyl quantisation a little bit to allow other representations of canonical commutator relations (2.1), based on the following observations. Suppose P and Q are two operators, satisfying the canonical relations (for $n = 1$). Consider the formal (by power series) exponentiation of these operators:

$$U(s) = e^{isP}, \; V(t) = e^{itQ}, \; s,t \in \mathbb{R}; \qquad (2.9)$$

if P and Q are self adjoint, then from formal computations we see that $U(s)$, $V(t)$ are unitary and

$$U(s_1)U(s_2) = U(s_1 + s_2) \qquad (2.10)$$

and the same for $V(t)$. Besides this we know $[P^n, Q] = nP^{n-1}$, so reasoning by induction we can find the commutators $[P^n, Q^m]$ for all m, n, for instance:

$$\begin{aligned}[P^n, Q^2] &= nP^{n-1}Q + nQP^{n-1} = 2nP^{n-1}Q - n[P^{n-1}, Q] \\ &= 2nP^{n-1}Q - n(n-1)P^{n-2}.\end{aligned}$$

In the end we get the *Weyl commutation relations*:

$$U(s)V(t) = e^{ist}V(t)U(s). \qquad (2.11)$$

If $n > 1$, one should consider the multi-parameter families $U(x) = e^{i(\sum_k x^k P_k)}$, $(x^1, \ldots, x^n) \in \mathbb{R}^n$ and $V(f) = e^{i(\sum_k f_k Q^k)}$, $(f_1, \ldots, f_n) \in (\mathbb{R}^n)^*$ (recall that $*$ denotes the dual of a vector space, i.e. the space of linear functionals; here we use the usual Euclidean structure on $\mathbb{R}^n$ for dualisation). Then the properties of $U(s)$, $V(t)$ including the formula (2.11) are straightforwardly generalised to this situation. On the other hand, if $U(a)$, $a \in \mathbb{R}$ is a 1-parameter group of unitary operators (i.e. $U(a+b) = U(a)U(b)$), which is *strong continuous*, i.e. such that

$$\lim_{t \to t_0} U(t)\varphi = U(t_0)\varphi, \forall \varphi \in \mathcal{H}, \qquad (2.12)$$

then by *Stone theorem*, the formula

$$\lim_{t \to 0} \frac{U(t)\varphi - \varphi}{it} = A\varphi \qquad (2.13)$$

determines an essentially self-adjoint operator, although unbounded in a general situation (this shall not make us shy however due to Wintner's theorem). Generalising this we come up with the following axiomatic description.

Let L be a real vector space, L^* its dual; let $\mathcal{H}$ be a Hilbert space and $\mathscr{U}(\mathcal{H})$ the group of unitary operators on $\mathcal{H}$. We shall say that a *Weyl representation* is given if we have two maps $U : L \to \mathscr{U}(\mathcal{H})$, $V : L^* \to \mathscr{U}(\mathcal{H})$, which are strong continuous (i.e. condition (2.12) holds), satisfy (2.10) and also the following equation is satisfied

$$U(x)V(f) = e^{if(x)}V(f)U(x), \qquad (2.14)$$

where $f(x)$ is the pairing of L and L^*.

In this case it is convenient to introduce the space $\mathcal{V} = L \oplus L^*$ and consider the operator

$$W(z) = e^{\frac{1}{2}if(x)}U(x)V(f) \qquad (2.15)$$

for any $z = (x, f) \in \mathcal{V}$. In this notation the commutation relation (2.14) is equivalent to

$$W(z + z') = e^{-\frac{1}{2}i\omega(z,z')}W(z)W(z'), \qquad (2.16)$$

where ω is the canonical nondegenerate antisymmetric form (canonical *symplectic form*) on $S = L \oplus L^*$:

$$\omega((x, f), (x', f')) = f(x') - f'(x), \qquad (2.17)$$

where $z = (x, f)$, $z' = (x', f')$.

The example which we considered before (with differential operators) corresponds to the following Weyl representation: we take $L = \mathbb{R}^n$, $\mathcal{H} = L_2(L)$, then

$$U : L \to \mathscr{U}(\mathcal{H}), \quad (U(x)\varphi)(y) = \varphi(y + x),$$
$$V : L^* \to \mathscr{U}(\mathcal{H}), \quad (V(f)\varphi)(y) = e^{if(y)}\varphi(y).$$

Equality (2.14) is readily checked. Observe that the operators $U(x)$, $V(f)$ are well-defined on the Hilbert space $\mathcal{H}$, although the operators P, Q such that $U(x) = e^{ixP}$, $V(f) = e^{ifQ}$ (scalar products xP, fQ are clear from the context) need not be determined for all $\varphi \in \mathcal{H}$ (for example, $P = i\frac{\partial}{\partial x}$ is well-defined only on smooth functions etc.). In effect, one can use *von Neumann's theorem* to show that Equations (2.9) determine a pair of essentially self-adjoint operators P, Q, well-defined on a dense subspace D in $\mathcal{H}$, such that the canonical commutator relations hold on D.

In the general case, using Stone and von Neumann's theorems we can define $R(z)$ as a formal logarithm of $W(z)$, i.e. write formally $W(z) = e^{iR(z)}$. It follows from (2.15) and strong continuity of $W(z)$ that $W(tz) = (W(z))^t$, so $R(tz) = tR(z)$, whenever it is well-defined. And since from (2.15) we get $W(z)W(z') = e^{\omega(z,z')}W(z')W(z)$, we compute (see exercises):

$$[R(z), R(z')] = -i\omega(z, z'). \tag{2.18}$$

Let us conclude this lecture by the following observation: if we have a Weyl representation, as described above, we can define the operator $a^W(P, Q)$ for any observable function $a(p, q)$; namely, if $\mathcal{F}(a)(x, f) = \mathcal{F}(a)(z)$ is the (inverse) Fourier transform of a, then (up to a constant factor)

$$Op^W(a)(Q, P) = \int_{\mathcal{V}=\mathbb{R}^{2n}} \mathcal{F}(a)(z)W(z)dz. \tag{2.19}$$

Since $(W(z))^\dagger = W(-z)$ and $\overline{\mathcal{F}(a)}(z) = \mathcal{F}(\bar{a})(-z)$, we see that this formula also produces self-adjoint operators from real functions. Some details on the construction given in this section can be further found in Chapter 9 of [EMS04].

3 Moyal quantisation

As we explained in the previous section, the basic principles of Quantum Mechanics and other physical theories related to it ask one to replace the commutative algebra of observable functions by a suitable noncommutative algebra (of operators on a Hilbert space). In particular, we need to find a construction that would allow one to replace commutative algebras by noncommutative ones so that the basic commutator relation (2.1) and its corollaries would be satisfied. This can be phrased in the framework of the purely algebraic *deformation theory* (see [GS88] or [Ge64]). This approach eventually lead to the formulation of the deformation quantisation problem by Flato and others in 1970s. In this section we will discuss Moyal product, which was the first example of formal deformations in Quantum Physics.

3.1 Asymptotic expansions

Let a, b be two observable functions on $\mathbb{R}^{2n}$ with coordinates $(x^1, \ldots, x^n, \xi_1, \ldots, \xi_n)$. We can use the canonical quantisation formula (2.7) to define the operators $Op(a)$, $Op(b)$; it is not difficult to see that the composition of these two operators is again an operator of the same sort: $Op(a) \circ Op(b) = Op(a \circ b)$, where

$$(a \circ b)(x, \xi) = \iint_{\mathbb{R}^n \times \mathbb{R}^n} e^{-2\pi i y \cdot \eta} a(x, \xi + \eta) b(x + y, \xi) dy d\eta; \quad (3.1)$$

here and below we assume that all the integrals are absolutely convergent. Similarly, one can show that the same is true for Weyl quantisation formula: using either (2.8) or (2.19), we get $Op^W(a) \circ Op^W(b) = Op^W(a \star b)$, where for $z \in \mathbb{R}^{2n} = \mathbb{R}^n \times (\mathbb{R}^n)^*$ we have

$$(a \star b)(z) = 2^{2n} \iint_{\mathbb{R}^{2n} \times \mathbb{R}^{2n}} e^{-4\pi i \omega(z - z', z - z'')} a(z') b(z'') dz' dz''. \quad (3.2)$$

Recall that ω is the canonical nondegenerate antisymmetric form on $\mathbb{R}^{2n} = \mathbb{R}^n \times (\mathbb{R}^n)^*$, given by the formula (2.17); if we introduce a basis $e_1, \ldots, e_n$ in $\mathbb{R}^n$ and use the dual basis $f^1, \ldots, f^n$ in $(\mathbb{R}^n)^*$, then the matrix of ω in basis $e_1, \ldots, e_n, f^1, \ldots, f^n$ is

$$I_{2n} = \left(\begin{array}{c|c} 0 & E_n \\ \hline -E_n & 0 \end{array} \right). \tag{3.3}$$

We shall denote by $\omega_{ij} = -\omega_{ji}$ the elements of the matrix representing ω in some basis and by $\pi^{ij} = -\pi^{ji}$ the elements of the inverse matrix, i.e. $\pi^{ik}\omega_{kj} = \delta^i_j = \omega_{jk}\pi^{ki}$.

In the theory of pseudodifferential operators one also describes the symbol $a \circ b$ of the composition $Op(a) \circ Op(b)$ of two operators in terms of the *asymptotic decomposition*. Namely, using the notation from Section 2.1.3, we can write:

$$a \circ b = \sum_{|I| \leq N} \frac{1}{I!} D^I_\xi a \, \partial^I_x b + r_N(a, b), \tag{3.4}$$

where for a multi-index $I = (i_1, \ldots, i_n)$ we put

$$D^I_\xi = D^{i_1}_{\xi_1} \ldots D^{i_n}_{\xi_n}, \quad D_{\xi_k} = \frac{1}{2\pi i} \frac{\partial}{\partial \xi_k}, \quad k = 1, \ldots, n$$

$$\partial^I_x = \partial^{i_1}_{x^1} \ldots \partial^{i_n}_{x^n}, \quad \partial_{x^k} = \frac{\partial}{\partial x^k}, \quad k = 1, \ldots, n$$

and $I! = i_1! \ldots i_n!$. The residue term r_N denotes the symbol of operator of degree greater[6] than N.

A similar asymptotic expansion now can be given for the Weyl quantisation, i.e. for the product $a \star b$, determined by (3.2): using the notation we have just introduced, we have

$$a \star b = \sum_{k=0}^{N} \frac{1}{2^k} \sum_{|I|+|J|=k} \frac{(-1)^{|J|}}{I! J!} D^I_\xi \partial^J_x a \, D^J_\xi \partial^I_x b + r_N(a, b), \tag{3.5}$$

where the second sum is over all multi-indices I, J such that $|I| + |J| = k$.

It is not difficult to write down the first couple of terms in this expansion: if $k = 0$, there are no partial derivations at all, so the

---
[6] In the theory of pseudodifferential operators degree of a symbol is determined by its behaviour when $\|(x, \xi)\| \to \infty$. We shall not need this in our lectures, as we are only interested in the algebraic side of the theory here.

first term is just $\frac{1}{2^0} \frac{(-1)^0}{0!} ab = ab$. And if $k = 1$, then either I, or J should be equal to $(0,\ldots,0)$ and the other multi-index is of the form $(0,\ldots,0,1,0,\ldots,0)$ where 1 can take any place in the string of zeroes, so the second term in expansion (i.e. for $k = 1$) is

$$\frac{1}{2}\sum_{j=1}^n \frac{1}{2\pi i}\left(\frac{\partial a}{\partial \xi_k}\frac{\partial b}{\partial x^k} - \frac{\partial a}{\partial x^k}\frac{\partial b}{\partial \xi_k}\right) = \frac{1}{4\pi i}\{a,b\}, \qquad (3.6)$$

here $\{,\}$ is the Poisson bracket on $C^\infty(\mathbb{R}^{2n})$ with coordinates (x,ξ) playing the role of (q,p) (c.f. the formula (1.6)). The fact that the Poisson bracket appears in this expansion is quite natural, taken into consideration the formula (2.18), since $\omega(z,z')$ is equal to the Poisson bracket of the linear functions on $\mathbb{R}^{2n}$, determined by z, z' with respect to the Euclidean pairing.

3.2 Moyal deformation series

In general, the series on the right-hand side of formulas (3.4) and (3.5) need not converge: these are just asymptotic series, which determine pseudodifferential operators. On the other hand, they give a nice algebraic approach to the quantisation problem. It is this algebraic point of view, which will pretty soon bring us to the *deformation quantisation* theory.

First of all we restore the Planck constant $\hbar$ at its place in the exponent of formula (3.2), i.e. the integrating kernel now is $e^{4\pi i\hbar\omega(z-z',z-z'')}$ so in asymptotic expansion the operator of degree k is multiplied by $\hbar^k$. Next we rescale $\hbar$ so that it will absorb the factor $\frac{1}{2\pi i}$. And finally, we note that formula (1.6) can be written as follows in the terms of matrices (ω_{ij}), (π^{ij}) (the matrix of symplectic form and its inverse):

$$\{f,g\} = \sum_{1\leq i,j\leq n}\pi^{ij}\frac{\partial f}{\partial z^i}\frac{\partial g}{\partial z^j}. \qquad (3.7)$$

Here $(z^1,\ldots,z^{2n})$ are the generic linear coordinates in $\mathbb{R}^{2n}$; in particular, if $z = (x,\xi)$, then $(\pi^{ij}) = (\omega_{ij}) = I_{2n}$ (see formula (3.3)) so this formula coincides with the standard one. Moreover, in this case all the coefficients of the differential operators in (3.5) can be written down in terms of the coefficients of these matrices, since $\pi^{ij} = -\omega_{ij} = \delta_{i-n,j} - \delta_{i,j-n}$

$(-1)^q \partial_{\xi_{i_1}} \ldots \partial_{\xi_{i_p}} \partial_{x_{j_1}} \ldots \partial_{x_{j_q}} a \, \partial_{x_{i_1}} \ldots \partial_{x_{i_p}} \partial_{\xi_{j_1}} \ldots \partial_{\xi_{j_q}} b$
$= \pi^{i'_1 i_1} \ldots \pi^{i'_p i_p} \pi^{j_1 j'_1} \ldots \pi^{j_q j'_q} \partial_{i'_1} \ldots \partial_{i'_p} \partial_{j_1} \ldots \partial_{j_q} a \, \partial_{i_1} \ldots \partial_{i_p} \partial_{j'_1} \ldots \partial_{j'_q} b.$

Here and below we shall usually abbreviate ∂_{z^k} to ∂_k, where $(z^1, \ldots, z^{2n})$ are the coordinates (x, ξ), and $i' = i + n$, $j' = j + n$. Summing up, we can now rewrite the asymptotic expansion (3.5) as the following *formal* power series, called *Moyal deformation series*, or *Moyal product*:

$$a \star b = \sum_{k=0}^{\infty} \frac{\hbar^k}{k! 2^k} \sum_{\substack{1 \leq i_1, \ldots, i_k \leq 2n \\ 1 \leq j_1, \ldots, j_k \leq 2n}} \pi^{i_1 j_1} \pi^{i_2 j_2} \ldots \pi^{i_k j_k} \partial_{i_1} \ldots \partial_{i_k} a \, \partial_{j_1} \ldots \partial_{j_k} b.$$

(3.8)

This formula, now known as the Moyal product was in effect first discovered by Hilbrand Groenewold in his thesis [Gr46]. However, it obtained popularity after the work of Jose Moyal [Mo49], published in 1949, where he (having discovered these formulas independently) makes use of the commutators with respect to the operation $\star$.

As before, we cannot speak about the convergence of any analytic kind in the formula (3.8). Instead, from this moment on we shall say, that in this formula $\hbar$ plays the role of *formal variable* and so the whole right-hand side of the expression (3.8) is a *formal power series in variable* $\hbar$.

Recall that for any commutative ring of coefficients R, *the ring of formal power series in a variable* t *with coefficients in* R *is*

$$R[[t]] = \left\{ \sum_{k \geq 0} a_k t^k \mid a_k \in R, \ k = 0, 1, \ldots \right\}.$$

For instance in the formula (3.8) we are dealing with the coefficient ring $R = C^\infty(\mathbb{R}^{2n})$. The sum and the product of two such expressions $F(t) = \sum_k a_k t^k$ and $G(t) = \sum_k b_k t^k$ are given by the standard formulas:

$$(F + G)(t) = \sum_k (a_k + b_k) t^k$$

and $(F \cdot G)(t) = \sum_k c_k t^k$ where

$$c_k = \sum_{p=0}^{k} a_p b_{k-p}.$$

The space $R[[t]]$ with these operations becomes a commutative ring. Some properties of this ring are listed in the exercises.

3.2.1 Moyal products

We can now study the formula (3.8) independently of all the previous considerations, including the integral formula (3.2). Then (3.8) can be regarded as a map, which associates to any pair f, g of smooth functions on $\mathbb{R}^{2n}$ a formal power series in $C^\infty(\mathbb{R}^{2n})[[\hbar]]$. This point of view gives us much freedom to extend this approach to the situations, where the analytic side of the story is very restricted or even non-existent.

First of all, we observe that the same formula (3.8) defines a formal series $(A \star B)(\hbar)$ for any pair of formal series $A = A(\hbar)$, $B = B(\hbar)$, if we assume that the operation $\star$ is bi-linear with respect to $\hbar$; i.e. we put

$$(A \star B)(\hbar) = \sum_{k \geq 0} \hbar^k c_k, \text{ where } c_k = \sum_{p=0}^{k} a_p \star b_{k-p}.$$

In fact every c_k in this formula is a formal power series and in general infinite sum of formal series is not well-defined, but since we multiply each c_k by $\hbar^k$, there will be only finite number of terms with any given power of $\hbar$.

Next, we observe that if we make a linear change of coordinates[7] $z^{i'} = C^{i'}_i z^i$ on $\mathbb{R}^{2n}$ for a constant invertible matrix $C = (C^{i'}_i)$ then the matrix π transforms as $\pi^{i'j'} = \pi^{ij} C^{i'}_i C^{j'}_j$ and partial derivations are related by the chain rule: $\partial_{i'} = C^i_{i'} \partial_i$, where $(C^i_{i'}) = C^{-1}$, so that

---

[7]Here and below we use Einstein's summation notation, i.e. $C^{i'}_i z^i = \sum_{i=1}^{2n} C^{i'}_i z^i$, $\pi^{ij} C^{i'}_i C^{j'}_j = \sum_{i,j=1}^{2n} \pi^{ij} C^{i'}_i C^{j'}_j$, etc.: whenever we have a pair of reappearing indices, the summation is mutely understood.

the formula (3.8) in new coordinates looks precisely the same, even though the matrix π need not look as before.

All these observations justify the following definition:

Definition 3.1. *Let V be a vector space (real or complex), and let π be a bilinear function on V^* (not necessarily symmetric or skew-symmetric). Then the π-**Moyal product** on V is the map $C^\infty(V)[[\hbar]]^{\otimes 2} \to C^\infty(V)[[\hbar]]$, given for any two functions $f, g \in C^\infty(V)$ by formula (3.8) and extended $\hbar$-linearly to $C^\infty(V)[[\hbar]]$ as explained above.*

We shall denote the π-Moyal product by $\star_\pi$ or just $\star$ if there's no ambiguity. The formula (3.8) is often abbreviated to the following

$$f \star_\pi g = m \circ \exp\left(\frac{\hbar}{2}\pi^{ij}(\partial_i \otimes \partial_j)\right)(f \otimes g), \qquad (3.9)$$

where $m : C^\infty(V) \otimes C^\infty(V) \to C^\infty(V)$ denotes the usual product of functions and the exponential function of the partial differential operator $\frac{\hbar}{2}\pi^{ij}\partial_i \otimes \partial_j$ is understood in the sense of formal power series, in which left and right tensor legs of this operator are composed separately. The purpose of using $\frac{\hbar}{2}\pi$ and not $\hbar\pi$ in the formula (3.9) is to preserve the identity: $f \star_\pi g - g \star_\pi f = \hbar\{f, g\} + o(\hbar)$, when $\pi = \omega^{-1}$ as in the original example. From algebraic point of view, this choice is not very important and we could have written

$$f \star_\pi g = m \circ \exp\left(\hbar\pi^{ij}(\partial_i \otimes \partial_j)\right)(f \otimes g), \qquad (3.9')$$

instead (as we will often do). In Physics textbooks, the formula (3.9') is sometimes further abbreviated to

$$f \star_\pi g = \exp(\hbar\pi(\overleftarrow{\partial} \otimes \overrightarrow{\partial}))(f, g). \qquad (3.10)$$

The arrows over the partial derivations denote the "direction, in which they act", i.e. whether they are applied to the left or right legs of the tensor square. Observe, that we make no assumptions on the form of the matrix π in this definition; this matrix can be degenerate and still the formula will be well-defined; for example if $\pi = 0$, then $a \star_\pi b = ab$.

3.2.2 Properties of the Moyal product

Proposition 3.2. *Let (V, π) be a vector space and a bilinear function on V^*, then the Moyal product $\star_\pi$ (see Definition 3.1) satisfies the following conditions:*

1. $f \star_\pi g = \sum_{k=0}^{\infty} \hbar^k B_k(f, g)$ for some bidifferential operators B_k;
2. $f \star_\pi g = fg + \frac{\hbar}{2}\pi(df, dg) + o(\hbar)$, where $\pi(df, dg) = \pi^{ij}\partial_i f \partial_j g$;
3. $f \star_\pi 1 = f = 1 \star_\pi f$;
4. if the bilinear function π is anti-symmetric then $B_k(f, g) = (-1)^k B_k(g, f)$;
5. if f and g are polynomials in linear coordinates on V, then $f \star_\pi g$ is equal to a finite sum of polynomials $\sum_{k=0}^{N} \hbar^k p_k$;
6. the product $\star_\pi$ is associative, i.e. for all f, g and h in $C^\infty(V)[[\hbar]]$
$$(f \star_\pi g) \star_\pi h = f \star_\pi (g \star_\pi h).$$

Recall that *bidifferential operator* B on V is a map $C^\infty(V)^{\otimes 2} \to C^\infty(V)$, which is equal to a differential operator of either of its arguments, when the other one is fixed; equivalently, $B(f, g)$ can be represented as a sum of the products of (iterated) partial derivatives of f and g with functional coefficients.

Proof. Proving all the properties of $\star_\pi$ except for the last one is easy, we leave it as an exercise to the readers. As for the associativity of the $\star_\pi$, it can be proved as follows: use formula (3.9) and the following version of Leibniz rule, expressed in terms of m and ∂_i:
$$\partial_i \circ m = m \circ (\partial_i \otimes 1 + 1 \otimes \partial_i),$$
then (omitting the subscript π for the sake of brevity) we get
$$f \star (g \star h)$$
$$= m \circ \exp\left(\frac{\hbar}{2}\pi^{ij}\partial_i \otimes \partial_j\right)\left(f \otimes m \circ \exp\left(\frac{\hbar}{2}\pi^{kl}\partial_k \otimes \partial_l\right)(g \otimes h)\right)$$
$$= m(1 \otimes m) \circ \exp\left(\frac{\hbar}{2}\pi^{ij}(\partial_i \otimes \partial_j \otimes 1 + \partial_i \otimes 1 \otimes \partial_j)\right)$$
$$\left(1 \otimes \exp\left(\frac{\hbar}{2}\pi^{kl}\partial_k \otimes \partial_l\right)\right)(f \otimes g \otimes h)$$

and hence
$$f \star (g \star h)$$
$$= m(1 \otimes m) \circ \exp\left(\frac{\hbar}{2}\pi^{ij}(\partial_i \otimes \partial_j \otimes 1 + \partial_i \otimes 1 \otimes \partial_j + 1 \otimes \partial_i \otimes \partial_j)\right)$$
$$(f \otimes g \otimes h).$$

Here we used the equation $e^A e^B = e^{A+B}$ for any commuting operators A and B (in this case A and B are partial derivatives), and the equality $e^{A \otimes 1} = e^A \otimes 1$. Now, similar computation shows that

$$f \star (g \star h)$$
$$= m(m \otimes 1) \circ \exp\left(\frac{\hbar}{2}\pi^{ij}(\partial_i \otimes \partial_j \otimes 1 + \partial_i \otimes 1 \otimes \partial_j + 1 \otimes \partial_i \otimes \partial_j)\right)$$
$$(f \otimes g \otimes h).$$

Since $m(1 \otimes m) = m(m \otimes 1)$ by associativity of m, these expressions are equal. □

3.2.3 Equivalence of Moyal $\star$-products

In previous paragraph we introduced the operation $\star_\pi$ for any bilinear function on V^*; we call it *product*, because this map satisfies the associativity condition and is clearly distributive, so that $(C^\infty(V)[[\hbar]], \star_\pi)$ is a ring. Thus we get an infinite family of ring structures on $C^\infty(V)[[\hbar]]$. As it is usual in Algebra, when we have two rings, we might ask if they are isomorphic or not, so it is natural to wonder when the algebra structures induced by $\star_\pi$ and $\star_{\pi'}$ are isomorphic. The following proposition gives a partial answer to this question.

Proposition 3.3. *Suppose $\alpha = (\alpha^{ij})$ is a symmetric bilinear function on V^*. Consider the linear map $T : C^\infty(V)[[\hbar]] \to C^\infty(V)[[\hbar]]$, given by $T = \exp\left(\frac{\hbar}{2}\alpha^{ij}\partial_i\partial_j\right)$, i.e.*

$$T(f) = f + \frac{\hbar}{2}\alpha^{ij}\partial_i\partial_j f + \frac{\hbar^2}{2^2 2!}\alpha^{ij}\alpha^{kl}\partial_i\partial_j\partial_k\partial_l f + \cdots.$$

Then T is well-defined and invertible and induces an isomorphism

$$T : (C^\infty(V)[[\hbar]], \star_\pi) \to (C^\infty(V)[[\hbar]], \star_{\pi+2\alpha}).$$

Proof. It is evident that T is well-defined since the degree of $\hbar$ grows; the invertibility of T follows from general principles of formal series (see Exercise E.1.11), but also in this case it is easy to find the explicit expression for the inverse map: $T^{-1} = \exp(-\frac{\hbar}{2}\alpha^{ij}\partial_i\partial_j)$.

Next the Leibniz rule yields

$$(\partial_i\partial_j) \circ m = m \circ (\partial_i\partial_j \otimes 1 + \partial_i \otimes \partial_j + \partial_j \otimes \partial_i + 1 \otimes \partial_i\partial_j).$$

Further we compute,

$$T(f \star_\pi g) = \exp\left(\frac{\hbar}{2}\alpha^{ij}\partial_i\partial_j\right) \circ \left(m \circ \exp\left(\frac{\hbar}{2}\pi^{kl}\partial_k \otimes \partial_l\right)(f \otimes g)\right)$$

$$= m \circ \exp\left(\frac{\hbar}{2}\left(\alpha^{ij}(\partial_i\partial_j \otimes 1 + \partial_i \otimes \partial_j + \partial_j \otimes \partial_i\right.\right.$$

$$\left.\left. + 1 \otimes \partial_i\partial_j) + \pi^{kl}\partial_k \otimes \partial_l\right)\right)(f \otimes g)$$

$$= m \circ \exp\left(\frac{\hbar}{2}(2\alpha^{ij} + \pi^{ij})(\partial_i \otimes \partial_j)\right)$$

$$\circ \exp\left(\frac{\hbar}{2}\alpha^{ij}(\partial_i\partial_j \otimes 1 + 1 \otimes \partial_i\partial_j)\right)(f \otimes g)$$

$$= m \circ \exp\left(\frac{\hbar}{2}(2\alpha^{ij} + \pi^{ij})(\partial_i \otimes \partial_j)\right) \circ (T \otimes T)(f \otimes g)$$

$$= Tf \star_{\pi+2\alpha} Tg.$$

Here we again used the properties of the exponents of commuting operators. $\square$

Below we shall call isomorphic $\star$-products $\star_\pi$ and $\star_{\pi+2\alpha}$ *equivalent*, and T their equivalence relation. Varying α we get a vast class of equivalent $\star$-products; in particular, for every matrix π we can take $-\alpha$ to be half of its symmetric part, i.e. $2\alpha^{ij} = -\frac{1}{2}(\pi^{ij} + \pi^{ji})$. In this case, $\pi^{ij} + 2\alpha^{ij} = \frac{1}{2}(\pi^{ij} - \pi^{ji})$ is skew-symmetric and this brings us to the following conclusion:

Corollary 3.4. *Every Moyal product is equivalent to a product with antisymmetric π, i.e. one can assume that $\pi^{ij} = -\pi^{ji}$.*

Observe that if π is skew-symmetric, then so is the bracket $\{f,g\}_\pi = \pi(df, dg)$, associated with it (see Exercise E.1.13).

Remark 3.5. In effect one can consider more general linear isomorphisms $T : C^\infty(V)[[\hbar]] \to C^\infty(V)[[\hbar]]$, associated with any series of partial differential operators:

$$T(f) = T_0(f) + \hbar T_1(f) + \hbar^2 T_2(f) + \cdots.$$

Such operator is invertible, iff T_0 is, so we can ask if two $\star$-products $\star_1$ and $\star_2$ are equivalent by T, i.e. if $T(f \star_1 g) = T(f) \star_2 T(g)$. We will not discuss this question here.

3.2.4 The complex case

As we have already mentioned complex conjugation plays an important role in physical applications of quantisation. So let us suppose that V is *complex*, i.e. the field of scalars we use is $\mathbb{C}$. Alternatively, one says that a real vector space V over $\mathbb{R}$ bears *a complex structure*, if there's a map $J : V \to V$, $J^2 = -1$. This map plays the role of multiplication by $\sqrt{-1}$: when J is given, then for any $v \in V$ we can define its product with $z = a + b\sqrt{-1} \in \mathbb{C}$ by the formula $z \cdot v = av + bJ(v)$ for all $a, b \in \mathbb{R}$.

Now in this case it is natural to assume that the $\star_\pi$-product of complex-valued functions on V satisfies the standard condition that the complex conjugation is anti-linear anti-isomorphism of the algebra $(C^\infty(V)[[\hbar]], \star)$:

$$\overline{f \star_\pi g} = \bar{g} \star_\pi \bar{f}. \tag{3.11}$$

This means that the complex structure on V is in accordance with the bilinear map π. Namely, comparing the coefficients at various powers of $\hbar$ in the series (3.8), we see that *equality (3.11) holds iff the matrix π is self-adjoint*, i.e. $\bar{\pi}^{ji} = \pi^{ij}$ or $\pi^\dagger = \pi$.

This happens for instance, if π comes from a Hermitian structure on V. Namely, let h be a (nondegenerate) Hermitian structure on V (with constant coefficients). Then the matrix of h is self-adjoint, and hence the same holds for $\pi = h^{-1}$.

Indeed remark, that if h is Hermitian, then $\omega = \Im(h)$ is real skew-symmetric, and $g = \Re(h)$ is real symmetric:

$$g(a,b) + \sqrt{-1}\omega(a,b) = h(a,b) = \overline{h(b,a)} = g(b,a) - \sqrt{-1}\omega(b,a).$$

These bilinear forms are related via the complex structure operator J:

$$g(J(a), J(b)) + \sqrt{-1}\omega(J(a), J(b)) = h(J(a), J(b)) = h(\sqrt{-1}a, \sqrt{-1}b)$$
$$= -(\sqrt{-1})^2 h(a,b) = h(a,b) = g(a,b) + \sqrt{-1}\omega(a,b),$$

so $g(J(a), J(b)) = g(a,b)$, $\omega(J(a), J(b)) = \omega(a,b)$. Similarly

$$g(J(a), b) + \sqrt{-1}\omega(J(a), b) = h(\sqrt{-1}a, b) = -\sqrt{-1}h(a,b)$$
$$= -\sqrt{-1}(g(a,b) + \sqrt{-1}\omega(a,b)) = \omega(a,b) - \sqrt{-1}g(a,b),$$

so $g(J(a), b) = \omega(a,b)$, $\omega(J(a), b) = -g(a,b)$.

If finally ω and g are invertible, then the matrix $\pi = \frac{1}{2}(g^{-1} + \sqrt{-1}\omega^{-1})$ is again self-adjoint:

$$2\pi^\dagger = (g^{-1})^\dagger - \sqrt{-1}(\omega^{-1})^\dagger = g^{-1} + \sqrt{-1}\omega^{-1} = 2\pi,$$

since g^{-1} and ω^{-1} are real symmetric and skew-symmetric respectively. Thus, we see that if one takes π to define the $\star$-product, then the condition (3.11) will hold.

3.3 The integral formula

Let us conclude this lecture with an attempt to justify the formula (3.8) by deriving it formally from the integral expression (3.2). To this end we shall use the following remark: the Fourier transform of the Dirac's δ-function is given by

$$\mathcal{F}(\delta)(x) = \int e^{-2\pi i x \cdot \xi} \delta(\xi) d\xi = 1$$

by definition of the δ. Indeed, computing formally we have:

$$\delta(x) = \mathcal{F}^{-1}(\mathcal{F}(\delta))(x) = \int e^{2\pi i x \cdot \xi} d\xi. \tag{3.12}$$

This agrees with the usual formula for Fourier transform:

$$f(x) = \int \delta(y-x) f(y) dy = \iint e^{2\pi i (y-x)\cdot \xi} f(y) d\xi dy = [\text{integrate by } y]$$
$$= \int e^{-2\pi i x \cdot \xi} \mathcal{F}^{-1}(f)(\xi) d\xi = \mathcal{F}(\mathcal{F}^{-1}(f))(x) = \mathcal{F}^{-1}(\mathcal{F}(f))(x).$$
$$\tag{3.13}$$

Using these observations we compute for $V = \mathbb{R}^{2n}$ with nondegenerate skew-symmetric form ω, as in the original definition (we have changed the coefficients a little for the sake of convenience):

$$\frac{1}{(2\pi\hbar\det(\omega))^{2n}} \iint_{\mathbb{R}^{2n}\times\mathbb{R}^{2n}} e^{-\frac{1}{i\hbar}\omega(z-z',z-z'')} a(z')b(z'')dz'dz''$$

$$= \frac{1}{(2\pi)^{2n}} \iiint_{(\mathbb{R}^{2n})\times 3} \delta(\zeta + z - z'') e^{\frac{i}{\hbar}\omega(\zeta,z-z')} a(z')b(z'')d\zeta dz'dz''$$

$$= \frac{1}{(2\pi)^{2n}} \iiint_{(\mathbb{R}^{2n})\times 3} \delta(\hbar\pi^{\sharp}(k) + z - z'') e^{ik\cdot(z-z')} a(z')b(z'') dk dz'dz''$$

$$= \frac{1}{(2\pi)^{4n}} \iiiint_{(\mathbb{R}^{2n})\times 4} \underbrace{e^{i(\hbar\pi^{\sharp}(k)+z-z'')\cdot q}}_{\delta(\hbar\pi^{\sharp}(k)+z-z'')} e^{ik\cdot(z-z')} a(z')b(z'') dq dk dz'dz''$$

$$= \frac{1}{(2\pi)^{4n}} \iiiint_{(\mathbb{R}^{2n})\times 4} e^{i\hbar\pi(k,q)} \underbrace{e^{ik\cdot(z-z')} a(z')}_{\mathcal{F}^{-1}(\mathcal{F}(a)(k))(z)} \underbrace{e^{iq\cdot(z-z'')} b(z'')}_{\mathcal{F}^{-1}(\mathcal{F}(b)(q))(z)} dq dk dz'dz''$$

$$= m \circ \exp(\hbar\pi^{ij}\partial_i \otimes \partial_j)(a \otimes b) = a \star_\pi b.$$

Here in the first line we formally introduced δ-function, so that the determinant and the sign $(i^{-1} = -i)$ are absorbed into the measure $\delta(\zeta)d\zeta$; then in the second line we replaced $\zeta \in \mathbb{R}^{2n}$ with the variable $k \in (\mathbb{R}^{2n})^*$, so that $\pi = \omega^{-1}$ determines an isomorphism $\pi^{\sharp} : (\mathbb{R}^{2n})^* \to \mathbb{R}^{2n}$. Next we replaced the δ-function by its Fourier transform (see (3.12)), and observed that $\pi^{\sharp}(k)\cdot q = \pi^{ij}k_i q_j = \pi(k,q)$; finally we noticed that the expressions, involving a and b would have looked as the composition of direct and inverse Fourier transform (see (3.13)), if it were not for the interference of $e^{i\hbar\pi(p,q)}$. Recalling, that the Fourier transform intertwines multiplication by k_i (or q_j) with partial derivation ∂_i, we get the result. It is important to remark, that all this reasoning was purely formal, as we have freely changed the order of integration and never asked for any convergence condition.

Exercises 1: General quantisation principles

E.1.1 Find the eigenvalues and eigenfunctions of the quantum harmonic oscillator, see Section 1.3.1.

E.1.2 Let $\mathcal{F}(f)(\xi)$ denote the Fourier transform of a (Schwarz) function $f(x)$ on $\mathbb{R}^n$. Recall that for any $a = a(x,\xi) \in C(\mathbb{R}^{2n})$ (with suitable decay conditions) we defined
$$(Op(a)f)(x) = \frac{1}{\sqrt{2^n \pi^n}} \int_{\mathbb{R}^n} a(x,\xi) \mathcal{F}(f)(\xi) e^{2i\pi x \cdot \xi} d\xi,$$
see formulas (2.6) and (2.7). Find $Q_i = Op(x_i)$, $P_i = Op(\xi_i)$ and $Op(x \cdot \xi)$.

E.1.3 Recall that for any $a = a(x,\xi) \in C(\mathbb{R}^{2n})$ (with suitable decay conditions) we defined *Weyl quantisation* of a by the formula (2.8)
$$(Op^W(a)(u))(x) = \iint_{\mathbb{R}^{2n}} e^{2\pi i(x-y) \cdot \xi} a\left(\frac{x+y}{2}, \xi\right) u(y) dy d\xi.$$
Find $Op^W(x_i)$, $Op^W(\xi_i)$ and $Op^W(x \cdot \xi)$.

E.1.4 Let $n = 3$. Find the eigenvalues and eigenfunctions of the momentum operator $P = \sum_{j=1}^{3} P_j$ in standard representation (i.e. by Op). Same for $Q = \sum_{j=1}^{3} Q_j$.

E.1.5 In previous notation, let $n = 3$. Find the expression for the angular momentum $L = Q \cdot P = \sum_{j=1}^{3} Q_j P_j$ and its square in spherical coordinates.

E.1.6 *Legendre's polynomials.* Let $P_l(x)$ be determined by the formula
$$(1 - 2xt + t^2)^{-\frac{1}{2}} = \sum_l P_l(x) t^l.$$
Prove the following properties of $P_l(x)$:

(a) $(l+1)P_{l+1}(x) - (2l+1)xP_l(x) + lP_{l-1}(x) = 0$;

(b) $P'_l(x) - 2xP'_{l-1}(x) + P'_{l-2}(x) = P'_{l-1}(x)$;

(c) $P_l(x) = \frac{1}{2^l l!} \frac{d^l}{dx^l} (x^2 - 1)^l$ (Rodriguez recurrence formula);

(d)
$$(1 - x^2) \frac{d^2}{dx^2} P_n(x) - 2x \frac{d}{dx} P_n(x) + n(n+1) P_n(x) = 0$$

(Legendre differential equation).

E.1.7 Let $P_n^m(x) = (1 - x^2)^{\frac{m}{2}} \frac{d^m}{dx^m} P_n(x)$ be the *associated Legendre polynomial*. Prove that it satisfies the associated Legendre differential equation:

$$(1 - x^2) \frac{d^2}{dx^2} P_n(x) - 2x \frac{d}{dx} P_n(x) + \left(n(n+1) - \frac{m^2}{1 - x^2} \right) P_n(x) = 0.$$

E.1.8* Use previous results to compute the eigenvalues and eigenfunctions of the square of the angular momentum L^2 and of the z-component of the angular momentum operator L_z.

E.1.9 Prove the commutation relation: $e^{isP} e^{itQ} = e^{ist} e^{isQ} e^{itP}$ for any two operators P, Q that satisfy the canonical commutation relations.

E.1.10 Let $\mathcal{H}$ be a Hilbert space, let L be any (real) vector space, L^* its dual and $U : L \to \mathcal{U}(\mathcal{H}), V : L^* \to \mathcal{U}(\mathcal{H})$ be two representations of the Abelian groups in $\mathcal{H}$ such that

$$V(f)U(x) = e^{if(x)} U(x) V(f).$$

This is called a *Weyl representation of L*. In what follows, we assume that it is fixed.

(a) Check the formula

$$W(z + z') = e^{-\frac{1}{2}\omega(z, z')} W(z) W(z')$$

for $z = (x, f), z' = (x', f') \in V = L \oplus L^*$ with standard symplectic structure, where $W(z) = e^{\frac{1}{2} if(x)} U(x) V(f)$.

(b) Let $R(z)$ be the "logarithm" of $W(z)$, i.e. $W(z) = e^{iR(z)}$. Check that
$$[R(z), R(z')] = -i\omega(z, z').$$

(c) Let $P(x)$ be the infinitesimal generator of $t \mapsto U(tx)$, $x \in L$; let $Q(x)$ be the infinitesimal generator of $t \mapsto V(tF(x))$, where $F : L \to L^*$ is an isomorphism, determined by a choice of basis. Show that the operators
$$a_i = \frac{1}{\sqrt{2}}(Q(e_i) + iP(e_i)), \quad a_i^\dagger = \frac{1}{\sqrt{2}}(Q(e_i) + iP(e_i)),$$
where $e_1, \ldots, e_n$ is the basis used before, satisfy the relations:
$$[a_j, a_k] = 0, \quad [a_j^\dagger, a_k^\dagger] = 0, \quad [a_j, a_k^\dagger] = \delta_{jk}.$$

E.1.11 Prove that in the ring $R[[t]]$ of formal power series the following statements hold:

(a) For any $F(t) = \sum_k a_k t^k$ with invertible a_0, there exists a unique inverse element $\frac{1}{F(t)} \in R[[t]]$;

(b) For any two formal power series $F(t), G(t)$ one can define their composition $F(G(t)) \in R[[t]]$;

(c) If $a_0 = 0$, $a_1 \neq 0$ and R is a field, one can find "inverse function" $F^{-1}(t) \in R[[t]]$, i.e. such element that $F^{-1}(F(t)) = t = F(F^{-1}(t))$.

E.1.12 Prove the properties 1.-5. of the Moyal product (see Proposition 3.2).

E.1.13 Prove that for any anti-symmetric $n \times n$-matrix $\pi = (\pi^{ij})$, the bracket $\{f, g\}_\pi = \pi^{ij} \partial_i f \partial_j g$ on $C^\infty(\mathbb{R}^n)$ satisfies all the properties of Poisson bracket (1.6).

4 Symplectic and Poisson structures

Before we can formulate the deformation quantisation problem, we need to establish the framework, in which it can be done in a sufficiently general way. This framework is that of symplectic and Poisson structures on manifolds, which we explain in this section.

4.1 The deformation quantisation problem in $\mathbb{R}^n$

The Moyal product (see Definition 3.1, Equations (3.8), (3.9)) and its properties (see Proposition 3.2) play important role in the quantisation theory: this product is the first example of a purely algebraic approach to the quantisation. In effect, let us consider the *Moyal algebra*, i.e. the algebra of formal power series of the variable $\hbar$ with coefficients in functions $C^\infty(\mathbb{R}^n)[[\hbar]]$, equipped with Moyal product. This is a noncommutative associative algebra. Let $n = 2k$ and let the skew-symmetric matrix π that determines the $\star$-product, be invertible, so we can choose a linear change of coordinates, such that π is given by the block form (3.3): $\pi = I_{2n}$.

Let us also restrict our attention from all smooth functions to just polynomials in the chosen coordinates on $\mathbb{R}^{2n}$. Then the resulting algebra can be identified with the algebra of polynomial differential operators on $\mathbb{R}^{2n}$, see Exercise E.2.1. Thus in the general case, the Moyal algebra can be regarded as a functorial way to associate a differential operator to any function on $\mathbb{R}^{2n}$. The main flaw of this idea is that we have no actual representation of the Moyal algebra on wave functions, we don't even have the convergence of the formal power series, that defines the product. This is not a problem from the point of view of pure Algebra though; it only means that we have abandoned the domain of analysis and only care about the algebraic properties of the formulas. This approach eventually leads to the deformation quantisation problem, see page 50, and the whole area related with this idea is called *deformation quantisation theory*.

The Moyal product formula is the first example of deformation quantisation. Its main defect is that it is extremely coordinates-sensitive: unless the matrix $\pi = (\pi^{ij})$ is constant (does not depend on the point in V), one cannot expect that the product $\star_\pi$, given by the formula (3.9) will be associative. It is not difficult to see that the associativity proof (see Proposition 3.2) fails unless $\partial_i \pi = 0$ for all i. On the other hand, nonlinear changes of variables turn constant bilinear function π and the bracket $\{f, g\}_\pi = \pi(df, dg)$ into generic one.

This brings forward the following problem, which we shall call *the general deformation quantisation problem* in $\mathbb{R}^n$:

> *For any matrix-valued function $\pi = \pi(z)$ (where $z = (z^1, \ldots, z^n)$ is some coordinate system) on $V = \mathbb{R}^n$, find a formula analogous to (3.8) or (3.9), which will turn the space of formal power series $C^\infty(V)[[\hbar]]$ into an associative algebra with $\hbar$-linear $\star$-product depending on π, $\star_\pi$. Describe different products $\star_\pi$ up to an equivalence.*

In this form the problem looks a bit too abstract (in effect, later we shall see that assumptions we made about π can also be refined and reformulated), so one usually adds the following important conditions that specify the form and properties of the $\star_\pi$-product:

> *Put $\{f, g\}_\pi = \pi(df, dg) = \sum_{i,j} \pi^{ij} \partial_i f \partial_j g$; then the associative $\star_\pi$-product should satisfy the conditions 1.–4. of Proposition 3.2, i.e.*
>
> 1. *$f \star_\pi g = \sum_{k=0}^\infty \hbar^k B_k(f, g)$ for some bidifferential operators B_k, $k = 0, 1, 2, \ldots$;*
> 2. *$f \star_\pi g = fg + \frac{\hbar}{2}\{f, g\}_\pi + o(\hbar)$, where $o(\hbar)$ are the terms of degree 2 and higher in $\hbar$;*
> 3. *$f \star_\pi 1 = f = 1 \star_\pi f$;*
> 4. *$B_k(f, g) = (-1)^k B_k(g, f)$.*

We have omitted the property 6 (which is just associativity) here because we had postulated it in the very beginning; as for the property 5, it deals with the polynomial functions, and the notion of polynomial function is not well-defined, if we allow generic coordinate systems on V. Moreover, condition 4 is rather technical and in many cases it is either removed at all or reformulated, and condition 3 means that the constant functions act as scalars if we substitute them in $\star_\pi$. This assumption is quite natural; it also can be reformulated in a more general way, but we shall not need it. From the point of view of bidifferential operators B_k it means that these operators have strictly positive bidegrees if $k > 0$.

It turns out that operation $\{,\}_\pi$ that we use here should satisfy some additional properties, if we want it to generate an associative product $\star_\pi$. Indeed, let us write down the associativity condition: expanding the equation $(f \star_\pi g) \star_\pi h = f \star_\pi (g \star_\pi h)$ in powers of $\hbar$ we get:

$$fgh + \frac{\hbar}{2}(\{f,g\}_\pi h + \{fg,h\}_\pi) + \hbar^2 \left(B_2(f,g)h \right.$$
$$\left. + \frac{1}{4}\{\{f,g\}_\pi, h\}_\pi + B_2(fg, h) \right)$$
$$= fgh + \frac{\hbar}{2}(\{f,gh\}_\pi + f\{g,h\}_\pi) + \hbar^2 \left(B_2(f,gh) \right.$$
$$\left. + \frac{1}{4}\{f, \{g,h\}_\pi\}_\pi + fB_2(g,h) \right).$$

If we compare terms with different degrees of $\hbar$, we see that the coefficients at $\hbar^0$ coincide automatically, and the coefficients at $\hbar$ are equal because the operation $\{,\}_\pi$ satisfies Leibniz rules in both variables (we shall usually omit the subscript π from $\{,\}_\pi$ if this causes no ambiguity):

$$\{f,g\}h + \{fg,h\}$$
$$= \{f,gh\} - \{f,h\}g + f\{g,h\} + g\{f,h\} = \{f,gh\} + f\{g,h\}.$$

Eventually the equation, involving $\hbar^2$ can be written as

$$\frac{1}{4}(\{\{f,g\},h\} - \{f,\{g,h\}\})$$
$$= fB_2(g,h) - B_2(fg,h) + B_2(f,gh) - B_2(f,g)h.$$

Let us rewrite the left-hand side as $\frac{1}{4}(\{\{f,g\},h\} + \{\{g,h\},f\})$ (this is possible due to the assumption we made: $\pi^{ij} = -\pi^{ji}$) and antisymmetrise this equation in f, g and h, i.e. consider all possible permutations of these symbols and take the sum of the corresponding expressions with the signs of permutations; this will give:

$$\{\{f,g\},h\} + \{\{h,f\},g\} + \{\{g,h\},f\} = 0 \qquad (4.1)$$

where on the left we took advantage of equation $\{f,g\} = -\{g,f\}$ and on the right we used condition 4. from above $(B_2(f,g) = B_2(g,f))$ and the commutativity of product of functions.

We have already met this equation before: this is the Jacobi identity for the Poisson bracket, see (1.9). That time it was a property of the standard Poisson bracket, associated with the Hamilton treatment of Newton's mechanics. Now this property is a natural condition, that should hold for the matrix-valued function π if we want to use it in the definition of a $\star$-product. It turns out, nonconstant matrix-valued functions with this property exist not only on Euclidean spaces, but also on manifolds[8] and play important role in Geometry and Differential equations; this allows one to extend the quantisation problem to nonlinear spaces. This is important from the point of view of applications to Physics, when one has to construct theories which work in curved spaces. We shall briefly discuss main properties of these structures here.

4.2 Symplectic structures on manifolds

Symplectic structures (see Definition 4.1) play an important role in the applications of Mathematics to various physical and mechanical problems, and had been studied extensively even before the term was actually coined. The ideas and methods that are somehow related with the study of symplectic structures are usually referred to as the *Symplectic Geometry* as a whole. There exist plenty of books on this subject, dealing with various aspects of these studies, all these books can be used as a good reference for the introduction to it. For example we can recommend the classical book by Arnold [Ar89] cited earlier,

---

[8]We assume that the reader is familiar with the Differential Geometry of manifolds at least on the level of basic definitions, such as local coordinate charts, tangent and cotangent bundles, differential forms, de Rham operator, vector fields and their commutators etc. There are many good introductory textbooks on Differential Geometry, any of which can be used as a reference; see for instance [KN63], or the first chapters of [He78].

Arnold and Givental's review [AG85] or the introductory chapters of the book by Ana Cannas da Silva [CdS08].

4.2.1 Definition and examples

Let M be a smooth manifold, not necessarily compact, but without boundary.

Definition 4.1. *One says that a **symplectic structure** on manifold M is given, if there's chosen a nondegenerate 2-form $\omega \in \Omega^2(M)$ with $d\omega = 0$; here and below d is de Rham differential operator on forms. In this case, manifold M with fixed ω is called **symplectic manifold** (M, ω), and the form ω is referred to as the symplectic form on M. Diffeomorphism $F : M_1 \to M_2$ between two symplectic manifolds (M_1, ω_1) and (M_2, ω_2), such that $F^*(\omega_2) = \omega_1$ is called **symplectomorphism**.*

Nondegeneracy of a form ω just means that the skew-symmetric matrix $\omega = (\omega_{ij})$, $\omega_{ij} = -\omega_{ji}$, representing the form in some local coordinate system on M is invertible (hence, the same is true for any local coordinate system). As one knows (see exercises) this is only possible if $\dim M$ is even. So from now on we shall assume that $\dim M = 2n$. Observe that if ω is such 2-form on M, then $\omega^n \in \Omega^{2n}(M)$ is a nowhere vanishing top degree form on M (see Exercise E.2.3). Thus any symplectic manifold M is automatically orientable; in particular if M is compact its top degree de Rham cohomology is not zero.

Let us begin by giving few standard examples of symplectic manifolds

Example 4.1. Let $M = \mathbb{R}^{2n}$ with coordinates $(p_1, \ldots, p_n, q^1, \ldots, q^n)$; consider the form

$$\omega = dp_1 \wedge dq^1 + \cdots + dp_n \wedge dq^n. \tag{4.2}$$

Clearly $d\omega = 0$; the matrix of ω is given by Equation (3.3): $I_{2n} = \begin{pmatrix} 0 & E_n \\ \hline -E_n & 0 \end{pmatrix}$. It is evidently invertible so that $I_{2n}^{-1} = -I_{2n}$.

Example 4.2. Let X be a smooth n-dimensional manifold, and T^*X its cotangent bundle. For any local coordinate system $(x^1, \ldots, x^n)$

on an open set $U \subset X$, we have the corresponding coordinates $(x^1, \ldots, x^n, \xi_1, \ldots, \xi_n)$ on $T^*U \subset T^*X$: $\xi_1, \ldots, \xi_n$ are defined as the unique coordinates on T^*U such that any covector $\varphi \in T_x^*U$, $x \in U$ is represented as $\xi_1(\varphi)dx^1 + \cdots + \xi_n(\varphi)dx^n$ for a fixed set of numbers $\xi_1(\varphi), \ldots, \xi_n(\varphi)$. Then we put:

$$\alpha = \xi_1 dx^1 + \cdots + \xi_n dx^n \in \Omega^1(T^*U).$$

When the coordinates on X are changed to $x^{1'}, \ldots, x^{n'}$ we have: $dx^k = \frac{\partial x^k}{\partial x^{k'}} dx^{k'}$, $\xi_{k'} = \frac{\partial x^{k'}}{\partial x^k} \xi_k$, so α does not depend on the choice of coordinates in X. Finally we put:

$$\omega = d\alpha = d\xi_1 \wedge dx^1 + \cdots + d\xi_n \wedge dx^n \in \Omega^2(T^*X).$$

Then clearly $d\omega = 0$ and from the coordinate expression it follows that ω is nondegenerate.

Example 4.3. Recall that a complex manifold M is called Kähler, if it is equipped with a Hermitian structure whose imaginary part is closed. Since, imaginary part of a Hermitian structure is always skew-symmetric (see Section 3.2.4), so it represents a differential 2-form on M. Since the Hermitian structure is positive definite, the form ω is nondegenerate. Thus any Kähler manifold is *par excellence* symplectic. In particular, any complex curve (complex manifold of complex dimension 1) is symplectic, because there are no 3-forms on X in this case.

4.2.2 Poisson brackets on symplectic manifolds

For every function f on symplectic manifold M we can define its *skew-gradient* or *Hamilton vector field* by the following rule:

$$\omega(X_f, Y) = -Y(f), \text{ for all vector fields } Y. \quad (4.3)$$

Observe that since ω is nondegenerate, the field X_f is uniquely defined: it is enough to put $X_f^i = \pi^{ij} \frac{\partial f}{\partial x^j}$, where $\pi = \omega^{-1}$. Indeed:

$$\omega(X_f, Y) = \omega_{ij} X_f^i Y^j = \omega_{ij} \pi^{ik} \frac{\partial f}{\partial x^k} Y^j = -\omega_{ji} \pi^{ik} \frac{\partial f}{\partial x^k} Y^j$$

$$= -\delta_j^k \frac{\partial f}{\partial x^k} Y^j = -\frac{\partial f}{\partial x^k} Y^k = -Y(f).$$

The fact that ω is a closed form has important corollaries. One of them is:

Proposition 4.2. *The form ω is invariant with respect to any Hamiltonian field X_f:*
$$\mathcal{L}_{X_f}\omega = 0,$$
where $\mathcal{L}_X$ is the Lie derivative of ω.

Proof. We use the *Cartan's "magic" derivation formula*: since $d\omega = 0$, we have
$$\mathcal{L}_X\omega = i_X d\omega + d i_X \omega = d i_X \omega,$$
where $i_X\omega$ is the contraction of ω and the field X; this is a 1-form such that
$$i_X\omega(Y) = \omega(X,Y), \text{ for any vector field } Y.$$

On the other hand, for a 1-form φ and any two fields Y, Z, the Cartan's formula for the de Rham differential gives
$$d\varphi(Y,Z) = Y(\varphi(Z)) - Z(\varphi(Y)) - \varphi([Y,Z]),$$
so in our case we have
$$(\mathcal{L}_{X_f}\omega)(Y,Z) = d(i_{X_f}\omega)(Y,Z)$$
$$= Y(\omega(X_f, Z)) - Z(\omega(X_f, Y)) - \omega(X_f, [Y,Z])$$
$$= -Y(Z(f)) + Z(Y(f)) + [Y,Z](f) = 0. \quad \square$$

Now one can define the *Poisson bracket*, associated with symplectic structure:

Definition 4.3. *Let f, g be two smooth functions on symplectic manifold M; then we put*
$$\{f, g\} = -X_f(g). \tag{4.4}$$

Proposition 4.4. *The bracket (4.4) satisfies all the properties of the Poisson bracket (see Section 1.2): it is skew-symmetric, linear over scalars, satisfies Leibniz rule and Jacobi identity.*

Proof. Let us rewrite the formula (4.4) a little:

$$\{f,g\} = -X_f(g) = -X_f^i \frac{\partial g}{\partial x^i}$$
$$= -\pi^{ij} \frac{\partial f}{\partial x^j} \frac{\partial g}{\partial x^i} = \pi^{ij} \frac{\partial f}{\partial x^i} \frac{\partial g}{\partial x^j} = \pi(df, dg),$$

since $\pi^{ij} = -\pi^{ji}$. Now the first three properties (skew-symmetry, linearity over scalars and Leibniz rule) are evident. As for the Jacobi identity, we observe that since $\mathcal{L}_{X_f}\omega = 0$ and $\pi = \omega^{-1}$, then $\mathcal{L}_{X_f}\pi = 0$, so using the properties of Lie derivatives and the skew symmetry of π we compute:

$$\{f, \{g,h\}\} = -X_f(\pi(dg, dh))$$
$$= -\mathcal{L}_{X_f}(\pi)(dg, dh) - \pi(\mathcal{L}_{X_f}(dg), dh) - \pi(dg, \mathcal{L}_{X_f}(dh))$$
$$= \pi(dh, d\mathcal{L}_{X_f}(g)) - \pi(dg, d\mathcal{L}_{X_f}(h))$$
$$= \{\{f,g\},h\} + \{g,\{f,h\}\},$$

which is precisely the Jacobi identity. □

Observe that we wrote Jacobi identity in the form

$$\{f, \{g,h\}\} = \{\{f,g\},h\} + \{g,\{f,h\}\}, \qquad (4.5)$$

which has a transparent algebraic meaning: *Poisson bracket satisfies Leibniz rule with respect to itself.*

4.2.3 Darboux' theorem

Symplectic manifolds have many important geometric and topological properties; in particular, one can develop the Hamiltonian mechanics on any symplectic manifold (M,ω) and obtain results similar to the results in $\mathbb{R}^{2n}$ with the standard form (4.2): $\omega = \sum_{i=1}^n dp_i \wedge dq^i$. It turns out, that in fact every symplectic structure on $2n$-dimensional manifold is locally isomorphic to this one. Namely

Theorem 4.5 (Darboux' theorem). *Let (M^{2n}, ω) be a symplectic manifold; then for every $x_0 \in M$ there exists in a an open neighbourhood of x_0 a local coordinate system $(p_1, \ldots, p_n, q^1, \ldots, q^n)$ centered*

at x_0, in which ω takes the form $\omega = \sum_{i=1}^{n} dp_i \wedge dq^i$. In other words, there exists a local symplectomorphism between M and $\mathbb{R}^{2n}$ equipped with the standard form. The coordinates $(p_1, \ldots, p_n, q^1, \ldots, q^n)$ are called **(local) Darboux coordinate system** in M.

We are not going to prove this theorem here in full generality, let us just sketch the main steps of the

Proof. We begin with choosing any function f in the neighbourhood of x_0 for which $f(x_0) = 0$ and $X_f(x_0) \neq 0$. This is possible, since ω is nondegenerate, so this just means that $df(x_0) \neq 0$. We put $p_1 = f$ and take N^{2n-1} to be any hypersurface that passes through x_0 transversally to X_f. Let $g_t : M \to M$ be the (local) 1-parameter group of diffeomorphisms, for which $\frac{dg_t}{dt} = X_f$, i.e. g_t is the flow of the Hamiltonian field X_f. Let q^1 be the function (defined in a neighbourhood of x_0) such that

$$q^1(x) = t, \text{ if and only if } g_t(x) \in N^{2n-1}.$$

Then q^1 is a smooth function in vicinity of x_0, such that $q_1(x_0) = 0$ and for any x near x_0

$$\{p_1, q^1\}(x) = -(X_f(q^1))(x) = -\frac{dq^1(g_t(x))}{dt} = -\frac{d}{dt}(t_0 - t) = 1.$$

Now we can go on reasoning by induction on dimension of M: we take $\tilde{M}^{2n-2}$ to be the joint level set $p_1 = q^1 = 0$ and check that restriction of ω to $\tilde{M}$ is a symplectic form; then by inductive hypothesis we can find the canonical coordinates $p_2, \ldots, p_n, q^2, \ldots, q^n$ on $\tilde{M}$, and then extend them in a natural way to a small open subset in M so that $(p_1, \ldots, p_n, q^1, \ldots, q^n)$ are the canonical coordinates there. □

In spite of the fact that this result is very important from the point of view of the general theory, it can hardly be used in practice to describe the structure of a symplectic manifold as one has very little control over the way canonical coordinates on open subsets of M interact. However, it means that locally we can always find a Moyal product on functions in $C^\infty(M)$; thus the general problem in this case consists of gluing together Euclidean pieces, so that the formulas for $\star$-product match on their intersections.

4.3 Poisson structures on manifolds

Symplectic structures and their properties are an important subject of study in Differential Geometry, and constitute a natural context for the deformation quantisation. However this is not the most general situation when this problem exists: in many naturally arising situations the condition of nondegeneracy of the matrix ω is too restrictive. It often happens that a dynamical system can be formulated in Hamiltonian way, while there is no global symplectic structure at sight. These observations justify the following definition:

Definition 4.6. *Let M be a smooth manifold (possibly noncompact). One says that M is **Poisson manifold** or is equipped with a **Poisson structure**, if there is given a bilinear over scalars operation $\{,\} : C^\infty(M) \otimes C^\infty(M) \to C^\infty(M)$ on functions on M, called **Poisson bracket**, which satisfies for all functions $f, g, h \in C^\infty(M)$ the following three conditions:*

(i) *Skew symmetry:* $\{f, g\} = -\{g, f\}$;

(ii) *Leibniz rule:* $\{fg, h\} = f\{g, h\} + \{f, h\}g$;

(iii) *Jacobi identity, see Section* 1.2 *and Equation* (4.5):

$$\{\{f, g\}, h\} + \{\{g, h\}, f\} + \{\{h, f\}, g\} = 0.$$

A map $F : M \to N$ between two Poisson manifolds is called a **Poisson morphism** if $\{F^*(f), F^*(g)\} = F^*(\{f, g\})$ for all $f, g \in C^\infty(N)$.

Of course, every symplectic manifold is *par excellence* Poisson manifold, but the contrary is not at all true; in effect it follows from the definition of the bracket associated with a symplectic form, that the Hamilton vector field $X_f = -\{f, \cdot\}$ cannot vanish in a point, where $df \neq 0$. Hence the simplest counterexample is any smooth manifold M equipped with trivial Poisson bracket: $\{f, g\} = 0$ for all f, g. A slightly more elaborate example is $\mathbb{R}^2$ with coordinates (x, y) and the bracket, determined by equality:

$$\{x, y\} = a(x, y)$$

for some function $a \in C^\infty(\mathbb{R}^2)$. Indeed, it follows from conditions above that in this case for all functions f, g on $\mathbb{R}^2$ we will have

$$\{f,g\} = a(x,y)(\partial_x f \partial_y g - \partial_y f \partial_x g),$$

and the Jacobi identity follows by direct computation (we leave it to the reader, see Exercise E.2.17). Then clearly the operation $\{f,\cdot\}$ vanishes in all points, where $a(x,y) = 0$.

The general theory of Poisson structures on manifolds and the study of their properties is the subject of *Poisson Geometry*, a recently emerged branch of Mathematics. In this book, we will use only very basic results from this field; our exposition here is basically self-contained. The reader interested in further acquaintance with the field can look into the book [CFM21] for instance for more details; a shorter and somewhat less formal text is in the lecture notes by Ciccoli [Ci06].

4.3.1 Polyvector fields

In order to understand deeper the algebraic properties and geometry behind a Poisson structure, it is convenient to modify Definition 4.6 as follows: since every map $X : C^\infty(M) \to C^\infty(M)$ that satisfies Leibniz rule is in effect given by differentiation along a vector field, we conclude from the properties (i) and (ii) that in any local coordinate system $(x^1, \ldots, x^n)$ on M Poisson bracket is given by the formula

$$\{f,g\} = \sum_{i,j=1}^n \pi^{ij} \frac{\partial f}{\partial x^i} \frac{\partial g}{\partial x^j} = \pi(df, dg), \tag{4.6}$$

where $\pi = (\pi^{ij})$ is an antisymmetric tensor field of type $(2,0)$. In other words, *π is a section of the bundle $\wedge^2 TM$, the exterior square of the tangent bundle of M*. The bundles $\wedge^k TM$ play important role in deformation quantisation, their sections can be regarded as dualisation of the differential forms. Namely:

Definition 4.7. *Let M be a smooth manifold. Then sections of the exterior powers $\wedge^k TM$, $k = 0, 1, \ldots, n$ of the tangent bundle of M, are called **polyvector fields** or simply **polyvectors** on M; we shall* denote the space of polyvector fields of degree k by $\mathcal{T}^k(M)$; in particular $\mathcal{T}^0(M) = C^\infty(M)$, $\mathcal{T}^1(M) = Vect(M)$ (the space of vector fields on M). Sections of $\mathcal{T}^2(M)$ are usually called **bivectors**.

Similarly to differential forms, any polyvector field $\varphi \in \mathcal{T}^p(M)$ in local coordinates $(x^1, \ldots, x^n)$ can be written in a unique way as

$$\varphi = \sum_{1 \leq i_1 < i_2 < \cdots < i_p \leq n} \varphi_{i_1 i_2 \ldots i_p}(x^1, \ldots, x^n) \partial_{i_1} \wedge \partial_{i_2} \wedge \cdots \wedge \partial_{i_p}.$$

Here $\partial_i = \frac{\partial}{\partial x^i}$ and $\varphi_{i_1 i_2 \ldots i_p}(x^1, \ldots, x^n)$ are smooth functions. Sometimes it is more convenient to regard φ as a sum of "decomposable" polyvectors:

$$\varphi = \sum_{k=1}^{N} X_1^k \wedge X_2^k \wedge \cdots \wedge X_p^k, \tag{4.7}$$

for some vector fields X_i^p; here $\wedge$ denotes the exterior product of vector fields. It is clear, that this product naturally extends to the graded product of polyvectors: $\wedge : \mathcal{T}^p(M) \otimes \mathcal{T}^q(M) \to \mathcal{T}^{p+q}(M)$, such that for $\varphi = X_1 \wedge \cdots \wedge X_p, \psi = Y_1 \wedge \cdots \wedge Y_q$

$$\varphi \wedge \psi = X_1 \wedge \cdots \wedge X_p \wedge Y_1 \wedge \cdots \wedge Y_q.$$

Obviously this product is *graded commutative*: $\varphi \wedge \psi = (-1)^{pq} \psi \wedge \varphi$.

Using the observation, that any differentiation of the algebra $C^\infty(M)$ is given by a vector field, it is easy to show that *every multilinear anti-symmetric map $D : C^\infty(M)^{\otimes p} \to C^\infty(M)$, satisfying the Leibniz rule in all arguments is given by a unique polyvector field $\varphi \in \mathcal{T}^p(M)$ by the formula*

$$D(f_1 \ldots, f_p) = \varphi(df_1, \ldots, df_p)$$
$$= \sum_{1 \leq i_1 < i_2 < \cdots < i_p \leq n} \varphi_{i_1 i_2 \ldots i_p}(x^1, \ldots, x^n) \partial_{i_1}(f_1) \wedge \cdots \wedge \partial_{i_p}(f_p).$$

Let us apply this observation to the left-hand side of Jacobi identity, see condition (*iii*) of Definition 4.6:

$$J_\pi(f, g, h) = \{\{f, g\}, h\} + \{\{g, h\}, f\} + \{\{h, f\}, g\}. \tag{4.8}$$

An easy computation (see Exercise E.2.8) shows that J_π is antisymmetric and satisfies the Leibniz rule. We shall denote by γ_π the corresponding polyvector. Clearly, *the bracket induced by a bivector satisfies Jacobi identity iff $\gamma_\pi = 0$*. Our purpose now is to identify γ_π.

4.3.2 Schouten bracket

Although of course the representation (4.7) is not unique, it is convenient for many purposes. For us the main advantage of using this formula is the following theorem, or rather *definition-theorem* in the sense of Grothendieck:

Theorem 4.8 (Definition of the Schouten bracket). *For every smooth manifold M there exists a unique bilinear operation*

$$[,] : \mathcal{T}^p(M) \otimes \mathcal{T}^q(M) \to \mathcal{T}^{p+q-1}(M) \text{ for all } p, q = 0, \ldots, n,$$

such that, the following three conditions (suitable graded versions of the conditions satisfied by Poisson bracket) hold:

(i) *Graded skew symmetry:* $[\alpha, \beta] = -(-1)^{(p-1)(q-1)}[\alpha, \beta];$

(ii) *Graded Leibniz rule with respect to the wedge product:*

$$[\alpha \wedge \beta, \gamma] = \alpha \wedge [\beta, \gamma] + (-1)^{q(r-1)}[\alpha, \gamma] \wedge \beta;$$

(iii) *Graded Jacobi identity:*

$$(-1)^{(p-1)(r-1)}[[\alpha, \beta], \gamma] + (-1)^{(q-1)(p-1)}[[\beta, \gamma], \alpha]$$
$$+ (-1)^{(r-1)(q-1)}[[\gamma, \alpha], \beta] = 0,$$

for all $\alpha, \beta, \gamma \in \mathcal{T}^*(M)$ of degrees p, q and r respectively, and such, that

(iv) $[f, g] = 0$ *for all functions $f, g \in C^\infty(M) = \mathcal{T}^0(M);$*

(v) $[X, \varphi] = \mathcal{L}_X \varphi$ *for any vector field X and polyvector φ.*

*This operation is called the **Schouten bracket** (or the **Schouten–Nijenhuis bracket**).*

We are not going to prove this theorem here in all details. Just observe, that the axioms (i)–(v) without the Jacobi identity uniquely determine this operation on polyvectors written in the form (4.7); indeed for any $\varphi = X_1 \wedge \cdots \wedge X_p, \psi = Y_1 \wedge \cdots \wedge Y_q$ we shall have

$$[\varphi,\psi] = \sum_{\substack{1\leq i\leq p \\ 1\leq j\leq q}} (-1)^{i+j+p-1}[X_i,Y_j]\wedge X_1 \wedge \cdots \wedge \widehat{X_i} \wedge \cdots \wedge X_p \wedge$$

$$\wedge Y_1 \wedge \cdots \wedge \widehat{Y_j} \wedge \cdots \wedge Y_q, \quad (4.9)$$

where $\widehat{}$ denotes the element missing from the row. We leave it to the reader to check that this formula is well-defined and satisfies all the properties of Theorem 4.8 (see Exercise E.2.9).

4.3.3 Poisson bivectors

Let now π be a bivector on M. It is our purpose to prove the following proposition:

Proposition 4.9. *The 3-vector field γ_π, such that*

$$J_\pi(f,g,h) = \gamma_\pi(df,dg,dh)$$

is equal to the half of the Schouten bracket $[\pi,\pi]$.

Observe that in general situation the bracket $[\pi,\pi]$ needs not be equal to 0 unlike the commutator of a vector field with itself: that is due to the graded nature of skew symmetry, satisfied by the Schouten bracket!

Our proof of Proposition 4.9 is based on the following expression due to Lichnerowicz, which represents Schouten bracket in terms of the natural pairing of polyvectors and differential forms and other natural operations on polyvectors and polycovectors:

$$\langle \omega, [P,Q]\rangle = (-1)^{(p-1)(q-1)}\langle di_Q\omega, P\rangle - \langle di_P\omega, Q\rangle + (-1)^p \langle d\omega, P\wedge Q\rangle,$$

where $\langle,\rangle$ is the natural pairing between forms and polyvector fields and $i_P\omega$ is the contraction of a form and a polyvector, generalising that of a form and a vector field. We leave the proof of this formula to the reader, see Exercise E.2.10.

Now we can take the 3-form $\omega = df \wedge dg \wedge dh$ and check that $\langle \omega, [\pi,\pi]\rangle = \frac{1}{2}J_\pi(f,g,h)$:

$$\langle \omega, [\pi,\pi]\rangle = -2\langle di_\pi\omega, \pi\rangle$$

since $d\omega = 0$. But

$$di_\pi\omega = \frac{1}{2}(d\pi(df,dg)dh + d\pi(dg,dh)df + d\pi(dh,df)dg)$$
$$= \frac{1}{2}(d\{f,g\} \wedge dh + d\{g,h\} \wedge df + d\{h,f\} \wedge dg),$$

so

$$\langle di_\pi\omega, \pi\rangle = -\frac{1}{4}(\pi(d\{f,g\},dh) + \pi(d\{g,h\},df) + \pi(d\{h,f\},dg))$$
$$= -\frac{1}{4}J_\pi(f,g,h).$$

Definition 4.10. *Bivectors $\pi \in \mathcal{T}^2(M)$, which satisfy the condition $[\pi,\pi] = 0$ are called **Poisson bivectors**; so we see that **choosing a Poisson structure on a manifold is equivalent to choosing a Poisson bivector on it**.*

4.3.4 Weinstein's theorem

The geometry of Poisson structures on manifolds is much more complicated than that of symplectic structures. The main problem is that the matrix $\pi = (\pi^{ij})$ can be very singular. Namely, let $rk\,\pi(x)$ be the rank of the skew symmetric matrix $(\pi^{ij}(x))$. Clearly π is symplectic iff $rk\,\pi(x) = 2n$ in all x. However $rk\,\pi(x)$ in generic case is a nonconstant semicontinous function, taking even positive values. Let us briefly describe the structure of Poisson manifolds depending on $rk\,\pi(x)$. An interested reader can look for details in Weinstein's paper.

The first important particular case, is when $rk\,\pi = 2k < 2n$ is constant. In this case, we can consider in every point $x \in M$ the subspace $V_x(\pi) \subset T_xM$, $\dim V_x = 2k$ on which π is nondegenerate. Then one can prove that $V(\pi)$ is an integrable distribution; its leaves are called *symplectic leaves* of the Poisson structure. In general case, when the function $rk\,\pi$ is not constant, M is stratified by sets of constant rank, which are further foliated by symplectic leaves of different dimensions.

The local structure of any Poisson manifold is described by the following statement, similar to the Darboux' Theorem 4.5:

Theorem 4.11 (Weinstein's theorem [Wei83]).

(*i*) *For any point* $x \in M$, $\dim M = n$ *there exists a local diffeomorphism of Poisson manifolds between an open neighbourhood* U *of* x *and the cartesian product of Poisson manifolds*[9] $\mathbb{R}^{2k} \times \mathbb{R}^{n-2k}$, *which sends* x *to* 0, *such that* $\mathbb{R}^{2k}$ *is symplectic space with standard symplectic structure and* $\mathbb{R}^{n-2k}$ *is equipped with a Poisson bivector* π^s *for which* $\pi^s(0) = 0$;

(*ii*) *In particular if the rank* $\mathrm{rk}\,\pi = 2k$ *is constant in a neighbourhood of* x, *then the structure* π^s *above can be taken identically equal to* 0, *so in this case one can find local coordinates* $(p_1, \ldots, p_k, q^1, \ldots, q^k, y^1, \ldots, y^{n-2k})$ *around* x *in which the Poisson bracket on* M *takes the form* $\{p_i, q^j\} = \delta_i^j$ *and all the other brackets of these coordinates vanish.*

4.4 Deformation quantisation problem

Now we can formulate the most general version of the deformation quantisation problem, that first was articulated in [BFFLS78₁] and [BFFLS78₂] and also in [Fl82]:

Problem 1. Let M be a Poisson manifold, π the corresponding Poisson bivector. Find an associative $\star$-product on $C^\infty(M)[[\hbar]]$, such that

1. $f \star_\pi g = \sum_{k=0}^{\infty} \hbar^k B_k(f, g)$ for some bidifferential operators B_k, $k = 0, 1, 2, \ldots$;

2. $f \star_\pi g = fg + \frac{\hbar}{2}\{f, g\}_\pi + o(\hbar)$.

Classify all $\star$-products on M up to an isomorphism, similar to the relation from Section 3.2.3, see Proposition 3.3: *two* $\star$-*products* $\star$ *and* $\star'$ *are called* **equivalent** *if there exists a formal power series* T *of differential operators beginning with the identity operator, such that the map* $T: C^\infty(M)[[\hbar]] \to C^\infty(M)[[\hbar]]$, *i.e.*

$$T(f) = f + \hbar T_1(f) + \hbar^2 T_2(f) + \cdots + \hbar^n T_n(f) + \ldots \quad (4.10)$$

induces isomorphism of algebras:

$$T(f \star g) = T(f) \star' T(g). \quad (4.11)$$

---
[9]In terms of bivectors, this means that the bivector on the product space is equal to the sum of bivectors on factors.

Sometimes in addition to the conditions 1 and 2 one also adds the following technical conditions:

3. $f \star_\pi 1 = f = 1 \star_\pi f$;

4. $B_k(f, g) = (-1)^k B_k(g, f)$,

which prove handy in many situations, but we shall often not take them into account. Also one can relax the condition that π is a Poisson structure and consider $\star$-products associated with more general *formal Poisson structures* i.e. formal power series in $\hbar$ with coefficients bivectors $\pi = \pi_0, \pi_1, \pi_2, \ldots$, such that

$$\Pi = \pi_0 + \hbar \pi_1 + \hbar^2 \pi_2 + \cdots$$

satisfies the condition $[\Pi, \Pi] = 0$ where $[,]$ is the formal extension of Schouten brackets to the power series. The purpose of considering formal structures will be clear from our discussion of Kontsevich's theorem, see Sections 17 and 18.

The remaining part of this course is dedicated to various methods that are used to solve the deformation quantisation problem. The main theorem here is the well-known *Kontsevich's theorem*, which says that the problem always has a solution, and two such solutions are equivalent, if the corresponding Poisson structures are (formally) equivalent. However before we address Kontsevich's construction we shall spend some time dealing with few simpler particular cases of this problem, first of all the cases when π is in fact generated by a symplectic form, or when $M = \mathbb{R}^n$ is a Euclidean space.

5 Deformations and Lie algebras: A nontrivial example

The material, which we discuss in this section is related with the basic properties of Lie algebras and Lie groups. There are plenty of introductory monographs, surveys and textbooks on this subject, of which we can recommend [He78], or in case you prefer a more algebraic approach, the well-known lecture notes by J.P. Serre [Se65].

5.1 Lie algebras and Poisson structures

One of the simplest but important nontrivial examples of Poisson structures is the following construction, related with Lie algebras, so we begin with recalling the main definitions of the Lie theory. Recall that *Lie algebra* is a vector space $\mathfrak{g}$ equipped with bilinear skew-symmetric bracket operation:

$$[,]: \mathfrak{g} \otimes \mathfrak{g} \to \mathfrak{g}, \ [X,Y] = -[Y,X], \ \forall X, Y \in \mathfrak{g}$$

called *the Lie bracket*. This operation should satisfy only one condition, Jacobi identity:

$$[[X,Y],Z] + [[Y,Z],X] + [[Z,X],Y] = 0, \ \forall X, Y, Z \in \mathfrak{g}.$$

Lie algebras appear naturally in various situations in Mathematics and Physics; examples are vector fields on a manifold with respect to the commutator operation, functions on a Poisson manifold with respect to Poisson bracket, or square matrices with respect to the matrix commutator: $[A, B] = AB - BA$. Finite-dimensional Lie algebras naturally arise from the study of *Lie groups*, they were extensively studied and classified in the XX-th century.

So let $\mathfrak{g}$ be a finite-dimensional real or complex Lie algebra. In physical literature it is tradition to write down the Lie bracket on $\mathfrak{g}$ in terms of the *structure constants*: if $e_1, \ldots, e_n$ is a basis in $\mathfrak{g}$, then the structure constants C_{pq}^r, $p, q, r = 1, \ldots, n$ are the (real or complex) numbers determined by equation:

$$[e_p, e_q] = C_{pq}^r e_r. \tag{5.1}$$

Anti-symmetry of the bracket now just means that $C_{pq}^r = -C_{qp}^r$ and Jacobi identity amounts to the following system of equations

$$C_{pq}^i C_{ir}^s + C_{qr}^i C_{ip}^s + C_{rp}^i C_{iq}^s = 0, \quad \forall p, q, r, s = 1, \ldots, n. \tag{5.2}$$

The following construction of Poisson structure is often referred to as the *Kirillov–Kostant* or *Kirillov–Kostant–Souriau structure*, but most probably it has been known much earlier to Sophus Lie and other XIX-th century mathematicians, and therefore it is called the *Lie-Poisson structure*. Let $\mathfrak{g}$ be a finite-dimensional Lie algebra and $\mathfrak{g}^*$ its dual space. Let $e_1, \ldots, e_n$ and $e^1, \ldots, e^n$ be the dual basis in $\mathfrak{g}$ and $\mathfrak{g}^*$ respectively, and let $x_1, \ldots, x_n$ be the coordinates in $\mathfrak{g}^*$, corresponding to this basis.

Definition 5.1. *The Kirillov–Kostant structure on $\mathfrak{g}^*$ is the Poisson structure, defined by the formula*

$$\{f, g\} = x_r C_{pq}^r \partial_p f \partial_q g \text{ for all } f, g \in C^\infty(\mathfrak{g}^*),$$

where the summation over the repeating indices is implied and $\partial_i = \frac{\partial}{\partial x_i}$.

The fact that this expression is independent of the choices made follows directly from the definitions, if we apply the tensor laws to the basis change from $e_1, \ldots, e_n$ to $e_{1'}, \ldots, e_{n'}$ (see also Remark 5.2 below). The Jacobi identity for this structure is equivalent to the corresponding identity for $\mathfrak{g}$, see Equation (5.2): we compute

$$\{\{f, g\}, h\} = x_s C_{ir}^s \partial_i (x_t C_{pq}^t \partial_p f \partial_q g) \partial_r h = x_s C_{ir}^s C_{pq}^i \partial_p f \partial_q g \partial_r h + \ldots,$$

where ... denote terms with quadratic derivatives; we do not take them into account since the left-hand side of Jacobi identity is always represented by a polyvector, see Section 4.3.1 and Equation (4.8). Now similar computations will give:

$$\{\{g, h\}, f\} = x_s C_{ir}^s C_{pq}^i \partial_p g \partial_q h \partial_r f + \ldots,$$
$$\{\{h, f\}, g\} = x_s C_{ir}^s C_{pq}^i \partial_p h \partial_q f \partial_r g + \ldots.$$

Summing up these three equations and eliminating the quadratic derivatives, we get:

$$\{\{f, g\}, h\} + \{\{g, h\}, f\} + \{\{h, f\}, g\}$$
$$= x_s (C_{ir}^s C_{pq}^i + C_{ip}^s C_{qr}^i + C_{iq}^s C_{rp}^i) \partial_p f \partial_q g \partial_r h,$$

where the right-hand side contains the factor, identical to Equation (5.2).

Remark 5.2. If we restrict the functions on $\mathfrak{g}^*$ to polynomials, we will be able to introduce the Kirillov–Kostant bracket $\{,\}$ in a coordinate-free manner: since by definition every linear function on $\mathfrak{g}^*$ is an element $X \in \mathfrak{g}$, we can identify the algebra of polynomial functions on $\mathfrak{g}^*$ with symmetric algebra of the vector space $\mathfrak{g}$, $S(\mathfrak{g})$. Now by Leibniz rule we see that any Poisson bracket on $S(\mathfrak{g})$ is uniquely determined by its values on $\mathfrak{g}$. So we assume that on $\mathfrak{g}$ the Poisson bracket coincides with the Lie bracket then the values of the bracket on the elements of $S(\mathfrak{g})$ are uniquely defined; for example on monomials $f = X_1 X_2 \ldots X_p$, $g = Y_1 Y_2 \ldots Y_q \in S(\mathfrak{g})$ we get the formula

$$\{f,g\} = \sum_{\substack{1 \leq i \leq p \\ 1 \leq j \leq q}} [X_i, Y_j] X_1 \ldots \widehat{X_i} \ldots X_p Y_1 \ldots \widehat{Y_j} \ldots Y_q,$$

c.f. the formula (4.9).

Another approach to this construction consists of the following observation: for all functions $f, g \in C^\infty(\mathfrak{g}^*)$ and every point $x \in \mathfrak{g}^*$ the differentials $df(x), dg(x)$ of these functions at x lie in the cotangent space $T_x^* \mathfrak{g}^*$ which is naturally isomorphic to $\mathfrak{g}$. Hence we can put:

$$\{f,g\}(x) = \langle x, [df(x), dg(x)] \rangle,$$

where $\langle, \rangle$ denotes the natural pairing between $\mathfrak{g}^*$ and $\mathfrak{g}$. It is easy to see that all these formulas induce the same Poisson structure.

5.2 Universal enveloping algebras and PBW theorem

One of the main algebraic constructions connected with Lie algebras is the universal enveloping algebra. It is applicable to all Lie algebras over arbitrary fields or even rings of coefficients, but we will restrict our attention to the real or complex finite-dimensional case. An interested reader wanting to learn more about the universal enveloping algebras and their properties can consult the wonderful book by Dixmier [Dix77].

So let $\mathfrak{g}$ be a finite-dimensional Lie algebra over $\Bbbk = \mathbb{R}$ or $\mathbb{C}$. Recall that homomorphism of Lie algebras is any linear map between them that preserves the Lie bracket, and that for every associative algebra A we can introduce the Lie bracket in it as the commutator of the elements in A: $[a, b] = ab - ba$. We shall denote this Lie algebra by $A_\mathcal{L}$

Definition 5.3. *Associative algebra $U\mathfrak{g}$ is called **universal enveloping algebra of** $\mathfrak{g}$, if there exists a homomorphism of Lie algebras $u : \mathfrak{g} \to U\mathfrak{g}_\mathcal{L}$ such that for any associative algebra A and any homomorphism of Lie algebras $f : \mathfrak{g} \to A_\mathcal{L}$ there exists a unique homomorphism of associative algebras $f_u : U\mathfrak{g} \to A$ such that $f = f_u \circ u : \mathfrak{g} \to A_\mathcal{L}$.*

The existence and uniqueness of $U\mathfrak{g}$ is proved by standard methods; for our purposes it is enough to know that the following construction gives universal enveloping algebras for all finite-dimensional real and complex Lie algebras $\mathfrak{g}$:

$$U\mathfrak{g} = T^\otimes \mathfrak{g} / \langle X \otimes Y - Y \otimes X - [X, Y] \mid X, Y \in \mathfrak{g} \rangle. \tag{5.3}$$

Here $T^\otimes \mathfrak{g} = \bigoplus_{k \geq 0} \mathfrak{g}^{\otimes k}$ is the tensor algebra of $\mathfrak{g}$, and $\langle \cdot \rangle$ denotes the ideal generated by the expression in the angular brackets.

The algebra $U\mathfrak{g}$ has lots of various properties. For us the most important one will be the fact that it is linearly isomorphic to the symmetric algebra $S(\mathfrak{g})$ of $\mathfrak{g}$. More accurately, let $e_1, \ldots, e_n$ as above be a basis in $\mathfrak{g}$; we observe that since $\mathfrak{g}^{\otimes k}$ is linearly spanned by the tensor products $e_{i_1} \otimes \cdots \otimes e_{i_k}$, the algebra $U\mathfrak{g}$ is spanned by the images of these expressions under the natural projection; using the commutator relation

$$e_p e_q = e_q e_p + C_{pq}^r e_r \tag{5.4}$$

inductively, we see that linearly $U\mathfrak{g}$ is generated by the linear combinations of the following form

$$\sum_{I=(i_1,\ldots,i_n)} a_I e_1^{i_1} \ldots e_n^{i_n}, \quad a_I \in \mathbb{k}, \quad i_k = 0, 1, \ldots. \tag{5.5}$$

Then the claim we made above is a consequence of the next theorem, called the *Poincaré–Birkhoff–Witt theorem* or *PBW theorem* for short:[10]

---

[10]The reader should be warned that in different books the statement of Theorem 5.4 can be formulated in different equivalent ways.

Theorem 5.4 (Poincaré–Birkhoff–Witt theorem). *Monomials $e_I = e_1^{i_1} \ldots e_n^{i_n}$ are linearly independent in $U\mathfrak{g}$ for different I.*

This means, that the representation (5.5) of elements in $U\mathfrak{g}$ by linear combinations of these monomials is in fact unique, i.e. that monomials $e_1^{i_1} \ldots e_n^{i_n}$ give a basis in $U\mathfrak{g}$. There exist many different proofs of this theorem, some of which are applicable to Lie algebras over general coefficient rings and even to Lie algebroids.[11] The proof that we shall sketch here for real or complex Lie algebras is also applicable to other ground fields.

Proof. Recall that basis in $T^\otimes \mathfrak{g}$ consists of the elements $e_{i_1} \otimes \cdots \otimes e_{i_p}$ for all $p \in \mathbb{Z}_{\geq 0}$ and all sets of indices $i_1, \ldots, i_p = 1, \ldots, n$. We shall now prove the linear independence of the elements $e_1^{i_1} \ldots e_n^{i_n} \in U\mathfrak{g}$ by constructing a map:

$$L : T^\otimes \mathfrak{g} \to T^\otimes \mathfrak{g}$$

such that

(i) $L(e_{i_1} \otimes e_{i_2} \otimes \cdots \otimes e_{i_p}) = e_{i_1} \otimes e_{i_2} \otimes \cdots \otimes e_{i_p}$, if $i_1 \leq i_2 \leq \cdots \leq i_p$;

(ii) $L(\mathcal{R}) = 0$, where $\mathcal{R} = \langle X \otimes Y - Y \otimes X - [X, Y] \mid X, Y \in \mathfrak{g} \rangle$ is the ideal, generating the algebra $U\mathfrak{g}$; in terms of the basis elements, this is equivalent to the condition

$$\begin{aligned} L(e_{i_1} &\otimes e_{i_2} \otimes \cdots \otimes e_{i_k} \otimes e_{i_{k+1}} \otimes \cdots \otimes e_{i_p}) \\ &= L(e_{i_1} \otimes e_{i_2} \otimes \cdots \otimes e_{i_{k+1}} \otimes e_{i_k} \otimes \cdots \otimes e_{i_p}) \\ &\quad + L(e_{i_1} \otimes e_{i_2} \otimes \cdots \otimes [e_{i_k}, e_{i_{k+1}}] \otimes \cdots \otimes e_{i_p}) \end{aligned} \quad (5.6)$$

for all sets of indices $i_1, \ldots, i_p$; observe that this equation is symmetric in the sense that it is the same for the sequences $i_1, \ldots, i_k, i_{k+1}, \ldots, i_p$ and $i_1, \ldots, i_{k+1}, i_k, \ldots, i_p$.

This would mean that the subspace $\mathcal{I} = Span(e_{i_1} \otimes e_{i_2} \otimes \cdots \otimes e_{i_p} \mid i_1 \leq i_2 \leq \cdots \leq i_p) \subset T^\otimes \mathfrak{g}$ trivially intersects with $\mathcal{R}$, so it is projected isomorphically onto $U\mathfrak{g}$.

---

[11] *Lie algebroids* are a generalisation of the Lie algebra of vector fields on a manifold; we will not discuss them in this course.

The map L is constructed on the basis of $T^{\otimes}\mathfrak{g}$ by induction on p and on the number of *inversions* in the sequence $i_1, i_2, \ldots, i_p$, i.e. the number of pairs i_k, i_l in this sequence, for which $k < l$ and $i_k > i_l$. Namely, if $p = 1$ or there are no inversions in the sequence (and hence $e_{i_1} \otimes e_{i_2} \otimes \cdots \otimes e_{i_p} \in \mathcal{I}$), we put L to be the identity map.

Now if $p > 1$ and the number of inversions is nonzero, there should exist a pair $i_k > i_{k+1}$; so we can use Equation (5.6): both terms in the right are already defined before, because the first of them has less inversions, and in the second the number of tensor legs is $p - 1$.

It only remains now to check that this procedure will not bring us to a contradiction: in general the resulting map L can depend on the way we chose pairs of inverted indices while constructing the map. To this end we consider two possibilities: we first exchange the order of e_{i_k} and $e_{i_{k+1}}$ and then of the pair e_{i_l} and $e_{i_{l+1}}$ where $l > k+1$, or we do it in the opposite order: first the pair e_{i_l} and $e_{i_{l+1}}$ and then e_{i_k} and $e_{i_{k+1}}$. In this case we clearly get the same result, since these pairs do not "interact".

A more interesting situation is when $l = k + 1$, so we have to deal with adjacent inversions: let $i_k > i_{k+1} > i_{k+2}$, we can bring it to the regular order i_{k+2}, i_{k+1}, i_k in two different ways. To see that the choice of order of transpositions here does not change the final result we can assume that $p = 3$; let X, Y, Z be the elements in the tensor product. We compute:[12]

$$L(X,Y,Z) = L(Y,X,Z) + L([X,Y],Z)$$
$$= L(Y,Z,X) + L(Y,[X,Z]) + L([X,Y],Z)$$
$$= L(Z,Y,X) + L([Y,Z],X) + L(Y,[X,Z]) + L([X,Y],Z)$$

or in the second case

$$L(X,Y,Z) = L(X,Z,Y) + L(X,[Y,Z])$$
$$= L(Z,X,Y) + L([X,Z],Y) + L(X,[Y,Z])$$
$$= L(Z,Y,X) + L(Z,[X,Y]) + L([X,Z],Y) + L(X,[Y,Z]).$$

---

[12]Here and below we shall often replace the tensor product signs by commas for the sake of brevity.

By inductive hypothesis L is well-defined for tensors with less than p legs and satisfies the condition (ii), so we have:

$$L(Z,[X,Y]) + L([X,Z],Y) + L(X,[Y,Z])$$
$$= L([X,Y],Z) + L(Y,[X,Z]) + L([Y,Z],X)$$
$$+ L([Z,[X,Y]]) + L([[X,Z],Y]) + L([X,[Y,Z]]),$$

and the third line of this equality vanishes due to the Jacobi identity, which proves the claim. □

5.3 PBW quantisation

Theorem 5.4 shows that as linear space $U\mathfrak{g}$ is isomorphic to the symmetric algebra $S(\mathfrak{g})$.[13] Indeed both algebras are the isomorphic images of the subspace $\mathcal{I} \subset T^\otimes \mathfrak{g}$ so they share the same basis $e_1^{i_1} \ldots e_n^{i_n}$ (or rather there's a canonical 1-1 correspondence between the elements of the basis' in $U\mathfrak{g}$ and $S(\mathfrak{g})$).

A more abstract and coordinate-free way of formulating this property is the following: let

$$\Bbbk = \mathcal{F}_0 U\mathfrak{g} \subseteq \mathcal{F}_1 U\mathfrak{g} \subseteq \mathcal{F}_2 U\mathfrak{g} \subseteq \cdots \subseteq \mathcal{F}_k U\mathfrak{g} \subseteq \mathcal{F}_{k+1} U\mathfrak{g} \subseteq \ldots \quad (5.7)$$

be the increasing filtration on $U\mathfrak{g}$ where $\mathcal{F}_k U\mathfrak{g}$ is the image of the subspace $\mathfrak{g}^{\leq k} = \bigoplus_{p \leq k} \mathfrak{g}^{\otimes p}$ under the natural projection. Let

$$Gr_\mathcal{F} U\mathfrak{g} = \bigoplus_{k \geq 0} Gr_\mathcal{F}^k U\mathfrak{g} = \bigoplus_{k \geq 0} \mathcal{F}_{k+1} U\mathfrak{g} / \mathcal{F}_k U\mathfrak{g}$$

be the associated graded algebra of this filtration. It is easy to see that for any two elements $a \in \mathcal{F}_p U\mathfrak{g}$, $b \in \mathcal{F}_q U\mathfrak{g}$ their commutator lies in $\mathcal{F}_{p+q-1} U\mathfrak{g}$: for any two basis elements $e_{i_1} \otimes \cdots \otimes e_{i_p}$, $i_1 \leq i_2 \leq \cdots \leq i_p$, $e_{j_1} \otimes \cdots \otimes e_{j_q}$, $j_1 \leq j_2 \leq \cdots \leq j_q$ of $\mathcal{I}$, their product in either order in $T^\otimes \mathfrak{g}$ can be brought by relations (5.4) to the fully ordered form $e_{k_1} \otimes \cdots \otimes e_{k_{p+q}} \in \mathcal{I}$, $k_1 \leq \cdots \leq k_{p+q}$ so that the difference will belong to the subspace $\mathfrak{g}^{\leq p+q-1} = \bigoplus_{t \leq p+q-1} \mathfrak{g}^{\otimes t}$. Hence, the algebra $Gr_\mathcal{F} U\mathfrak{g}$ will be commutative.

---

[13]In effect, one can strengthen this statement a little bit: one can make the isomorphism between $S(\mathfrak{g})$ and $U\mathfrak{g}$ to preserve the adjoint action of the simply-connected group G associated with $\mathfrak{g}$ and to intertwine the natural cocommutative coproducts on both sides.

On the other hand, consider the linear map

$$S(\mathfrak{g}) \xrightarrow{\tilde{\sigma}} U\mathfrak{g} \to Gr_{\mathscr{F}} U\mathfrak{g}$$

where in terms of the basis $\tilde{\sigma}$ is the linear isomorphism, given by the equation

$$\tilde{\sigma}(e_1^{i_1} \ldots e_n^{i_n}) = e_1^{i_1} \ldots e_n^{i_n}, \tag{5.8}$$

and the second map is the natural projection.[14] It follows from the same considerations and from Theorem 5.4 that this composition induces linear isomorphisms $S^k(\mathfrak{g}) \cong Gr_{\mathscr{F}}^k U\mathfrak{g}$ in all degrees and commutes with the products, since all commutators in $U\mathfrak{g}$ fall in lower filtrations. Hence, we can reformulate Theorem 5.4 as follows: *the graded algebra $Gr_{\mathscr{F}} U\mathfrak{g}$ of $U\mathfrak{g}$ is isomorphic to the symmetric algebra $S(\mathfrak{g})$*. One can use this isomorphism to relate the Poisson bracket on $S(\mathfrak{g})$ with the product in $U\mathfrak{g}$.

For our purposes it is convenient to modify the constructions of $U\mathfrak{g}$ a little bit: we put

$$\widehat{U}_\hbar \mathfrak{g} = T^\otimes \mathfrak{g}[[\hbar]]/\langle X \otimes Y - Y \otimes X - \hbar[X,Y] \mid X,Y \in \mathfrak{g}\rangle. \tag{5.3'}$$

We will also replace the map $\tilde{\sigma}$ by the following coordinate-free (i.e. independent of the choice of the basis) construction:

$$\sigma(X_1 \ldots X_p) = \frac{1}{p!} \sum_{s \in S_p} X_{s(1)} \cdot \ldots \cdot X_{s(p)}, \tag{5.8'}$$

where the sum is taken over all permutations s of order p and on the left we regard $X \in \mathfrak{g}$ as an element of $S(\mathfrak{g})$ while on the right it is the corresponding element of $U\mathfrak{g}$, so that $\cdot$ is the product in $U\mathfrak{g}$; the fact that σ is a linear isomorphism follows from the PBW theorem. Of course, σ needs not coincide with $\tilde{\sigma}$ (but the compositions of these two maps with the natural projection onto the graded algebra do coincide). The map σ is usually referred to as *the symmetrisation map*.

---

[14] The projection $U\mathfrak{g} \to Gr_{\mathscr{F}} U\mathfrak{g}$ is **not** a linear map; however, when precomposed with $\tilde{\sigma}$ it induces a linear isomorphism, independent on the choices we make.

It is clear that σ can be extended to an $\hbar$-linear isomorphism (also denoted by σ):

$$\sigma : S(\mathfrak{g})[[\hbar]] \to \widehat{U}_\hbar \mathfrak{g}. \tag{5.9}$$

Then we have the following proposition, which we shall prove below.

Proposition 5.5. *The formula*

$$f \star g = \sigma^{-1}(\sigma(f) \cdot \sigma(g)), \ f, g \in S(\mathfrak{g}), \tag{5.10}$$

where $\cdot$ stands for the product in $\widehat{U}_\hbar \mathfrak{g}$, determines an associative $\star$-product on $C^\infty(\mathfrak{g}^*)$, associated with the Kirillov–Kostant structure, see Definition 5.1.

This $\star$-product is called sometimes the **Poincaré–Birkhoff–Witt quantisation** (or **the PBW quantisation** for short). One also uses the name **Campbell–Baker–Hausdorff quantisation** (**CBH-quantisation** for short) due to its close relation with the Campbell–Baker–Hausdorff formula, which we don't discuss here (see [Se65] or [He78]). This is the simplest nontrivial example of deformation quantisation of a Poisson manifold; although the manifold in question is topologically trivial, $\mathfrak{g}^* \cong \mathbb{R}^n$, the Poisson structure on it is not. In effect it is quite interesting: it has nonconstant rank and its symplectic leaves are orbits of the coadjoint representation of the Lie group G associated with $\mathfrak{g}$. Later we shall give other constructions leading to this quantisation.

5.3.1 Proof of Proposition 5.5

The associativity of the product (5.10) on $S(\mathfrak{g})$ and the fact that 1 plays the role of unit in the new multiplication follows directly from the definition; hence by continuity principle this product will be automatically associative on $C^\infty(\mathfrak{g})$ if we can extend it there. To this end, we need to establish the existence of the bidifferential operators B_k (see the Problem 1, page 50), which would determine this product on $S(\mathfrak{g})$ independently of the formula (5.10); in our proof we follow the exposition in the paper of Gutt [Gu83].

We begin by computing the first term: for any two elements $X, Y \in \mathfrak{g} \subset U\mathfrak{g}$ we have $\sigma(X) \cdot \sigma(Y) = X \cdot Y$ and

$$\sigma(XY) = \frac{1}{2}(X \cdot Y + Y \cdot X) = \frac{1}{2}(2X \cdot Y + Y \cdot X - X \cdot Y) = X \cdot Y - \frac{\hbar}{2}[X, Y].$$

So

$$\sigma^{-1}(\sigma(X) \cdot \sigma(Y)) = \sigma^{-1}(\sigma(XY) + \frac{\hbar}{2}\sigma([X,Y])) = XY + \frac{\hbar}{2}[X,Y].$$

Using this formula, the fact that the commutator of elements of an associative algebra satisfies Leibniz rule and Remark 5.2 we can prove by induction that $f \star g = fg + \frac{\hbar}{2}\{f, g\} + o(\hbar)$ for all $f, g \in S(\mathfrak{g})$ (compare the reasoning below).

In order to complete the proof, we will need the following lemma, due to Berezin [Be67]:

Lemma 5.6. *Let* $X \in \mathfrak{g}$, $X_1 X_2 \ldots X_p \in S(\mathfrak{g})$ *then*

$$X \cdot \sigma(X_1 X_2 \ldots X_p)$$
$$= \sigma(X X_1 \ldots X_p) + \sum_{j=1}^{p} \frac{(-1)^j \hbar^j}{j!} B_j$$
$$\sum_{1 \leq r_1 < \cdots < r_j \leq p} \sigma([[\ldots [X, X_{r_1}], \ldots], X_{r_j}] X_1 \ldots \widehat{X}_{r_1} \ldots \widehat{X}_{r_j} \ldots X_p).$$

Here the second sum is taken over all j-tuples of indices $r_1 < r_2 \cdots < r_j$ and B_j is **the jth Bernoulli number**.

Recall that *Bernoulli numbers* are rational numbers determined by the following generating series:

$$\frac{t}{e^t - 1} = \sum_n B_n \frac{t^n}{n!}.$$

They appear in many situations throughout Mathematics. Below we shall use the following recurrence relation, which with equality $B_1 = 1$ uniquely determines these numbers (see Exercise E.2.13):

$$\sum_{k=0}^{n-1} \binom{n}{k} B_k = 0. \tag{5.11}$$

Proof. In our proof we follow the original paper by Berezin. For the sake of brevity, we shall omit $\hbar$ from our equations. Let $k(t) = tX + t_1X_1 + \cdots + t_pX_p$, $t, t_1, \ldots, t_p \in \mathbb{R}$ be a real linear function with values in $\mathfrak{g} \subseteq U\mathfrak{g}$. Then

$$\frac{d}{dt}(k(t))^{p+1} = X(k(t))^p + k(t)X(k(t))^{p-1} + \cdots + (k(t))^pX.$$

On the other hand, by induction one can prove the following formula for any two elements $X, k \in \mathfrak{g} \subseteq U\mathfrak{g}$ and all $r \geq 0$

$$k^r X = \sum_{s=0}^{r} \binom{r}{s} \underbrace{[k,[k,\ldots[k,X]\ldots]]}_{s \text{ times}} k^{r-s},$$

so we continue previous equation:

$$\frac{d}{dt}(k(t))^{p+1} = \sum_{r=0}^{p} \left(\sum_{s=0}^{r} \binom{r}{s} \underbrace{[k,[k,\ldots[k,X]\ldots]]}_{s \text{ times}} k^{r-s} \right) k^{p-r}$$

$$= \sum_{s=0}^{r} \binom{p+1}{s+1} \underbrace{[k,[k,\ldots[k,X]\ldots]]}_{s \text{ times}} k^{p-s}, \quad (5.12)$$

where we used the well-known equality

$$\sum_{r=0}^{p} \binom{r}{s} = \binom{p+1}{s+1},$$

proved by induction (see Exercise E.2.12). On the other hand, clearly for any m we have

$$(k(t))^m = \sum_{q+\sum_{i=1}^{p} q_i = m} \frac{m!}{q!q_1!\ldots q_p!} t^q t_1^{q_1} \ldots t_p^{q_p} \sigma(X^q X_1^{q_1} \ldots X_p^{q_p}). \quad (5.13)$$

Plugging this into (5.12) and setting $t = 0$ we get

$$\frac{d}{dt}(k(t))^{p+1}\Big|_{t=0} = (p+1)! \Bigg\{ X \sum_{\sum_{i=1}^{p} q_i = N} \frac{t_1^{q_1} \ldots t_p^{q_p}}{q_1!\ldots q_p!} \sigma(X_1^{q_1} \ldots X_p^{q_p})$$

$$+ \frac{1}{2!} \sum_{k=1}^{p} t_k[X_k, X] \sum_{\sum_{i=1}^{p} q_i = N-1} \frac{t_1^{q_1} \ldots t_p^{q_p}}{q_1!\ldots q_p!} \sigma(X_1^{q_1} \ldots X_p^{q_p}) + \ldots \Bigg\}.$$

(5.14)

Deformations and Lie algebras   63

On the other hand, applying $\frac{d}{dt}|_{t=0}$ to (5.13), comparing the coefficients of this expression and (5.14) at equal powers of $t_1, \ldots, t_p$ and finally setting $t_1 = t_2 = \cdots = t_p = 1$, we get

$$\sigma(XX_1 \ldots X_p) = X\sigma(X_1 \ldots X_p) + \frac{1}{2!} \sum_i [X_i, X]\sigma(X_1 \ldots \widehat{X_i} \ldots X_p)$$
$$+ \frac{1}{3!} \sum_{i,j} [X_i, [X_j, X]]\sigma(X_1 \ldots \widehat{X_i} \ldots \widehat{X_j} \ldots X_p) + \ldots$$
(5.15)

Reasoning inductively from Equation (5.15) we conclude that there exist rational coefficients B_j such that

$$X \cdot \sigma(X_1 X_2 \ldots X_p)$$
$$= \sigma(XX_1 \ldots X_p) + \frac{B_1}{1!} \sum_i \sigma([X_i, X]X_1 \ldots \widehat{X_i} \ldots X_p)$$
$$+ \frac{B_2}{2!} \sum_{i,j} \sigma([X_i, [X_j, X]]X_1 \ldots \widehat{X_i} \ldots \widehat{X_j} \ldots X_p) + \ldots$$

Moreover, it is easy to show that $B_1 = 1$ and that these coefficients satisfy Equation (5.11), hence they are Bernoulli numbers. □

Now we are finally able to show that there exists a sequence of bidifferential operators, which determine the PBW $\star$-product, and even to find explicit expressions for few first elements of this sequence. Indeed, Lemma 5.6 means that the $\star$-product of a linear function on $\mathfrak{g}^*$ given by $p \in S(\mathfrak{g})$ and a monomial $Q = p_1 \ldots p_m$ takes the form

$$p \star (p_1 \ldots p_n)$$
$$= pQ + \sum_r \frac{\hbar^r(-1)^r}{r!} \sum_{j_1, \ldots, j_r} \{\ldots \{p, p_{j_1}\}, \ldots\}, p_{j_r}\} \frac{\partial^r}{\partial p_{j_1} \ldots \partial p_{j_r}} Q.$$

We suppose that the sequence of coordinate functions $p_1, \ldots p_m$ can contain repetitions. Thus, the same formula holds for all polynomials Q.

Now if P is another monomial, $P = p_1 \ldots p_k \in S(\mathfrak{g})$, we have

$$\sigma(P) = \frac{1}{k!} \sum_{s \in S_k} p_{s(1)} \cdots p_{s(k)}$$

$$= \frac{1}{k} \sum_{j=1}^{k} \frac{1}{(k-1)!} \sum_{\substack{s \in S_k \\ s(1)=j}} p_j p_{s(2)} \cdots p_{s(k)}$$

$$= \frac{1}{k} \sum_{j=1}^{k} p_j \cdot \sigma\left(\frac{\partial}{\partial p_j} P\right).$$

Applying σ^{-1} to this equality, we get

$$P = \frac{1}{k} \sum_{j=1}^{k} p_j \star \frac{\partial}{\partial p_j} P.$$

Now we can reason as follows: for all Q consider the product $P \star Q$:

$$P \star Q = \left(\frac{1}{k} \sum_{j=1}^{k} p_j \star \frac{\partial}{\partial p_j} P\right) \star Q = \frac{1}{k} \sum_{j=1}^{k} p_j \star \left(\frac{\partial}{\partial p_j} P \star Q\right),$$

where we used the associativity of the PBW $\star$-product on polynomials. By inductive hypothesis, we conclude the existence of bidifferential operators here too: it is sufficient to observe that a differential operator on Euclidean space is uniquely determined by its values on polynomials. This completes the proof of Theorem 5.4.

Remark 5.7. It is also possible to use this method to prove the skew symmetry of the bidifferential operators in the PBW deformation series. We are not going to do it here.

6 Crash course in homological algebra, I. Examples

This and the next lectures are introduced into the notes with the main purpose of making the exposition as self-contained as possible. Here we only give a very brief list of major terms used in the field and formulate some basic results. The reader who is familiar with the main results, ideas and principles of the Homological algebra can skip it. On the other hand, if you would like to learn more about this important subject, there are plenty of excellent detailed textbooks on Homological algebra, of which we can recommend [Wei94] and [GM03].

6.1 General nonsense

One of the main ideas of modern Mathematics is the idea of invariants, i.e. of some construction associated with an object we study so that its result is not changed, when we modify the object in a certain way. These constructions often give a number, or more generally an element of some group as their value, but quite often the group itself is such invariant. Of course the problem to discriminate two non-Abelian groups can be quite tricky, so it is preferable to work with Abelian groups instead.

By far the most popular constructions that gives a sequence of Abelian groups, is that of *homology* or *cohomology groups*. Originating from Topology, the ideas and methods connected with these constructions are now often regarded as a separate branch of Mathematics, *Homological algebra*. In this section we give few basic definitions and examples of the objects, studied in this theory.

So let

$$\cdots \leftarrow C_{k-2} \xleftarrow{d} C_{k-1} \xleftarrow{d} C_k \xleftarrow{d} C_{k+1} \xleftarrow{d} C_{k+2} \leftarrow \cdots \qquad (6.1)$$

be a sequence of Abelian groups indexed by $k \in \mathbb{Z}$ and homomorphisms such that $d \circ d = 0$ at all stages. Then one says that $C_{\cdot} = (\{C_k\}, d)$ is *a chain complex*, elements $x \in C_k$ are called *k-chains*, or *k-dimensional chains of $C_{\cdot}$* and any of the maps d is called *differential* or *boundary map* of $C_{\cdot}$; one usually assumes that

$C_k = 0$ for $k < 0$ and $k \gg 1$. *Homology groups of $C.$* are defined to be the quotients:

$$H_k(C.) = \ker(d : C_k \to C_{k-1})/\mathrm{im}(d : C_{k+1} \to C_k), \quad k \in \mathbb{Z}. \quad (6.2)$$

Elements in $\ker(d : C_k \to C_{k-1})$ are called *closed k-chains* or *k-cycles*, and the elements in $\mathrm{im}(d : C_{k+1} \to C_k)$ are called *exact k-chains* or *k-boundaries*. Cycles x, $x' \in C_k$ which fall into the same homology class are called *homologous*.

A collection of homomorphisms $f_k : C_k \to C'_k$, $k \in \mathbb{Z}$ between the chain groups of two chain complexes, such that $f_{k-1} \circ d = d' \circ f_k$ for all k is called *chain map*. Such maps induce homomorphisms between homology groups $f_* : H_*(C.) \to H_*(C'.)$ for all integer values of $*$. An important relation on chain maps is the *chain homotopy*: chain maps $\{f_k\}, \{f'_k\} : C. \to C'.$ are called homotopic if there exists a collection of homomorphisms $h_k : C_k \to C'_{k+1}$ for which:

$$f_k - f'_k = h_{k-1} \circ d + d' \circ h_k, \text{ for all } k \in \mathbb{Z}.$$

The collection of maps $\{h_k\}$ is also called *chain homotopy*; chain homotopy is an equivalence relation on chain maps. Homotopic chain maps induce identical morphisms of the homology groups.

Remark 6.1. The notion of *cochain complexes, cochains, cohomology groups, cocycles* etc. can be obtained by merely changing the direction of all arrows in (6.1): if the differential of the complex raise all dimensions by 1,

$$\cdots \to C^{k-2} \xrightarrow{\partial} C^{k-1} \xrightarrow{\partial} C^k \xrightarrow{\partial} C^{k+1} \xrightarrow{\partial} C^{k+2} \to \cdots \quad (6.1')$$

then one says that we deal with a **co***chain complex* $(C^., \partial)$, whose **co***homology groups* are

$$H^k(C^.) = \ker(\partial : C^k \to C^{k+1})/\mathrm{im}(\partial : C^{k-1} \to C^k), \quad k \in \mathbb{Z}. \quad (6.2')$$

All other definitions are modified in a similar manner. In future we will not pay much attention to the differences between chain and cochain complexes: in effect, mere change of indices $m = -n$ turns chain complexes into cochain complexes and vice-versa.

Various properties of these and related constructions as well as their applications to Algebra, Geometry and Topology are widely studied by *Homological Algebra*. We shall use these results rather freely in our course, so the reader is supposed to have certain knowledge of this subject (see Exercise E.2.15*) although most time we are not going to use more than very basic properties of the homological constructions.

The primary purpose of using homological algebra in our course is that in many cases it is convenient to formulate the sufficient and/or necessary conditions of quantisation in terms of certain cohomology classes; in good cases one can check if these classes vanish or not, thus giving an answer to the quantisation problem.

6.2 Examples

Let us give few important examples of homology and cohomology theories, which will play important role in the rest of our course.

6.2.1 De Rham cohomology

If M is a smooth n-dimensional real manifold, one has a cohomological complex naturally associated with it:

$$\Omega^*(M) = (\{\Omega^k(M)\}, d):$$
$$0 \to C^\infty(M) = \Omega^0(M) \xrightarrow{d} \Omega^1(M) \xrightarrow{d} \ldots \xrightarrow{d} \Omega^n(M) \to 0$$

i.e. $C^{\cdot}$ is the space of differential forms on M with natural grading equipped with de Rham differential. Its cohomology groups (in fact vector spaces) $H^*_{dR}(M)$ are called *de Rham cohomology of M*. We hope that the reader has sufficient knowledge of this theory including its applications to Geometry and Topology of manifolds and its relation with other theories.

6.2.2 Dolbeault cohomology

If the manifold M bears in addition a complex structure, then the space of complex differential k-forms breaks down into the sum of bigraded components $\Omega^k(M) = \bigoplus_{p+q=k} \Omega^{p,q}(M)$, where in local complex coordinates $z^1, \ldots, z^n$, $z^i = x^i + \sqrt{-1}y^i$ on M we have

$$\Omega^{p,q}(M) = \left\{ \sum_{I,\bar{J}} a_{I,\bar{J}}(z,\bar{z}) dz^{i_1} \wedge \ldots \wedge dz^{i_p} \wedge d\bar{z}^{\bar{j}_1} \wedge \ldots \wedge d\bar{z}^{\bar{j}_q} | a_{I,\bar{J}}(z,\bar{z}) \in C^\infty(M) \right\},$$

where $I, \bar{J}$ are multiindices:

$$I = (1 \leq i_1 < \cdots < i_p \leq n), \ \bar{J} = (1 \leq \bar{j}_1 < \cdots < \bar{j}_q \leq n),$$

and $dz^i = dx^i + \sqrt{-1}dy^i$, $d\bar{z}^k = dx^k - \sqrt{-1}dy^k$, i.e. $\Omega^{p,q}(M)$ is the subspace spanned by p holomorphic and q anti-holomorphic forms. In this case, de Rham differential d is equal to the sum of two components: $d = \partial + \bar{\partial}$, where

$$\partial : \Omega^{p,q}(M) \to \Omega^{p+1,q}(M), \ \bar{\partial} : \Omega^{p,q}(M) \to \Omega^{p,q+1}(M).$$

In this case, *Dolbeault cohomology* $H^{p,q}_{\bar{\partial}}(M)$ is the q-th cohomology group of the complex

$$\Omega^{p,\cdot}(M) = (\{\Omega^{p,q}(M)\}, \bar{\partial}) :$$

$$0 \to \Omega^{p,0}(M) \xrightarrow{\bar{\partial}} \Omega^{p,1}(M) \xrightarrow{\bar{\partial}} \cdots \xrightarrow{\bar{\partial}} \Omega^{p,n-p}(M) \to 0.$$

This cochain complex and its cohomology plays crucial role in complex Differential Geometry; it is very sensitive to deformation of the complex structure on M. In fact $H^{p,q}_{\bar{\partial}}(M) = H^q(M, \Omega^p)$, where Ω^p is the *sheaf of holomorphic p-forms on M*. We will not speak much about sheaf cohomology in this course, although some particular examples thereof will appear. A reader, not familiar with this notion can skip them. An important particular case is when M is compact *Kähler* (i.e. the imaginary part of the Hermitian form on M is a closed 2-form); then we have $H^k_{dR}(M) = \bigoplus_{p+q=k} H^{p,q}_{\bar{\partial}}(M)$, so in this case Dolbeault cohomology groups are finite-dimensional. Kähler manifolds are by definition automatically symplectic manifolds and deformation quantisation of these manifolds is of particular interest, as the $\star$-product series contains information about the Kähler structure.

6.2.3 Lichnerowicz–Poisson cohomology

Let (M, π) be a Poisson manifold, i.e. $\pi \in \mathcal{T}^2(M)$ is a bivector, which satisfies the equation $[\pi, \pi] = 0$ for Schouten bracket $[,]$ (see Sections 4.3.2, 4.3.3). Consider the map

$$d_\pi : \mathcal{T}^k(M) \to \mathcal{T}^{k+1}(M), \ d_\pi(\varphi) = [\pi, \varphi],$$

for any k-polyvector $\varphi \in \mathcal{T}^k(M)$. Then

Proposition 6.2. $d_\pi^2 = 0$.

Proof. We compute by graded Jacobi identity:

$$d_\pi^2(\varphi) = [\pi, [\pi, \varphi]] = (-1)^k(-[[\pi, \varphi], \pi])$$
$$= (-1)^k((-1)^k[[\varphi, \pi], \pi] + (-1)^k[[\pi, \pi], \varphi]) = [[\varphi, \pi], \pi],$$

because $[\pi, \pi] = 0$. But, by graded skew symmetry

$$[[\varphi, \pi], \pi] = -(-1)^k[\pi, [\varphi, \pi]] = (-1)^{k+k-1}[\pi, [\pi, \varphi]] = -[\pi, [\pi, \varphi]],$$

so we have $d_\pi^2(\varphi) = -d_\pi^2(\varphi)$ and hence it vanishes. $\square$

Now we can define the *Lichnerowicz Poisson cohomology*, $H_\pi^*(M)$ of (M, π) as the cohomology of the complex:

$$\mathcal{T}^*(M) = (\{\mathcal{T}^k(M)\}, d_\pi):$$
$$0 \to C^\infty(M) = \mathcal{T}^0(M) \xrightarrow{d_\pi} \mathcal{T}^1(M) \xrightarrow{d_\pi} \ldots \xrightarrow{d_\pi} \mathcal{T}^n(M) \to 0.$$

It is easy to compute in natural terms $H_\pi^0(M)$ and $H_\pi^1(M)$:

$$H_\pi^0(M) = \ker(d_\pi : C^\infty(M) \to \mathcal{T}^1(M)),$$

since there are no "-1-dimensional cochains". On the other hand, $d_\pi(f) = X_f$, the Hamilton vector field of f: indeed, if $\pi = \sum_k X_k \wedge Y_k$, then by the properties of Schouten bracket (see Section 4.3.3), we have $d_\pi(f) = \sum_k (Y_k(f) X_k - X_k(f) Y_k)$, so for any smooth function g, we have

$$(d_\pi(f))(g) = \sum_k (Y_k(f) X_k(g) - X_k(f) Y_k(g))$$
$$= \pi(dg, df) = -\{f, g\} = X_f(g).$$

Thus, we conclude that $H^0_\pi(M) = Cas(M, \pi)$, the space of *Casimir functions on M*, i.e. of functions whose Hamilton vector fields vanish.

Similarly,

$$H^1_\pi(M) = \ker(d_\pi : \mathcal{T}^1(M) \to \mathcal{T}^2(M))/\operatorname{im}(d_\pi : C^\infty(M) \to \mathcal{T}^1(M)).$$

By the properties of Schouten bracket, $d_\pi(X) = [\pi, X] = -\mathcal{L}_X \pi$ so $d_\pi(X) = 0$ iff π is invariant under the action by the vector field X. Such vector fields are called *Poisson vector fields*, we denote the space of Poisson fields on M by $Pois(M, \pi)$. On the other hand by our previous observation $\operatorname{im}(d_\pi : C^\infty(M) \to \mathcal{T}^1(M))$ consists of Hamilton fields. So

$$H^1_\pi(M) = Pois(M, \pi)/Ham(M, \pi).$$

In general these cohomology groups can be rather hard to compute, see Exercise E.2.18*. An important particular case, when this problem can be solved is when $\pi = \omega^{-1}$, i.e. when the Poisson structure is induced from a symplectic form. Then the following is true:

Proposition 6.3. *If Poisson structure comes from a symplectic form then $H^*_\pi(M) = H^*_{dR}(M)$.*

Proof. Let $\pi = \omega^{-1}$, consider the maps $\pi^\sharp : \Omega^k(M) \to \mathcal{T}^k(M)$, $\omega_\flat : \mathcal{T}^k(M) \to \Omega^k(M)$, raising and lowering all the indices of tensors induced by the bivector π and the form ω respectively. These maps are mutually inverse linear isomorphisms.

We are going to show that these isomorphisms identify de Rham differential with d_π. To this end, we observe that by definition both maps commute with wedge products. On the other hand both differentials clearly satisfy the Leibniz rule with respect to the wedge products: it is a well-known fact for de Rham operator, and the identity for d_π follows from the graded Leibniz rule, satisfied by Schouten bracket: for $\alpha \in \mathcal{T}^p(M)$, $\beta \in \mathcal{T}^q(M)$

$$d_\pi(\alpha \wedge \beta) = [\pi, \alpha \wedge \beta] = [\pi, \alpha] \wedge \beta + (-1)^p \alpha \wedge [\pi, \beta]$$
$$= d_\pi(\alpha) \wedge \beta + (-1)^p \alpha \wedge d_\pi(\beta).$$

So it is enough to show that $\pi^\sharp \circ d \circ \omega_\flat = d_\pi$ for all $f \in C^\infty(M)$, but this is evident: $d_\pi(f) = X_f = \pi^\sharp(df)$. $\square$

Remark 6.4. In addition to the Lichnerowicz Poisson cohomology theory, we defined, there exists a somewhat dual construction of Poisson homology, introduced by Brylinski. We shall not use it in our course so we recommend the interested reader to read the original paper by that author.

6.2.4 Lie algebra homology and cohomology

Let $\mathfrak{g}$ be a Lie algebra over a field $\Bbbk$, and let $\mathfrak{M}$ be a (left) $\mathfrak{g}$-module, i.e. a $\Bbbk$-vector space, equipped with a bilinear map

$$\mathfrak{g} \otimes \mathfrak{M} \to \mathfrak{M}, \; X \otimes m \mapsto X \cdot m,$$

such that

$$X \cdot (Y \cdot m) - Y \cdot (X \cdot m) = [X, Y] \cdot m \text{ for all } m \in \mathfrak{M}, \; X, Y \in \mathfrak{g}.$$

Consider the following cochain complex, sometimes called *the Chevalley–Eilenberg complex of $\mathfrak{g}$ with coefficients in $\mathfrak{M}$*:

$$C^{\cdot}(\mathfrak{g}, \mathfrak{M}) = (\{\widetilde{\mathrm{Hom}}(\mathfrak{g}^{\times k}, \mathfrak{M})\}_{k \geq 0}, \delta),$$

where $\widetilde{\mathrm{Hom}}(\mathfrak{g}^{\times k}, \mathfrak{M})$ is the space of $\Bbbk$-multi-linear maps from $\mathfrak{g}^{\times k}$, $k \geq 0$ with values in $\mathfrak{M}$, completely anti-symmetric in all arguments. In particular, $C^0(\mathfrak{g}, \mathfrak{M}) = \mathfrak{M}$, $C^1(\mathfrak{g}, \mathfrak{M}) = \mathrm{Hom}(\mathfrak{g}, \mathfrak{M})$. If $\mathfrak{g}$ is finite-dimensional, then $C^k(\mathfrak{g}, \mathfrak{M}) = \mathrm{Hom}(\wedge^k \mathfrak{g}, \mathfrak{M}) = \wedge^k \mathfrak{g}^* \otimes \mathfrak{M}$, however our definition works in general case, so we prefer to skip this identifications for now.

The differential δ of this complex is given by the following formula: for all $\varphi \in C^p(\mathfrak{g}, \mathfrak{M})$ and all $X_0, \ldots, X_p \in \mathfrak{g}$

$$\delta\varphi(X_0, \ldots, X_p) = \sum_{i=0}^{p}(-1)^i X_i \cdot \varphi(X_0, \ldots, \widehat{X_i}, \ldots, X_p)$$
$$- \sum_{0 \leq i < j \leq p}(-1)^{i+j-1}\varphi([X_i, X_j], X_0, \ldots, \widehat{X_i}, \ldots, \widehat{X_j}, \ldots, X_p).$$

In this case the equation $\delta^2 = 0$ can be proved by direct computation: for the sake of brevity we shall only consider the case $p = 1$, the general case is proved by a similar, but lengthy calculation. Let $X, Y, Z \in \mathfrak{g}$ be arbitrary vectors, then

$$\begin{aligned}
\delta^2(\varphi)(X,Y,Z) &= X \cdot \delta\varphi(Y,Z) - Y \cdot \delta\varphi(X,Z) + Z \cdot \delta\varphi(X,Y) \\
&\quad - \delta\varphi([X,Y],Z) + \delta\varphi([X,Z],Y) - \delta\varphi([Y,Z],X) \\
&= X \cdot \Big(Y \cdot \varphi(Z) - Z \cdot \varphi(Y) - \varphi([Y,Z])\Big) \\
&\quad - Y \cdot \Big(X \cdot \varphi(Z) - Z \cdot \varphi(X) - \varphi([X,Z])\Big) \\
&\quad + Z \cdot \Big(X \cdot \varphi(Y) - Y \cdot \varphi(X) - \varphi([X,Y])\Big) \\
&\quad - [X,Y] \cdot \varphi(Z) + Z \cdot \varphi([X,Y]) + \varphi([[X,Y],Z]) \\
&\quad + [X,Z] \cdot \varphi(Y) - Y \cdot \varphi([X,Z]) - \varphi([[X,Z],Y]) \\
&\quad - [Y,Z] \cdot \varphi(X) + X \cdot \varphi([Y,Z]) + \varphi([[Y,Z],X]) \\
&= 0
\end{aligned}$$

because $X \cdot (Y \cdot \varphi(Z)) - Y \cdot (X \cdot \varphi(Z)) = [X,Y] \cdot \varphi(Z)$ and similarly for all other pairs of elements, and also $\varphi([[X,Y],Z]) - \varphi([[X,Z],Y]) + \varphi([[Y,Z],X]) = 0$ by Jacobi identity.

The cohomology of the Chevalley–Eilenberg complex $C^{\cdot}(\mathfrak{g}, \mathfrak{M})$ is often referred to as *the Chevalley–Eilenberg cohomology of $\mathfrak{g}$ with coefficients in $\mathfrak{M}$* or just as *the cohomology of $\mathfrak{g}$ with coefficients in $\mathfrak{M}$*, denoted by $H^*_{CE}(\mathfrak{g}, \mathfrak{M})$ or just $H^*(\mathfrak{g}, \mathfrak{M})$.

It is worth computing the first couple of groups in this cohomology theory: $H^0(\mathfrak{g}, \mathfrak{M}) = \ker(\delta : \mathfrak{M} \to C^1(\mathfrak{g}, \mathfrak{M}), \delta m(X) = X \cdot m$, so

$$H^0(\mathfrak{g}, \mathfrak{M}) = \mathfrak{M}_0 = \{m \in \mathfrak{M} \mid X \cdot m = 0 \ \forall X \in \mathfrak{g}\}.$$

If $\mathfrak{M}$ is equipped with trivial action of $\mathfrak{g}$ (i.e. $X \cdot m = 0$ for all $X \in \mathfrak{g}$, $m \in \mathfrak{M}$), then $\delta m = 0$ for all $m \in \mathfrak{M}$ and an element $\varphi \in C^1(\mathfrak{g}, \mathfrak{M})$ is closed iff $\varphi([X,Y]) = 0$ for all $X, Y \in \mathfrak{g}$. Linear maps $\mathfrak{g} \to \mathfrak{M}$ with this property are sometimes called *traces*; thus $H^1(\mathfrak{g}, \mathfrak{M}) = Trace(\mathfrak{g}, \mathfrak{M})$.

If $\mathfrak{g}$ is a finite-dimensional Lie algebra, one can somewhat dually define the *Chevalley–Eilenberg homological complex of $\mathfrak{g}$*: if $\mathfrak{M}$ is right $\mathfrak{g}$-module we put

$$C_{\cdot}(\mathfrak{g}, \mathfrak{M}) = (\{\mathfrak{M} \otimes \wedge^k \mathfrak{g}\}_{k \geq 0}, \partial_{CE}),$$

with differential

$$\partial_{CE}(m \otimes X_1 \wedge \cdots \wedge X_p) = \sum_{i=1}^{p}(-1)^{i-1} m \cdot X_i \otimes X_1 \wedge \ldots \wedge \widehat{X_i} \wedge \ldots \wedge X_p$$
$$- \sum_{1 \leq i < j \leq p}(-1)^{i+j-1} m \otimes [X_i, X_j] \wedge \cdots \wedge \widehat{X_i} \wedge \cdots \wedge \widehat{X_j} \wedge \cdots \wedge X_p.$$

The equation $\partial_{CE}^2 = 0$ is proved by a computation similar to what we did above. As before, one can compute the first couple of cohomology groups, for example,

$$H_0(\mathfrak{g}, \mathfrak{M}) = \mathfrak{M}/\mathfrak{M} \cdot \mathfrak{g} = \mathfrak{M}^0.$$

And if $\mathfrak{M} = \Bbbk$ with trivial action, then

$$H_1(\mathfrak{g}, \Bbbk) = \mathfrak{g}/[\mathfrak{g}, \mathfrak{g}].$$

Remark 6.5. A reader, familiar with Lie algebroids will easily see that these constructions can be applied in a word-for-word manner in that situation, thus yielding new examples of homology and cohomology theories. In some cases, these theories turn out to be something we have already discussed here; for instance if we consider the algebroid $\mathcal{T}^1(M)$ of vector fields on a smooth manifold M with identical anchor map, and consider the complex $C^{\cdot}(\mathcal{T}^1(M), C^\infty(M))$ of $C^\infty(M)$-linear maps, then Cartan derivation formula shows that this complex is in effect isomorphic to the de Rham complex of M.

One should be cautious so as *not to confuse the construction of algebroid cohomology with the cohomology of the infinite-dimensional Lie algebra of vector fields, where we only assume $\Bbbk$-linearity of the cochains, and not that they are $C^\infty(M)$-linear.* On the other hand, averaging over the group and Cartan formula show that *de Rham cohomology of compact Lie group G is isomorphic to the Lie algebra cohomology of the corresponding Lie algebra of left- or right-invariant vector fields on G with coefficients in trivial 1-dimensional module.* Thus the Chevalley–Eilenberg cohomology is closely related to the topology of Lie groups.

6.3 Algebraic structures on cohomology groups

One of the important features of various homology and cohomology theories are additional algebraic structures, such as multiplication, or Lie bracket etc., which one can introduce on the (co)homology groups. These structures are usually induced from analogous structures on the complexes; it is also usually supposed that chain maps commute with these structures, so that the whole algebraic construction is invariant.

Example 6.1. De Rham cohomology of any smooth manifold is equipped with the structure of graded algebra, induced from the wedge-product of differential forms. Indeed for $\alpha \in \Omega^p(M)$, $\beta \in \Omega^q(M)$, we have

$$d(\alpha \wedge \beta) = d\alpha \wedge \beta + (-1)^p \alpha \wedge d\beta,$$

so $d(\alpha \wedge \beta) = 0$ if both forms are closed; moreover, if one of them is exact, for example $\alpha = d\alpha'$ then $\alpha \wedge \beta = d(\alpha' \wedge \beta)$, so the cohomology class is not changed, when we replace a form by a cohomologous one. The same is clearly true for the Dolbeault cohomology.

Example 6.2. If (M, π) is a Poisson manifold, then its Lichnerowicz Poisson complex is equipped with wedge-product and the Schouten bracket. Both these structures survive on cohomology: the first follows, just as in Example 6.1, directly from the graded Leibniz rule that holds for the Schouten bracket (and hence for d_π too). The second follows in the same manner from graded Jacoby identity:

$$d_\pi([\alpha, \beta]) = [\pi, [\alpha, \beta]] = [[\pi, \alpha], \beta] + (-1)^{p-1}[\alpha, [\pi, \beta]]$$
$$= [d_\pi \alpha, \beta] + (-1)^{p-1}[\alpha, d_\pi \beta].$$

Example 6.3. De Rham complex of Poisson manifold bears in addition to the wedge-product structure, the structure of (graded) Lie algebra; it is uniquely determined by the graded Leibniz rule with respect to the exterior product and the following equation for 1-forms α and β:

$$\{\alpha, \beta\} = \mathcal{L}_{\pi^\sharp(\alpha)}\beta - \mathcal{L}_{\pi^\sharp(\beta)}\alpha - d\pi(\alpha, \beta).$$

However one can show that this bracket induces trivial (i.e. identically equal to 0) operation on the level of de Rham cohomology.

Example 6.4. De Rham cohomology of a compact connected Lie group G is naturally endowed with the map

$$\Delta : H^*_{dR}(G) \to H^*_{dR}(G) \hat{\otimes} H^*_{dR}(G),$$

where on the right $\hat{\otimes}$ denotes the *graded* tensor product. This map is induced from the group product $G \times G \to G$; one can show that this coproduct is in fact graded cocommutative (i.e. the signed swap of tensor legs does not change the result). In terms of the Lie algebra cohomology, see Remark 6.5, this map is induced from the identifications: $C^{\cdot}(\mathfrak{g}, \mathbb{R}) = \wedge^{\cdot} \mathfrak{g}^*$ and the formula

$$\Delta(X^*) = X^* \otimes 1 + 1 \otimes X^*.$$

Warning! This and the next sections give barely a glance at the eminent domain of Homological algebra. We are by no means intending to teach the reader even the basics of this subject, which is of tremendous importance for modern Mathematics. So the reader should not think that after completing these sections he or she will be familiar with the methods of Homological algebra. Our modest goal is to familiarise the reader with the terminology and introduce the main constructions that we will often encounter further in the lectures. We hope that the reader will be able to follow our reasonings, at least at the level of general definitions. On other hand we have omitted many important subjects, concerned with the methods and constructions used for the computations of homology, the most prominent of our omissions being the theory of spectral sequences. We would like to urge the reader who would like to master the material in this book and go further, to study these subjects on his or her own; the references are for instance [Wei94], or [MCl01].

Exercises 2: Poisson structures, quantisation and cohomology

E.2.1 Let $\omega = (\omega_{ij})$ be a constant nondegenerate anti-symmetric $2n \times 2n$ matrix; we define Weyl algebra as

$$W_\hbar(\mathbb{R}^{2n}, \omega) = \mathbb{R}[\hbar]\langle y_1, \ldots, y_{2n}\rangle / ([y_i, y_j] = \hbar \omega_{ij})$$

where $\mathbb{R}\langle y_1, \ldots, y_{2n}\rangle$ stand for the algebra of "noncommutative polynomials" in variables $y_1, \ldots, y_{2n}$ and $\hbar$ is a formal variable. Let $\sigma : \mathbb{R}[\hbar][y_1, \ldots, y_{2n}] \to W_\hbar(\mathbb{R}^n, \omega)$ be the $\hbar$-linear map from the space of (usual) polynomials $y_1, \ldots, y_{2n}$ with coefficients in $\mathbb{R}[\hbar]$ into the Weyl algebra, given by

$$\sigma(y_{i_1} \ldots y_{i_k}) = \frac{\hbar^k}{k!} \sum_{s \in S_k} y_{i_{s(1)}} \cdots y_{i_{s(k)}}.$$

(a) Prove, that σ is a $\hbar$-linear isomorphism.

(b) Find an expression for the $\star$-product in $\mathbb{R}[\hbar][y_1, \ldots, y_{2n}]$ induced by

$$f \star g = \sigma^{-1}(\sigma(f) \cdot \sigma(g)).$$

E.2.2 Use the integral formula (c.f. (3.2)) for the Moyal $\star$-product

$$a \star b = \frac{1}{(2\hbar\pi)^{2n}} \int_{\mathbb{R}^{2n} \times \mathbb{R}^{2n}} e^{\frac{1}{i\hbar}\omega(z-z', z-z'')} a(z') b(z'') dz' dz''$$

and prove the associativity of this product in symplectic case.

E.2.3 Prove that differential 2-form ω on a compact $2n$-dimensional manifold is nondegenerate iff the form ω^n is nonvanishing in all points.

E.2.4 Prove that every compact symplectic manifold has nontrivial cohomology in all even dimensions.

E.2.5 Let ξ be a vector field on a symplectic manifold M such that $\mathcal{L}_\xi \omega = 0$. Show that in a neighbourhood of every point in M there exists a 1-form α such that $d\alpha = 0$ and $\xi = \pi^\sharp(\alpha)$ (here $\pi = \omega^{-1}$). Such fields are called *locally Hamiltonian* or *Poisson*.

E.2.6 Show that if $H^1_{dR}(M) = 0$, where M is symplectic manifold, then every Poisson field on M is in fact Hamiltonian.

E.2.7 Show that locally Hamiltonian fields on a symplectic manifold form a Lie subalgebra; moreover, for any two locally Hamiltonian fields ξ, η their commutator is in fact Hamiltonian with Hamiltonian function equal to $\omega(\xi, \eta)$.

E.2.8 Let π be a bivector on M and $\{f, g\}_\pi = \pi(df, dg)$ is the associated bracket of functions; let $J_\pi : C^\infty(M)^{\otimes 3} \to C^\infty(M)$ be the map, given by

$$J_\pi(a, b, c) = \{\{a, b\}_\pi, c\}_\pi + \{\{b, c\}_\pi, a\}_\pi + \{\{c, a\}_\pi, b\}_\pi.$$

Prove that J_π is antisymmetric and satisfies Leibniz rule in all variables.

E.2.9 Prove that the Schouten bracket, defined by explicit formula (4.9) for decomposable polyvectors (i.e. for polyvectors equal to wedge-product of vector fields) is well defined on all polyvectors and satisfies the graded Jacobi identity.

E.2.10 Prove Lichnerowicz's formula:

$$\langle \omega, [P, Q] \rangle = (-1)^{(p-1)(q-1)} \langle di_Q \omega, P \rangle \\ - \langle di_P \omega, Q \rangle + (-1)^p \langle d\omega, P \wedge Q \rangle,$$

where $\omega \in \Omega^{p+q-1}(M)$, $P \in \Lambda^p(M)$, $Q \in \Lambda^q(M)$, $\langle , \rangle$ is the natural pairing between forms and polyvector fields and $i_P \omega$ is the convolution of form and polyvector. **Hint:** prove that the expression, determined by the right-hand side of this formula satisfies the conditions (i), (ii), (iv) and (v) of Theorem 4.8.

E.2.11 Prove that nondegenerate 2-form ω is closed iff ω^{-1} is Poisson bivector.

E.2.12 Prove the equality: $\sum_{r=0}^{p} \binom{r}{s} = \binom{p+1}{s+1}$. **Hint:** use induction and Pascal's rule.

E.2.13 *Bernoulli numbers* B_n are usually defined by the formula:

$$\frac{t}{e^t - 1} = \sum_n B_n \frac{t^n}{n!}.$$

Prove recurrence formula $\sum_{k=0}^{n-1} \binom{n}{k} B_k = 0$.

E.2.14 Prove that the bracket, which sends $a, b \in Gr_{\mathcal{F}} U\mathfrak{g}$ to the image in $Gr_{\mathcal{F}} U\mathfrak{g}$ of the commutator of $\bar{a}, \bar{b} \in U\mathfrak{g}$ (where a and b are images of $\bar{a}$ and $\bar{b}$ under the natural projection $U\mathfrak{g} \to Gr_{\mathcal{F}} U\mathfrak{g}$) is well-defined and check the conditions of Poisson brackets for it.

E.2.15* Open a book on homological algebra and read about the basic constructions, such as the exact sequences, chain maps, chain homotopies, 5-lemma, spectral sequences, etc.

E.2.16* Compute the Lichnerowicz–Poisson cohomology of $\mathfrak{g}^*$ with usual Poisson structure in terms of the Lie algebra cohomology of $\mathfrak{g}$.

E.2.17 Prove by direct computation that the formula

$$\{f, g\} = a(x, y)(\partial_x f \partial_y g - \partial_y f \partial_x g)$$

for any smooth function $a(x, y)$ determines a Poisson bracket on $C^\infty(\mathbb{R}^2)$.

E.2.18* Compute the Lichnerowicz–Poisson cohomology of $\mathbb{R}^2$ with the Poisson structure, given by equation $\{x, y\} = x^2 + y^2$.
Hint: see [Na97].

7 Crash course in homological algebra, II. Hochschild homology

Among many homology and cohomology theories, one that plays the crucial role in the deformation theory of algebras and rings is the *Hochschild cohomology*. This theory can be considered as a dualisation of the *Hochschild homology* theory, although one should exercise a certain degree of caution here, as the duality between Hochschild chains and cochains is not direct. In this (and the next) section, we describe basic constructions related with Hochschild (co)homology. Most of this material can be found in many books, including the earlier mentioned books by Weibel [Wei94] and Gelfand and Manin [GM03]. Other wonderful references are the Loday's and Witherspoon's books [Lo98], [Wi19] which contain much more than what we need here.

7.1 Definitions

Let us begin with the definitions of Hochschild homology and Hochschild cohomology. Consider an associative algebra with unit A over a field $\Bbbk$; let M be an A-*bi*module, i.e. a $\Bbbk$-vector space for which there exist both the left and the right actions of A, commuting with each other:

$$(a \cdot m) \cdot b = a \cdot (m \cdot b) \text{ for all } a, b \in A, \ m \in M.$$

Below we shall usually omit the multiplication sign $\cdot$.

Definition 7.1. *Hochschild cohomology* of A with coefficients in M is the cohomology of the complex

$$C^{\cdot}(A, M) = (\{\mathrm{Hom}(A^{\times k}, M)\}_{k \geq 0}, \delta),$$

where $\mathrm{Hom}(A^{\times p}, M)$ denotes the space of $\Bbbk$-multilinear maps (and as usual we put $C^0(A, M) = M$), and the differential (**Hochschild coboundary map**) δ is given by the formula: for any $\varphi \in C^p(A, M)$

and any $a_0, \ldots, a_p \in A$

$$\delta(\varphi)(a_0, \ldots, a_p) = a_0 \cdot \varphi(a_1, \ldots, a_p)$$
$$+ \sum_{i=1}^{p} (-1)^i \varphi(a_0, \ldots, a_{i-1}a_i, \ldots, a_p)$$
$$+ (-1)^{p+1} \varphi(a_0, \ldots, a_{p-1}) \cdot a_p.$$

These cohomology groups are denoted by $H^*(A, M)$. An important particular case of this construction is $A = M$ with the usual product as the bimodule structure; these groups are often denoted by $HH^*(A)$.

The fact that $\delta^2 = 0$ follows from direct, but a bit lengthy computation; for example, if $p = 2$

$$\begin{aligned}\delta^2(\varphi)(a,b,c,d) &= a\delta(\varphi)(b,c,d) - \delta(\varphi)(ab,c,d) + \delta(\varphi)(a,bc,d) \\ &\quad - \delta(\varphi)(a,b,cd) + \delta(\varphi)(a,b,c)d \\ &= a(b\varphi(c,d) - \varphi(bc,d) + \varphi(b,cd) - \varphi(b,c)d) \\ &\quad - (ab\varphi(c,d) - \varphi(abc,d) + \varphi(ab,cd) - \varphi(ab,c)d) \\ &\quad + (a\varphi(bc,d) - \varphi(abc,d) + \varphi(a,bcd) - \varphi(a,bc)d) \\ &\quad - (a\varphi(b,cd) - \varphi(ab,cd) + \varphi(a,bcd) - \varphi(a,b)cd) \\ &\quad + (a\varphi(b,c) - \varphi(ab,c) + \varphi(a,bc) - \varphi(a,b)c)d \\ &= 0.\end{aligned}$$

All the terms here cancel out because of the associativity of the product in A and commutativity of the left and right actions of A on M.

Example 7.1.

7.1.1 If $k = 0$, $C^0(A, M) = M$ and $\delta m(a) = a \cdot m - m \cdot a$ for all $a \in A$, $m \in M$, so $H^0(A, M) = Z_A(M)$, the center of the bimodule:

$$Z_A(M) = \{m \in M \mid a \cdot m = m \cdot a, \ \forall a \in A\}.$$

In particular, if $M = A$, $HH^0(A) = Z(A)$, the center of A.

7.1.2 If $k = 1$, $C^1(A, M) = \mathrm{Hom}(A, M)$, and for all $\xi \in C^1(A, M)$, $a, b \in A$

$$\delta\xi(a, b) = a\xi(b) - \xi(ab) + \xi(a)b.$$

Thus, $\delta\xi = 0$ iff ξ satisfies the *Leibniz rule*:
$$\xi(ab) = a\xi(b) + \xi(a)b.$$
Maps $\xi : A \to M$ with this property are called *derivations of A with values in A-bimodule M*, denoted $Der(A, M)$. The derivations, given by coboundaries δm, $m \in M$ are called *inner derivations* and denoted $In(A, M)$:
$$\delta(m)(a) = am - ma.$$
So we get the final equality
$$H^1(A, M) = Der(A, M)/In(A, M).$$
This space is often denoted by $Out(A, M)$. If $M = A$, we get $HH^1(A) = Out(A)$.

Although we will only very little use Hochschild homology in these lectures, it is worth giving the definition of this homology theory as well.

Definition 7.2. *Hochschild homology* of A with coefficients in M, denoted $H_*(A, M)$ is the homology of the complex
$$C_.(A, M) = (\{M \otimes A^{\otimes k}\}_{k \geq 0}, b),$$
where the differential $b : M \otimes A^{\otimes k} \to M \otimes A^{\otimes k-1}$ is given by the formula (we replace the tensor signs $\otimes$ by commas for brevity)
$$b(m, a_1, \ldots, a_k) = (ma_1, a_2, \ldots, a_k)$$
$$+ \sum_{i=1}^{k-1}(-1)^i(m, a_1, \ldots, a_ia_{i+1}, \ldots, a_k)$$
$$+ (-1)^k(a_km, a_1, \ldots, a_{k-1}).$$

If $M = A$, these homology groups are denoted by $HH_*(A)$.

As before, the equality $b^2 = 0$ is proved by a direct computation; we leave this to the reader as a good exercise.

Remark 7.3. One should not be fooled by the names: *in general, there is no duality pairing between Hochschild homology and Hochschild cohomology with the same coefficients module M for generic M.* In order to get such pairings, one needs to choose special coefficient modules, see Loday's book [Lo98] for details.

7.2 Derived tensor product and derived Hom_R functors

One of the major flaws of Definitions 7.1 and 7.2 is that they depend on a particular choice of complexes. Moreover the standard complex, that computes this (co)homology is rather big even if the algebra A is finite-dimensional, which is rarely the case. A standard way to circumvent this difficulty in homological algebra is by giving an abstract characterisation of homology in terms of *derived functors* or their generalisations. It turns out that similar constructions exist in the case of Hochschild homology and cohomology. We are going to explain them very briefly now; this description allows one in some situations to use smaller complexes to compute the (co)homology groups.

First of all, recall that a (right) module P over a ring R is called *projective*, if for any surjective module morphism $f : M \twoheadrightarrow N$ of R-modules and any morphism $g : P \to N$ there exists a map $h : P \to M$, such that $g = f \circ h$, see diagram (7.1):

$$\begin{array}{ccc} & & M \\ & \overset{\exists h}{\nearrow} & \downarrow f \\ P & \xrightarrow{g} & N \end{array} \qquad (7.1)$$

An important example of projective modules are *free* modules, i.e. modules of the form $V \otimes R$ for some vector space V (not necessarily finite-dimensional). Let now M be an arbitrary (right) R-module.

Definition 7.4.

(i) *A chain complex* $P_\cdot = (\{P_k\}_{k \geq 0}, d)$ *is called* **projective resolution** *of* M, *if for all* $k \geq 0$, P_k *is a projective R-module, the differential d is a morphism of R-modules (at all stages) and there exists a morphism* $\varepsilon : P_0 \to M$, *called* **augmentation**, *such that the complex*

$$0 \leftarrow M \xleftarrow{\varepsilon} P_0 \xleftarrow{d} P_1 \xleftarrow{d} P_2 \xleftarrow{d} P_3 \xleftarrow{d} \ldots$$

has trivial homology in all dimensions; such complexes are called **acyclic**.

(ii) If N is a left R-module and $P.$ a projective resolution of M then the **R-torsion groups of M and N of degree k** are defined as the homology in degree k of the complex

$$P. \otimes_R N = (\{P_k \otimes_R N\}_{k \geq 0}, d \otimes \mathbb{1}),$$

$$P_0 \otimes_R N \xleftarrow{d \otimes \mathbb{1}} P_1 \otimes_R N \xleftarrow{d \otimes \mathbb{1}} P_2 \otimes_R N \xleftarrow{d \otimes \mathbb{1}} \ldots.$$

They are denoted $Tor_k^R(M, N) = H_k(P. \otimes_R N)$.

(iii) If N is a right R-module and $P.$ a projective resolution of M then the **R-extension groups of M and N of degree k** are defined as the degree k cohomology of the complex

$$\mathrm{Hom}_R(P, N) = (\{\mathrm{Hom}_R(P_k, N)\}_{k \geq 0}, d^*),$$

$$\mathrm{Hom}_R(P_0, N) \xrightarrow{d^*} \mathrm{Hom}_R(P_1, N) \xrightarrow{d^*} \ldots.$$

They are denoted $Ext_R^k(M, N) = H^k(\mathrm{Hom}_R(P., N))$.

Remark 7.5. In literature the groups $Tor_k^R(M, N)$ are called *the derived tensor product groups of M and N over R* and $Ext_R^k(M, N)$ are called *the derived Hom_R groups of M and N*. They are basic examples of derived functors, associated with left- or right-exact functors on Abelian categories, see [Wei94], [GM03].

7.3 Hochschild homology and cohomology as derived functors

One of the first results in homological algebra is that *under mild conditions on M and R there always exist projective resolutions and the groups $Tor_k^R(M, N)$, $Ext_R^k(M, N)$ do not in effect depend on the choice of such resolutions.*[15] We are not going to prove this fact here, as it is pretty standard but technical. Instead, we shall reinterpret Hochschild homology and cohomology as higher torsions and extensions respectively.

---

[15] It is also worth mentioning that in order to compute $Tor_k^R(M, N)$ one can take projective resolution of N instead of a resolution of M; in case of , $Ext_R^k(M, N)$ one needs *injective* resolutions of N.

To this end, consider the *enveloping algebra* of A, given by definition as $A^e = A^o \otimes A$, where A^o is A with inverted product: $A^o = A$ as linear spaces, but $a \cdot_o b = b \cdot a$. Then it is evident, that *right and left A^e-modules can be naturally identified with A-bimodules.*

We are now going to construct a resolution of A by free (and hence projective) A^e-modules. To this end, we consider the complex

$$B.(A) = (\{A^{\otimes p+2}\}_{p \geq 0}, b'), \tag{7.2}$$

$$A \otimes A \xleftarrow{b'} A \otimes A \otimes A \xleftarrow{b'} A \otimes A^{\otimes 2} \otimes A \xleftarrow{b'} \cdots$$

where the differential b' is given by the formula

$$\begin{aligned} b'[x|a_1|\ldots|a_p|y] &= [xa_1|a_2|\ldots|a_p|y] \\ &+ \sum_{i=1}^{p-1}(-1)^i[x|a_1|\ldots|a_ia_{i+1}|\ldots|a_p|y] \\ &+ (-1)^p[x|a_1|\ldots|a_{p-1}|a_py]. \end{aligned} \tag{7.3}$$

Here $x, y, a_1, \ldots, a_p$ are elements in A and we use the bars | instead of the tensor product signs to save the space; for this reason this complex is often referred to as *the bar-construction* or *the bar-resolution*. It is evident that for all $p \geq 0$, $B_p(A)$ is free A^e-module.

Consider the augmentation $\varepsilon : B_0(A) = A \otimes A \to A$, $[a|b] \mapsto ab$; obviously this is a map of A^e-modules. We are going to show that $B.(A)$ is an acyclic resolution of A; to this end shall construct contracting chain homotopy, i.e. a collection of linear maps $h_p : B_p(A) \to B_{p+1}(A)$, $p \geq 0$, such that

$$h_{p-1} \circ b' + b' \circ h_p = \mathbb{1}_{B_p(A)}. \tag{7.4}$$

Indeed, if such $\{h_p\}$ exists, then for every cycle c, $b'(c) = 0$ in the augmented complex

$$c = \mathbb{1}(c) = (h_{p-1} \circ b' + b' \circ h_p)(c) = b'(h_p(c)),$$

so c is exact and its homology is trivial.

To define such chain homotopy, it is enough to put $h_p[x|a_1|\ldots|a_p|y] = [1|x|a_1|\ldots|a_p|y]$, where $1 \in A$ is the unit; in dimension -1 this gives the map $A \to B_0(A)$, $a \mapsto [1|a]$. We compute

$$h_{p-1}(b'[x|a_1|\ldots|a_p|y]) = [1|xa_1|a_2|\ldots|a_p|y]$$
$$+ \sum_{i=1}^{p-1}(-1)^i[1|x|a_1|\ldots|a_ia_{i+1}|\ldots|a_p|y]$$
$$+ (-1)^p[1|x|a_1|\ldots|a_{p-1}|a_py],$$
$$b'(h_p[x|a_1|\ldots|a_p|y]) = [x|a_1|\ldots|a_p|y] - [1|xa_1|a_2|\ldots|a_p|y]$$
$$- \sum_{i=1}^{p-1}(-1)^i[1|x|a_1|\ldots|a_ia_{i+1}|\ldots|a_p|y]$$
$$- (-1)^p[1|x|a_1|\ldots|a_{p-1}|a_py],$$

which gives the equality (7.4).

Now it is a matter of simple observation that Hochschild chain and cochain complexes coincide with the complexes, determined by this resolution:

$$M \otimes_{A^e} B.(A) \cong C.(A, M), \quad \mathrm{Hom}_{A^e}(B.(A), M) \cong C^{\cdot}(A, M),$$

so we have proved

Proposition 7.6.

$$H_k(A, M) = \mathrm{Tor}_k^{A^e}(A, M), \quad H^k(A, M) = \mathrm{Ext}_{A^e}^k(A, M).$$

7.4 Polynomial algebras

We are now going to take advantage of the freedom that comes from the possibility to choose projective resolutions, in order to compute the Hochschild homology of the polynomial algebra, or equivalently the symmetric algebra of a vector space. More examples of this sort can be found in exercises.

7.4.1 Computation in low dimensions

Let V be a finite-dimensional vector space, let

$$A = S(V) = \bigoplus_{k \geq 0} S^k(V).$$

We would like to compute $HH_*(A)$. It is not difficult to do this in low dimensions: for instance $HH_0(A) = A/[A,A] = A$, since A is commutative. Further $HH_1(A) = A \otimes A/\mathrm{im}(b)$, where

$$b(x \otimes y \otimes z) = xy \otimes z - x \otimes yz + zx \otimes y.$$

Due to the fact that A is commutative, this map is a homomorphism of left A-modules, where $A^{\otimes 3}$ is free A-module generated by $A^{\otimes 2}$. Image of this map in $A \otimes A$ is the A submodule, spanned by $-b(1 \otimes x \otimes y) = 1 \otimes xy - x \otimes y - y \otimes x$. We claim that there exists the following sequence of left A-modules:

$$0 \to \mathrm{im}(b) \to A \otimes A \xrightarrow{\chi} A \otimes V \to 0$$

where $\chi : A \to A \otimes V$, is given by the formula

$$\chi(a \otimes v_1 \ldots v_p) = \sum_{i=1}^{p} av_1 \ldots \widehat{v_i} \ldots v_p \otimes v_i,$$

for any monomial $v_1 \ldots v_p \in S^p(V)$ and $\chi(a \otimes 1) = 0$. An easy calculation shows that $\chi(\mathrm{im}(b)) = 0$. Moreover the natural inclusion $i : V = S^1(V) \to A = S(V)$ induces a natural splitting $i : A \otimes V \to A \otimes A$ of this sequence (i.e. $\chi \circ i = \mathrm{id}_{A \otimes V}$), and one can show that

$$1 \otimes v_1 \ldots v_p - i(\chi(1 \otimes v_1 \ldots v_p))$$
$$= 1 \otimes v_1 \ldots v_p - \sum_{i=1}^{p} v_1 \ldots \widehat{v_i} \ldots v_p \otimes v_i \in \mathrm{im}(b).$$

Indeed: if $p = 2$,

$$1 \otimes xy - x \otimes y - y \otimes x = -b(1 \otimes x \otimes y), \ x, y \in V.$$

If $p = 3$

$$1 \otimes xyz - yz \otimes x - xz \otimes y - xy \otimes x = 1 \otimes xyz - x \otimes yz - yz \otimes x$$
$$+ x \otimes yz - xz \otimes y - xy \otimes z$$
$$= -b(1 \otimes x \otimes yz) - b(x \otimes y \otimes z),$$

and so on, by induction. We conclude that $HH_1(A) = A \otimes V$.

7.4.2 Small resolution

In higher dimensions these computations become very cumbersome, and we resort to the homological constructions discussed earlier: we are going to describe a small resolution of $A = S(V)$ by free A^e-modules. To this end we consider the *exterior algebra of V*, $\Lambda^{\cdot}V = \bigoplus_{p=0}^{\dim V} \Lambda^p V$; this is a graded commutative algebra over $\Bbbk$.

Let $K^{\cdot} = K^{\cdot}(V) = S(V) \otimes \Lambda^{\cdot}V \otimes S(V)$ so that the grading in $K^{\cdot}$ is induced from $\Lambda^{\cdot}V$; this free A-bimodule is equal to the tensor product of algebras, so it has a natural structure of graded-commutative algebra. We introduce a differential $d : K^p \to K^{p-1}$ in $K^{\cdot}$ so that d satisfies the graded Leibniz rule and

$$d(f \otimes x \otimes g) = fx \otimes 1 \otimes g - f \otimes 1 \otimes xg, \quad \forall f, g \in S(V), \ x \in V = \Lambda^1 V. \tag{7.5}$$

Now $d^2 = 0$, because of the grading and commutativity of A, e.g,

$$\begin{aligned} d^2(f \otimes x \wedge y \otimes g) &= d(fx \otimes y \otimes g - f \otimes y \otimes xg \\ &\quad - fy \otimes x \otimes g + f \otimes x \otimes yg) \\ &= fxy \otimes 1 \otimes g - fx \otimes 1 \otimes yg \\ &\quad - fy \otimes 1 \otimes xg + f \otimes 1 \otimes yxg \\ &\quad - fyx \otimes 1 \otimes g + fy \otimes 1 \otimes xg \\ &\quad + fx \otimes 1 \otimes yg - f \otimes 1 \otimes xyg \\ &= 0. \end{aligned}$$

We are going to prove that $(K^{\cdot}(V), d)$ is in effect a resolution of A with augmentation

$$\varepsilon : K^0 = A \otimes \Bbbk \otimes A \to A$$

given by the multiplication.

To this end we observe that $K^{\cdot}(V \oplus W) = K^{\cdot}(V) \hat{\otimes} K^{\cdot}(W)$: in effect $S(V \oplus W) = S(V) \otimes S(W)$, $\Lambda^{\cdot}(V \oplus W) = \Lambda^{\cdot}V \hat{\otimes} \Lambda^{\cdot}W$. Now due to the *Künneth formula*[16] it is enough to show that $K^{\cdot}(V)$ is acyclic for $V = \Bbbk^1$.

In that case $A = S(V) = \Bbbk[t]$ and the augmented resolution takes form
$$A \xleftarrow{\varepsilon} K^1(V) \xleftarrow{d} K^1(V),$$
$$\Bbbk[t] \xleftarrow{m} \Bbbk[t] \otimes \Bbbk[t] \xleftarrow{d} \Bbbk[t] \otimes \Lambda^1[u] \otimes \Bbbk[t]$$

where u is the unique "skew variable", the generator of $\Lambda^{\cdot}\Bbbk^1$. Differential here is given by $d(f \otimes u \otimes g) = tf \otimes g - f \otimes tg$. Now acyclicity in degree 0 follows from the splitting of m, given for instance by $h_0(f) = 1 \otimes f$, and in degree 1 we use the map

$$h_1(t^p \otimes t^q) = \sum_{j=1}^{p} t^{p-j} \otimes u \otimes t^{q+j-1}.$$

Then

$$dh_1(t^p \otimes t^q) = \sum_{j=1}^{p}(t^{p-j+1} \otimes t^{q+j-1} - t^{p-j} \otimes t^{q+j}) = t^p \otimes t^q - 1 \otimes t^{p+q},$$

and hence $dh_1 + h_0 \varepsilon = \mathrm{id}$, so the complex is contractible.

7.4.3 The algebraic Hochschild–Kostant–Rosenberg theorem

Using the resolution $K^{\cdot}(V)$ it is now easy to compute the Hochschild homology and cohomology of $A = S(V)$: consider the tensor product $A \otimes_A^e K^{\cdot}(V) \cong S(V) \otimes \Lambda^{\cdot}(V)$. Since $K^{\cdot}(V)$ had the graded algebra structure, so does this product and since the differential d on $K^{\cdot}(V)$ satisfies Leibniz rule, so does the differential it induces on this tensor

---

[16]Künneth formula expresses the homology of the graded tensor product of two complexes in terms of homology of the components. We assume that the reader is familiar with it; however, in our case it is enough to observe that if every cocycle in complexes L and M is in fact a coboundary, and L, M are complexes of vector spaces, then so is every cocycle in $L \hat\otimes M$: this follows by direct computation. Moreover, contracting chain homotopies on tensor factors induce similar maps in the tensor product.

product. Hence in order to compute this differential it is enough to take $f \otimes v \in S(V) \otimes \Lambda^1(V)$. We compute

$$d(f \otimes v) = fv - vf = 0.$$

Similarly, $\mathrm{Hom}_{A^e}(K^{\cdot}(V), A) \cong \mathrm{Hom}(\Lambda^{\cdot}(V), A) \cong \Lambda^{\cdot}V^* \otimes S(V)$ and the differential is again trivial. Hence we obtain the following result:

Proposition 7.7. *Hochschild homology of $S(V)$ is given by $HH_k(S(V)) = S(V) \otimes \Lambda^k V$; similarly Hochschild cohomology of this algebra is $HH^k(S(V)) = S(V) \otimes \Lambda^k V^*$.*

In order to better understand the meaning of Proposition 7.7, consider the following construction. Let A be any *commutative* algebra. One can construct the following A-bimodule, called *Kähler differential 1-forms of A*; this bimodule is denoted $\Omega^1(A)$, it is the commutative bimodule,[17] spanned by the symbols $a\,db$ for $a, b \in A$ so that the symbol db is linear in b. This space is endowed with natural A-bimodule structure, i.e. $a' \cdot (a\,db) \cdot a'' = a'aa''db$, $a, a', a'', b \in A$ (in effect it is enough to consider only the left action since A is commutative), and factorised modulo the Leibniz rule:[18]

$$\Omega^1(A)$$
$$= \left\{ \sum_i a_i db_i, a_i, b_i \in A \right\} / \langle d(b+c) = db + dc, d(bc) = c\,db + b\,dc \rangle. \tag{7.6}$$

The map
$$d: A \to \Omega^1(A), \ d(a) = da, \tag{7.7}$$

satisfies the Leibniz rule in an evident way. It turns out that the construction of this *differential* $d: A \to \Omega^1(A)$ is universal in the following sense: *for every commutative A-bimodule M and every map $\delta : A \to M$, such that $\delta(ab) = a\delta(b) + b\delta(a)$, there exists a unique morphism of A-bimodules $u : \Omega^1(A) \to M$ such that $\delta = u \circ d$; then $\Omega^1(A)$ is uniquely determined by this property.*

---
[17]One calls an A-bimodule M commutative if $am = ma$ for all $a \in A$, $m \in M$.
[18]In order that these conditions lead us to no contradiction, the condition that A is commutative cannot be removed; however there exists a noncommutative version of such "differential forms", that plays an important role in cyclic homology theory and in Noncommutative Geometry, see exercises.

In the case $A = S(V)$, one can further simplify the definition of Kähler forms: it follows from the Leibniz rule that

$$d(v_1 \ldots v_p) = \sum_{i=1}^{p} v_1 \ldots \widehat{v_i} \ldots v_p \, dv_i. \qquad (7.8)$$

Let $M = S(V) \otimes V$ and let $\delta : S(V) \to M$ be given by

$$\delta(v_1 \ldots v_p) = \sum_{i=1}^{p} v_1 \ldots \widehat{v_i} \ldots v_p \otimes v_i. \qquad (7.9)$$

Then δ satisfies the Leibniz rule, so by (7.8), (7.9) and the universal property of $\Omega^1(A)$ we conclude that *there exist an isomorphism of $S(V)$-modules $\Omega^1(S(V)) \cong S(V) \otimes V$*.

We see that the result of Section 7.4.1 can be described as the isomorphism $HH_1(S(V)) \cong \Omega^1(S(V))$; moreover the map χ that determines this isomorphism is given by the formula $\chi(a \otimes b) = a\,db$. We can extend this observation to higher dimensions by setting for any commutative algebra A

$$\Omega^n(A) = \Lambda_A^n(\Omega^1(A))$$
$$= \underbrace{\Omega^1(A) \otimes_A \Omega^1(A) \otimes_A \ldots \otimes_A \Omega^1(A)}_{n \text{ times}} / \langle da \wedge db = -db \wedge da \rangle. \qquad (7.10)$$

The derivation d extends in an evident way to a graded differential $d : \Omega^n(A) \to \Omega^{n+1}(A)$ for all $n \geq 0$ (where $\Omega^0(A) = A$) and the complex $\Omega^{\cdot}(A) = (\{\Omega^n(A)\}_{n \geq 0}, d)$ is called *the complex of Kähler forms on A*.

If $A = S(V)$ then

$$\Omega^n(S(V)) = \Lambda_{S(V)}^n(\Omega^1(S(V))) = \Lambda_{S(V)}^n(S(V) \otimes V) = S(V) \otimes \Lambda^n V.$$

So the statement of Proposition 7.7 can be reinterpreted as the isomorphism $HH_n(S(V)) \cong \Omega^n(S(V))$.

It is easy to see that the map $\chi : A \otimes A \to \Omega^1(A)$ for $A = S(V)$ from Section 7.4.1 can be extended to all commutative A and all $n \geq 0$ to the map $\chi : C_n(A, A) \to \Omega^n(A)$ by the formula:

$$\chi : A \otimes A^{\otimes n} \to \Omega^n(A), \chi(f_0, f_1, f_2 \ldots, f_n) = f_0 df_1 \wedge df_2 \wedge \ldots \wedge df_n. \qquad (7.11)$$

Then $\chi \circ b = 0$, e.g.

$$\chi(b(a \otimes b \otimes c)) = \chi(ab \otimes c - a \otimes bc + ac \otimes b) = ab\,dc - a\,d(bc) + ac\,db = 0,$$

since $d(bc) = b\,dc + c\,db$. Thus if we introduce operator

$$\partial : \Omega^n(A) \to \Omega^{n-1}(A)$$

to be the zero map, then we see that

$$\chi : (C.(A,A), b) \to (\Omega^{\cdot}(A), \partial = 0)$$

is a chain map, and the result of Proposition 7.7 can be formulated as follows:

Proposition 7.8. *Let $A = S(V)$ then the chain map χ induces an isomorphism in homology: $HH_n(A) \cong \Omega^n(A)$ for all n.*

It turns out, that this statement holds for a vast class of commutative algebras, called *smooth algebras*. There exist various equivalent definitions of smooth algebras, an interested reader can consult Loday's book [Lo98]. Speaking a little loosely, *suppose that the finitely-generated commutative algebra A has form*

$$A = \Bbbk[x_1, \ldots, x_n]/\langle f_i(x_1, \ldots, x_n) = 0, \ i = 1, \ldots, m \rangle;$$

then A is smooth iff the equations $f_i(x_1, \ldots, x_n) = 0, \ i = 1, \ldots, m$ determine a smooth submanifold in $\Bbbk^n$. Then the following statement holds

Theorem 7.9 (Hochschild–Kostant–Rosenberg theorem). *Let A be a smooth algebra. Then the map $\chi : C.(A,A) \to \Omega^{\cdot}(A)$ induces an isomorphism $HH_n(A) \cong \Omega^n(A)$. The map χ is called* **Hochschild–Kostant–Rosenberg map.**

We are interested in smooth manifolds and algebras of smooth functions on them, which are not finitely-generated, so this statement cannot be applied directly to this case. Therefore we shall not prove this statement here (interested reader can find it in [Lo98]). Instead, in order to prove a "smooth analog" of this theorem we will need to modify the definitions of the Hochschild cohomology complex. This will be done in the next lecture.

8 Hochschild cohomology of C^∞-functions: Hochschild–Kostant–Rosenberg theorem

This section can be regarded as the third part of our "crash course" on Homological algebra. However, the subject that it deals with, namely, properties of the Hochschild cohomology of the algebra of smooth functions on a manifold, is less universal on one hand and on the other hand it is closely related to the deformation theory that we will develop later. Therefore we decided to treat it as a separate topic here.

One of the principal results discussed in this section is the "complete smooth version" of the Hochschild–Kostant–Rosenberg theorem (Theorems 8.4 and 8.9), which includes all additional structures such as the Gerstenhaber brackets and the ∪-product. Unlike the purely algebraic version, it is not so well documented in textbooks, or at least in the books that we are aware of. One can find a homological version of this theorem in [GVF01] for example; however the cohomological variant of this theorem is hard to find in books, so we give references to the original papers [CDG80], [Ge63] where these results first appeared.

8.1 Local and differentiable Hochschild complexes

Let M be a smooth manifold. It is our prime purpose here to compute the Hochschild cohomology of the algebra $C^\infty(M)$ of smooth functions on M.

To this end we observe that $A = C^\infty(M)$ is commutative so we can consider the Hochschild–Kostant–Rosenberg map (7.11),

$$\chi : C_\cdot(A, A) \to \Omega^\cdot(M),$$

where this time on the right-hand side we have the de Rham complex of M:

$$\chi(f_0, f_1, \ldots, f_n) = f_0 df_1 \wedge \cdots \wedge df_n.$$

Just like in the algebraic case, we see that $\chi \circ b = 0$ so χ induces a map χ_* of Hochschild homology into $\Omega^\cdot(M)$.

However, the map χ needs not induce an isomorphism in cohomology, at least the constructions from the previous section are not applicable since the algebra $C^\infty(M)$ is infinitely generated. In effect, one can say that the algebraic construction of Hochschild complex is not quite adequate when one studies such infinitely generated algebras; for instance, if we speak in the terms of Hochschild homology, then maximum that one can show in this case is that the Kähler differential forms are embedded into the Hochschild homology of $C^\infty(M)$ (see e.g. [GVF01] or [Ka87]). If we want to get an isomorphism, we should modify the construction of Hochschild complex so that it would take into account the topology of the function space or some other features of smooth functions.[19]

The same is true for Hochschild cohomology: here too in order to guarantee that χ_* is an isomorphism we need to modify the definition of cohomological Hochschild complex. One can do it by choosing appropriately completed tensor products in the definition of this complex so that this complex would turn into something that we are familiar with from Differential Geometry. However in these lecture notes we try to avoid analytic difficulties, in particular the difficulties related with completions of topological tensor products. So let us restrict our attention to Hochschild cohomology. Here one can do quite well by suitably restricting the class of multilinear maps from $C^\infty(M)$ to itself instead of completing the topological vector spaces. First, we give the following definition:

Definition 8.1. *Let $A = C^\infty(M)$, then*

(i) *a Hochschild cochain $\varphi \in C^p(A, A)$ is called **local**, if $\varphi(f_1, \ldots, f_p) \equiv 0$ on an open set $U \subseteq M$, whenever there exists $i = 1, \ldots, p$ such that $f_i(x)|_U \equiv 0$.*

(ii) *a Hochschild cochain $\varphi \in C^p(A, A)$ is called **differentiable**, if it is given by a p-differential operator.*

---

[19] Informally speaking, in order to take care of the topological properties of $C^\infty(M)$, one should choose such completed tensor product $\tilde\otimes$ of topological algebras that $C^\infty(M) \tilde\otimes C^\infty(M) \cong C^\infty(M \times M)$. In this case one can use a geometric construction to get a small resolution for $C^\infty(M)$ as topological algebra.

It is easy to see that local and differentiable cochains form subcomplexes in $C^{\cdot}(C^\infty(M), C^\infty(M))$; we shall denote these subcomplexes by $C^{\cdot}_{loc}(C^\infty(M), C^\infty(M))$ and $C^{\cdot}_{diff}(C^\infty(M), C^\infty(M))$, respectively. We shall denote by $HH_{loc}(C^\infty(M))$, $HH_{diff}(C^\infty(M))$ the corresponding cohomology.

Clearly, since any partial derivation of a function vanishes on the open set, where the function is identically equal to 0, we have $C^{\cdot}_{diff}(C^\infty(M), C^\infty(M)) \subseteq C^{\cdot}_{loc}(C^\infty(M), C^\infty(M))$. It turns out that the contrary is also true in some sense:

Theorem 8.2. *Every local cochain can be locally (in an open neighbourhood of any point) represented by a polydifferential operator.*

Observe that using partition of unity we can "sew" the local expressions together; in particular this means that *on compact manifold the complexes $C^{\cdot}_{loc}(C^\infty(M), C^\infty(M))$ and $C^{\cdot}_{diff}(C^\infty(M), C^\infty(M))$ coincide.* Thus in future we shall not pay much attention to the distinctions between local and differentiable cochains; in general, unless we need to underline the differences between various Hochschild complexes, we will omit the subscripts below, mutely assuming that only differentiable cochains are allowed whenever we deal with algebras of smooth functions.

We are not going to give a full proof of Theorem 8.2 here. Let us only sketch few important steps. First of all, we observe that the general case of p-cochain follows by induction on p, and the key step here is the proof for $p = 1$, i.e. the following statement, called *Peetre's theorem*:

Lemma 8.3 (Peetre's theorem). *If the linear (over constants) operator $P : C^\infty(M) \to C^\infty(M)$ is local, then it is locally (in a vicinity of any point) equal to a differential operator.*

This statement first appeared in the papers [Pe59], [Pe60]; its proof now can be found in many textbooks on Geometric Analysis, see for instance [He84]. Observe that it is enough to prove this statement for Euclidean space in a neighbourhood U of the origin; using locality of P, one can show that for every $a \in U$ there exists $m \in \mathbb{N}$, $C > 0$ and an open neighbourhood V of a such that $\|P(u)\|_0 \leq C\|u\|_m$ for all u; here $\|\cdot\|_k$ denotes the semi-norm of u

as $C^k(V)$-function. Then for a multiindex $\alpha = (\alpha_1, \ldots, \alpha_n)$, $|\alpha| \leq m$ (we use notation from Sections 2.1.3 and 3.1) we consider the smooth functions

$$a_\alpha(a) = P((x^1 - a^1)^{\alpha_1} \ldots (x^n - a^n)^{\alpha_n})(a), \ a = (a^1, \ldots, a^n) \in U.$$

We claim that for all $u \in C^\infty(U)$

$$Pu = \sum_{|\alpha| \leq m} \frac{a_\alpha}{\alpha!} \partial_x^\alpha u. \tag{8.1}$$

To this end we consider the difference

$$f_{m,a}(x) = u(x) - \sum_{|\alpha| \leq m} \frac{\prod_{i=1}^n (x^i - a^i)^{\alpha_i}}{\alpha!} \partial_x^\alpha u(a). \tag{8.2}$$

This function has trivial Taylor coefficients at a. Therefore using "hat" functions, we can approximate it in $C^m(V)$-norm by functions g_ν, which vanish identically near a and are equal to $f_{m,a}(x)$ also identically outside a small open neighbourhood of a. Then it follows from the inequality for the norms $\|\cdot\|_0$, $\|\cdot\|_m$ that

$$\|P(f_{m,a} - g_\nu)\|_0 \leq C\|f_{m,a} - g_\nu\|_m \to 0 \text{ as } \nu \to \infty.$$

Thus $Pf_{m,a}(a) = \lim_{\nu \to \infty} Pg_\nu(a) = 0$, since $g_\nu \equiv 0$ near a. Now the formula (8.1) follows by applying P to (8.2).

8.2 Smooth Hochschild–Kostant–Rosenberg theorem

8.2.1 The main theorem

Our primary purpose in this section it is to show that for differentiable cochains complex, an analog of Hochschild–Kostant–Rosenberg Theorem 7.9 holds:

Theorem 8.4 (Differentiable Hochschild–Kostant–Rosenberg theorem). *For any compact smooth manifold M, the following isomorphisms take place*

$$HH_{loc}^p(C^\infty(M)) = HH_{diff}^p(C^\infty(M)) \cong \mathcal{T}^p(M),$$

and the isomorphism on the right is induced by the map

$$\chi : \mathcal{T}^p(M) \to C^p_{diff}(C^\infty(M), C^\infty(M)),$$
$$\chi(\varphi)(f_1, \ldots, f_p) = \langle df_1 \wedge \cdots \wedge df_p, \varphi \rangle,$$

where $\langle , \rangle$ is the natural pairing of a p-differential form and the polyvector field $\varphi \in \mathcal{T}^p(M)$ (see Section 4.3.3). The map χ is called the **Hochschild–Kostant–Rosenberg map**.

Before this, let us observe that in polynomial case, i.e. for $A = S(V)$ we already had a similar identification: $HH^p(S(V)) \cong S(V) \otimes \Lambda^p V^*$ (see Proposition 7.7), where V^* is the dual space; in fact $S(V)$ plays the role of polynomial functions on V^*, and V^* can be regarded as the tangent space of itself at the origin (in fact, at any point). Thus $S(V) \otimes \Lambda^{\cdot} V^*$ can be interpreted as the space of polyvector fields with polynomial coefficients on V^*. Unfortunately, the proof we used there cannot be generalised directly to the smooth case.[20] So we first have to localise it to open coordinate neighbourhoods and then modify the reasoning somehow; this stratagem is used in this section. In the next section, we reproduce the original proof of Theorem 8.4 by Cahen, de Wilde and Gutt (see [CDG80]); their reasonings are rather technical so to save the space we shall only delineate the major steps at some point. The advantage of their reasoning on the other hand is that it allows a more detailed analysis of the local structure of the cocycles in the Hochschild complex; this will be of much help later in the construction of $\star$-product, see Section 12.

Proof. It is easy to see that $\delta(\chi(\omega)) = 0$ for all $\omega \in \mathcal{T}^{\cdot}(M)$. We shall show that for every differential p-cocycle c there exists $\omega \in \mathcal{T}^p(M)$ such that

$$c = \chi(\omega) + \delta E \qquad (8.3)$$

for some $p - 1$ differential cochain E. The first step of the proof is to reduce the statement to the local situation: let $U_1 \ldots, U_N$ be a

---
[20] As we have explained, that proof can still be modified to work well in the smooth case; in fact one can introduce a small resolution of $C^\infty(M)$, based on differential geometric considerations.

good open cover of M and $\{\varphi_k\}_{k=1}^N$, $\sum_k \varphi_k \equiv 1$ be the corresponding partition of unity. Then for every $k = 1, \ldots, N$ we have maps:

$$j_k : C_{\text{diff}}^p(C^\infty(M), C^\infty(M)) \to C_{\text{diff}}^p(C^\infty(M), C^\infty(U_k)),$$
$$j_k(c)(f_1 \ldots, f_p) = c(f_1 \ldots, f_p)|_{U_k},$$
$$\varphi_k^* : C_{\text{diff}}^p(C^\infty(M), C^\infty(U_k)) \to C_{\text{diff}}^p(C^\infty(M), C^\infty(M)),$$
$$\varphi_k^*(c)(f_1 \ldots, f_p) = \varphi_k c(f_1 \ldots, f_p).$$

In the first case we consider functions on U_k with natural bimodule structure over $C^\infty(M)$. These maps commute with differentials, since the algebra $C^\infty(M)$ is commutative. Then since $\sum_k \varphi_k \equiv 1$, we have $\sum_{k=1}^N \varphi_k^* \circ j_k = \text{id}$.

Further $C_{\text{diff}}^p(C^\infty(M), C^\infty(U_k))$ can be regarded as the space of p-differential operators on U_k, and as such it coincides with $C_{\text{diff}}^p(C^\infty(U_k), C^\infty(U_k))$, so if we know that the statement is true for all U_k then we can obtain the elements ω and E, satisfying (8.3) on M by using the partition of unity.[21] So it is enough to prove the theorem for $M = \mathbb{R}^n$.

Now the result follows from the previous theorem for the polynomial algebra. Indeed, on one hand we can consider the complex of differentiable Hochschild cochains of the polynomial algebra $S(\mathbb{R}^n)$ with coefficients in $C^\infty(\mathbb{R}^n)$; clearly

$$C_{\text{diff}}^{\cdot}(C^\infty(\mathbb{R}^n), C^\infty(\mathbb{R}^n)) = C_{\text{diff}}^{\cdot}(S(\mathbb{R}^n), C^\infty(\mathbb{R}^n)) \subseteq C^{\cdot}(S(\mathbb{R}^n), C^\infty(\mathbb{R}^n)).$$

On the other hand, it is easy to see that *every element* $\varphi \in C^{\cdot}(S(\mathbb{R}^n), C^\infty(\mathbb{R}^n))$ *can be described as a polydifferential operator of unbounded degree*. Since δ does not change the degree of a polydifferential operator we conclude that *if an element* $\alpha \in C_{\text{diff}}^{\cdot}(S(\mathbb{R}^n), C^\infty(\mathbb{R}^n))$ *is equal to a coboundary of an element* $\beta \in C^{\cdot}(S(\mathbb{R}^n), C^\infty(\mathbb{R}^n))$, *then there exists* $\beta' \in C_{\text{diff}}^{\cdot}(S(\mathbb{R}^n), C^\infty(\mathbb{R}^n))$ *such that* $\alpha = \delta\beta'$. Now since the Hochschild-Kostant-Rosenberg map χ sends $\mathcal{T}^{\cdot}(\mathbb{R}^n)$ into $C_{\text{diff}}^{\cdot}(S(\mathbb{R}^n), C^\infty(\mathbb{R}^n))$ and induces an isomorphism in cohomology for $C^{\cdot}(S(\mathbb{R}^n), C^\infty(\mathbb{R}^n))$ (to see this it is enough to use the small resolution of $S(\mathbb{R}^n)$ from Section 7.4.2 as

---

[21] A reader, familiar with the 5-lemma and Maier–Vietoris sequences will easily recognise them in this reasoning, thus making this proof more abstract.

in the proof of Proposition 7.7, since the result does not depend on the coefficients), we conclude that for every cocycle c there exist $\omega \in \mathcal{T}^p(\mathbb{R}^n)$ and $E \in C^\cdot_{diff}(S(\mathbb{R}^n), C^\infty(\mathbb{R}^n))$ for which Equation (8.3) holds. □

8.2.2 Symbol calculus

The proof we have just used except for the first step was based on the algebraic considerations, stemming from the algebraic case (polynomial algebra). For future references, it is worth having a more "hand-on" differential variant of this theorem. Here we shall sketch a construction taken from the paper [CDG80], based on the accurate study of symbols of the polydifferential operators.

Namely, if we fix the coordinates $x = (x^1, \ldots, x^n)$ in $\mathbb{R}^n$ we can define the *symbol* $\sigma(c)$ of a polydifferential operator c (we shall call it p-symbol): it is the polynomial in np variables $\xi_0 = (\xi_0^1, \ldots, \xi_0^n), \ldots, \xi_{p-1} = (\xi_{p-1}^1, \ldots, \xi_{p-1}^n)$ with coefficients in $C^\infty(\mathbb{R}^n)$ depending on x; namely, if

$$c(f_0 \ldots, f_{p-1}) = \sum_{\alpha^0, \ldots, \alpha^{p-1}} c_{\alpha^0, \ldots, \alpha^{p-1}}(x) \partial_x^{\alpha^0} f_0 \ldots \partial_x^{\alpha^{p-1}} f_{p-1},$$

where α^k are multiindices: $\alpha^k = (\alpha_1^k, \ldots, \alpha_n^k)$, $\alpha_j^k \in \mathbb{N}_0$, then

$$\sigma(c) = \sum_{\alpha^0, \ldots, \alpha^{p-1}} c_{\alpha^0, \ldots, \alpha^{p-1}}(x) \xi_0^{\alpha^0} \ldots \xi_{p-1}^{\alpha^{p-1}}.$$

The symbols are characterised by their (multi-) degree: *a symbol σ has degree $(r_0, \ldots, r_{p-1})$ if it is a polynomial of degree r_i in variables ξ_i*. It is clear that *a cochain c is the image of a polyvector field under the map χ iff its symbol is equal to a poly-linear totally skew-symmetric map in $\xi_0, \ldots, \xi_{p-1}$; in particular its multidegree is $(1, 1, \ldots, 1)$*. Let us denote the space of p-symbols on $\mathbb{R}^n$ by $Symb^p(\mathbb{R}^n)$.

From now on we shall only work with symbols. The differential in complex $C_{diff}(C^\infty(\mathbb{R}^n), C^\infty(\mathbb{R}^n))$ induces the map $\bar{\delta} : Symb^p(\mathbb{R}^n) \to Symb^{p+1}(\mathbb{R}^n)$:

$$\bar{\delta}(\sigma)(\xi_0,\ldots,\xi_p) = \sigma(\xi_1,\ldots,\xi_p) + \sum_{i=1}^{p}(-1)^i\sigma(\xi_0,\ldots,\xi_{i-1}+\xi_i,\ldots,\xi_p)$$
$$+ (-1)^{p+1}\sigma(\xi_0,\ldots,\xi_{p-1}). \tag{8.4}$$

Of course, c is cocycle iff $\bar{\delta}(\sigma(c)) = 0$. So the statement we want to prove will follow from the following: if $\bar{\delta}(\sigma) = 0$, then there a skew-symmetric symbol $w \in Symb^p(\mathbb{R}^n)$ and an element $E \in Symb^{p-1}(\mathbb{R}^n)$ for which $\sigma = w + \bar{\delta}E$.

Now let us consider the homogenous top-degree part of $\sigma \in Symb^p(\mathbb{R}^n)$, (the principal symbol of a polydifferential operator); we shall denote the operation of taking the top degree part of a polynomial by s. Clearly, $\bar{\delta}(\sigma) = 0$ iff $s(\bar{\delta}(\sigma)) = 0$.

It is our purpose to describe the principal symbol of $\bar{\delta}\sigma$ for homogenous $\sigma \in Symb^p(\mathbb{R}^n)$; observe that σ is homogenous iff $\sigma = s(\sigma)$. We consider the operations $\bar{\delta}_i : Symb^p(\mathbb{R}^n) \to Symb^{p+1}(\mathbb{R}^n)$, $i = 0,\ldots,p-1$:

$$\bar{\delta}_i(\sigma)(\xi_0,\ldots,\xi_p) = \sigma(\xi_0,\ldots,\widehat{\xi_i},\ldots,\xi_p) - \sigma(\xi_0,\ldots,\xi_i+\xi_{i+1},\ldots,\xi_p)$$
$$+ \sigma(\xi_0,\ldots,\widehat{\xi_{i+1}},\ldots,\xi_p). \tag{8.5}$$

Here as usual $\widehat{\cdot}$ denotes the missing element. Then

$$\bar{\delta}\sigma = \sum_{i=0}^{p-1}(-1)^i\bar{\delta}_i\sigma. \tag{8.6}$$

It is easy to see that for homogenous σ of multi-degree $(r_0,\ldots,r_{p-1})$

$$\bar{\delta}_i\sigma = 0 \text{ if } r_i = 1, \tag{8.7}$$
$$\bar{\delta}_i\sigma(\xi_0,\ldots,\xi_p) = \sigma(\xi_0,\ldots,\widehat{\xi_i},\ldots,\xi_p) \text{ if } r_i = 0. \tag{8.8}$$

In the general case, since for a homogenous σ its degree r_i is determined by Euler vector field in direction of ξ_i, $e_i = \sum_{k=1}^{n}\xi_i^k\frac{\partial}{\partial\xi_i^k}$ and since $\bar{\delta}_i$ does not change the degree of a symbol outside r_i we have the following characterisation of $s(\bar{\delta}_i\sigma)$:

$$s(\bar{\delta}_i\sigma)(\xi_0,\ldots,\xi_i,\xi_i,\xi_{i+2},\ldots,\xi_p) = (2-2^{r_i})\sigma(\xi_0,\ldots,\xi_i,\xi_{i+2},\ldots,\xi_p).$$

Let us dente by $T_j : Symb^p(\mathbb{R}^n) \to Symb^{p-1}(\mathbb{R}^n)$, $j = 0, 1, \ldots, p-2$ the map

$$T_j(\sigma)(\xi_0, \ldots, \xi_{p-2}) = \sigma(\xi_0, \ldots, \xi_{j-1}, \xi_j, \xi_j, \xi_{j+1}, \ldots, \xi_{p-2}).$$

Then, using the lexicographic order in the set of possible multi-degrees of p-differential operators, we can prove the following statements: let σ be a homogenous cocycle of multi-degree $(r_0, \ldots, r_p)$, then the following is true:

(i) The multi-degree of σ in ξ_{p-1} direction is $r_{p-1} \leq 1$.

(ii) If $r_i = r_{i+1} = \cdots = r_{p-1} = 1$ and $k \geq i$, let

$$\sigma'(\xi_0 \ldots, \xi_k, \xi_{k+1}, \ldots, \xi_{p-1}) = \frac{1}{2}(\sigma(\xi_0 \ldots, \xi_k, \xi_{k+1}, \ldots, \xi_{p-1})$$
$$- \sigma(\xi_0 \ldots, \xi_{k+1}, \xi_k, \ldots, \xi_{p-1}))$$

then

$$\sigma - \sigma' = \frac{(-1)^{k-1}}{2} \bar{\delta}(T_k(\sigma)).$$

(iii) If $r_i = \cdots = r_{p-1} = 1$ and σ' is the symbol σ symmetrised in the last $p - i$ variables, then there exists $E \in Symb^{p-1}(\mathbb{R}^n)$, such that $\sigma - \sigma' = \bar{\delta}E$.

(iv) If $r_i = r_{i+1} = \cdots = r_{j-1} = 0$, $r_j = r_{j+1} = \cdots = r_{p-1} = 1$ and σ is totally skew-symmetric in the last $p - j$ variables, then $j - i$ is even and $\sigma = (-1)^i \bar{\delta}(T_i(\sigma))$.

(v) If $r_i > 1$, $r_{i+1} = r_{i+2} = \cdots = r_{p-1} = 1$ and σ is skew-symmetric in the last $p - i - 1$ variables, then $\sigma = \frac{(-1)^{i+1}}{2^{r_i - 1}} \bar{\delta}(T_i(\sigma))$.

For instance the claim (i) follows from the fact that if $r_{p-1} > 1$, then the top-degree term of $\bar{\delta}\sigma$ will be $(r_0, \ldots, r_{p-1} - 1, 1)$ which cannot be "killed" by any other term in the coboundary. Some details can be found in the original paper by Cahen, Gutt and de Wilde.

Remark 8.5. Observe that the skew-symmetrisation map $SSym : C^p(C^\infty(M), C^\infty * (M)) \to C^p(C^\infty(M), C^\infty * (M))$,

$$SSym(c)(f_1, \ldots, f_p) = \frac{1}{p!} \sum_{s \in S_p} (-1)^s c(f_{s(1)}, \ldots, f_{s(p)}),$$

where S_p is the permutation group of p elements, sends coboundaries to zero and is equal to identity on the image of χ (this follows from the fact that the multiplication of functions is commutative, and on the other hand, every term in the formula for coboundary δ contains exactly one product of two functions, so that when we exchange these functions in the process of skew-symmetrisation these terms cancel out). Hence applying $SSym$ to both sides of (8.3) we see that for any p-cocycle c, its image in the Hochschild cohomology of $C^\infty(M)$ is represented by the element $SSym(c) = \chi(\omega)$.

8.3 Algebraic structures on Hochschild cohomology

The space of Hochschild cochains of an associative algebra bears two important binary operations: the cup-product (in many books it's written as $\cup$-product) and the Gerstenhaber bracket; while the first one is quite evident and simple to define, the definition of the second operation is less obvious and was first discovered by Murray Gerstenhaber in [Ge63].

8.3.1 The $\cup$-product

Let A be an algebra and $\varphi \in C^p(A, A)$, $\psi \in C^q(A, A)$ be two Hochschild cochains. Since A is an algebra, we can consider the following $(p+q)$-cochain:

$$\varphi \cup \psi(a_1, \ldots, a_{p+q}) = \varphi(a_1, \ldots, a_p)\psi(a_{p+1}, \ldots, a_{p+q}). \quad (8.9)$$

Then

$$\delta(\varphi \cup \psi)(a_1, \ldots, a_{p+q+1}) = a_1\varphi(a_2, \ldots, a_{p+1})\psi(a_{p+2}, \ldots, a_{p+q+1})$$
$$+ \sum_{i=1}^{p}(-1)^i\varphi(a_1, \ldots, a_ia_{i+1}, \ldots, a_{p+1})\psi(a_{p+2}, \ldots, a_{p+q+1})$$
$$+ (-1)^{p+1}\varphi(a_1, \ldots, a_p)a_{p+1}\psi(a_{p+2}, \ldots, a_{p+q+1})$$
$$- (-1)^{p+1}\varphi(a_1, \ldots, a_p)a_{p+1}\psi(a_{p+2}, \ldots, a_{p+q+1})$$
$$+ \sum_{j=1}^{q}(-1)^{p+j}\varphi(a_1, \ldots, a_p)\psi(a_{p+1}, \ldots, a_{p+j}a_{p+j+1}, \ldots, a_{p+q+1})$$
$$+ (-1)^{p+q+1}\varphi(a_1, \ldots, a_p)\psi(a_{p+1}, \ldots, a_{p+q})a_{p+q+1}.$$

So
$$\delta(\varphi \cup \psi)(a_1, \ldots, a_{p+q+1}) = (\delta(\varphi) \cup \psi + (-1)^p \varphi \cup \delta\psi)(a_1, \ldots, a_{p+q+1}).$$

It follows directly that the $\cup$-product is an associative graded product on the Hochschild complex $C^{\cdot}(A, A)$; it satisfies the graded Leibniz rule with respect to the Hochschild coboundary operation, and hence it induces an associative multiplication $HH^p(A) \otimes HH^q(A) \to HH^{p+q}(A)$ in cohomology. If $A = C^\infty(M)$ then we have the following equality: for all $\varphi \in \mathcal{T}^p(M)$, $\psi \in \mathcal{T}^q(M)$ and all smooth functions $f_1, \ldots, f_{p+q}$

$$\begin{aligned}
\chi(\varphi \wedge \psi)(f_1, \ldots, f_{p+q}) &= \langle df_1 \wedge \cdots \wedge df_{p+q}, \varphi \wedge \psi \rangle \\
&= \sum_{\sigma \in Sh(p,q)} (-1)^\sigma \langle df_{\sigma(1)} \wedge \cdots \wedge df_{\sigma(p)}, \varphi \rangle \langle df_{\sigma(p+1)} \wedge \cdots \wedge df_{\sigma(p+q)}, \psi \rangle \\
&= \sum_{\sigma \in Sh(p,q)} (-1)^\sigma (\chi(\varphi) \cup \chi(\psi))(f_{\sigma(1)}, \ldots, f_{\sigma(p+q)});
\end{aligned} \quad (8.10)$$

here $Sh(p, q)$ is the set of all (p, q)-*shuffles*, i.e. of all permutations $\sigma \in S_{p+q}$, such that $\sigma(1) < \sigma(2) < \cdots < \sigma(p)$ and $\sigma(p + 1) < \sigma(2) < \cdots < \sigma(p + q)$. On the other hand (see Remark 8.5), the cohomology class, induced by $\chi(\varphi) \cup \chi(\psi)$ is represented by the skew-symmetrisation of this cocycle, and it is easy to see that $Sym(\chi(\varphi) \cup \chi(\psi))$ is exactly the right-hand side of Equation (8.10), so on the level of cohomology

$$[\chi(\varphi \wedge \psi)] = [\chi(\varphi)] \cup [\chi(\psi)] \in HH^{p+q}(C^\infty(M)).$$

In other words, *Hochschild-Kostant-Rosenberg map induces an isomorphism of graded commutative algebras* $\chi: \mathcal{T}^*(M) \cong HH^*_{\mathit{diff}}(C^\infty(M))$, *although it is not a homomorphism of algebras itself.* In particular $HH^*_{\mathit{diff}}(C^\infty(M))$ is graded commutative with respect to the cup-product. It turns out that the $\cup$-product on $HH^*(A)$ is always graded commutative, even if A is not commutative. To prove this we shall need to introduce one more algebraic structure on Hochschild complex.

8.3.2 Gerstenhaber bracket

As we have shown earlier (see Example 7.1.2) Hochschild cohomology in dimension 1 of any algebra A naturally coincides with the space of all outer derivations of A, $Out(A) = Der(A)/Inn(A)$. If ξ, η are two derivations of A, then their commutator, determined by the rule

$$[\xi, \eta](f) = \xi(\eta(f)) - \eta(\xi(f)), \ \forall f \in A$$

is again a derivation of A. Moreover, it is not difficult to see that inner derivations make up a Lie ideal with respect to this commutator:

$$[\xi, ad_g](f) = ad_{\xi(g)}(f), \ \forall f, g \in A, \text{ where } ad_g(f) = gf - fg.$$

Thus this structure survives on $HH^1(A) = Out(A)$. It turns out that this construction can be extended further to cochains of arbitrary dimensions.

To this end we define for any two cochains $f^p \in C^p(A, A)$, $g^q \in C^q(A, A)$ and any index $i = 0, 1, \ldots, p-1$ the following new cochain $f^p \circ_i g^q \in C^{p+q-1}(A, A)$:

$$f^p \circ_i g^q(a_1, \ldots, a_{p+q-1})$$
$$= f^p(a_1, \ldots, a_i, g^q(a_{i+1}, \ldots, a_{i+q}), a_{i+q+1}, \ldots, a_{p+q-1}). \quad (8.11)$$

Of course, we can exchange the roles of f^p and g^q so that for all $j = 0, 1, \ldots, q-1$ we will obtain $g^q \circ_j f^p \in C^{p+q-1}(A, A)$.

Combining $\circ_i$ for various i, we obtain the map $f^p \circ g^q$:

$$f^p \circ g^q = \sum_{i=0}^{p-1} (-1)^{i(q-1)} f^p \circ_i g^q. \quad (8.12)$$

Finally we define *Gerstenhaber bracket of f^p and g^q*, denoted by $[f^p, g^q]$, as follows

$$[f^p, g^q] = f^p \circ g^q - (-1)^{(p-1)(q-1)} g^q \circ f^p. \quad (8.13)$$

It is clear, that this bracket is graded anti-commutative in the following sense:

$$[f^p, g^q] = -(-1)^{(p-1)(q-1)} [g^q, f^p],$$

c.f. part (i) of Theorem 4.8. It turns out, that it has many more good properties. First of all, we have the following statement, due to Gerstenhaber:

Proposition 8.6. *The bracket $[,]$ satisfies graded Jacobi identity, similar to part (iii) of Theorem 4.8:*

$$(-1)^{(p-1)(r-1)}[[f^p, g^q], h^r] + (-1)^{(q-1)(p-1)}[[g^q, h^r], f^p]$$
$$+ (-1)^{(r-1)(q-1)}[[h^r, f^p], g^q] = 0, \qquad (8.14)$$

for all Hochschild cochains f^p, g^q, h^r of degrees p, q, r, respectively.

Proof. The statement follows by careful inspection of definitions: first we observe that

$$(f^p \circ_i g^q) \circ_j h^r = \begin{cases} (f^p \circ_j h^r) \circ_{i+r} g^q, & 0 \leq j \leq i-1 \\ f^p \circ_i (g^q \circ_{j-i} h^r), & i \leq j \leq i+q-1 \\ (f^p \circ_{j-q} h^r) \circ_i g^q, & i+q \leq j \leq p+q-2, \end{cases} \qquad (8.15)$$

and similarly

$$f^p \circ_i (g^q \circ_j h^r) = (f^p \circ_i g^q) \circ_{i+j} h^r. \qquad (8.16)$$

Summing up these equations for all i, j with appropriate signs gives:

$$(f^p \circ g^q) \circ h^r - f^p \circ (g^q \circ h^r)$$
$$= \sum_{j=0}^{p+q-2} (-1)^{j(r-1)} \sum_{i=0}^{p-1} (-1)^{i(p-1)} (f^p \circ_i g^q) \circ_j h^r$$
$$- \sum_{i=0}^{p-1} (-1)^{i(p+r-1)} \sum_{j=0}^{q-1} (-1)^{j(r-1)} f^p \circ_i (g^q \circ_j h^r)$$
$$= \sum_{\substack{0 \leq j \leq i-1 \\ i+q \leq j \leq p+q-2}} (-1)^{i(p-1)+j(r-1)} (f^p \circ_i g^q) \circ_j h^r, \qquad (8.17)$$

since all the terms in the middle cancel out due to the equalities (8.15), (8.16). Next, abbreviating the sum on the right of this equation to $\sum'_{i,j}$ we get

$$f^p \circ [g^q, h^r] = f^p \circ (g^q \circ h^r) - (-1)^{(q-1)(r-1)} f^p \circ (h^r \circ g^q)$$
$$= (f^p \circ g^q) \circ h^r - \sum_{i,j}{}'(-1)^{i(p-1)+j(r-1)}(f^p \circ_i g^q) \circ_j h^r$$
$$- (-1)^{(q-1)(r-1)} \left((f^p \circ h^r) \circ g^q \right.$$
$$\left. - \sum_{i,j}{}'(-1)^{i(p-1)+j(q-1)}(f^p \circ_i h^r) \circ_j g^q \right)$$
$$= (f^p \circ g^q) \circ h^r - (-1)^{(q-1)(r-1)}(f^p \circ h^r) \circ g^q,$$

since all the terms in $\sum'_{i,j}$ cancel out due to (8.15). Thus we get

$$(f^p \circ g^q) \circ h^r - (-1)^{(q-1)(r-1)}(f^p \circ h^r) \circ g^q = f^p \circ [g^q, h^r]. \quad (8.18)$$

Finally, taking cyclic permutations of f^p, g^q and h^r in (8.18) and summing up with proper signs yields (8.14). □

This proposition and the methods used in its proof yield the following immediate consequence:

Proposition 8.7.

(i) *Hochschild coboundary operator δ satisfies the graded right Leibniz rule with respect to the Gerstenhaber bracket,[22] more accurately, we have:*

$$\delta([f^p, g^q]) = (-1)^{q-1}[\delta f^p, g^q] + [f^p, \delta g^q].$$

(ii) *The $\cup$-product and Gerstenhaber bracket are related by the following equality: for all cochains f^p, g^q:*

$$f^p \cup g^q - (-1)^{pq} g^q \cup f^p = (-1)^q (\delta(g^q \circ f^p)$$
$$- (-1)^{q-1} \delta g^q \circ f^p - (-1)^{p+q} g^q \circ \delta f^p).$$

---
[22]In order to obtain the usual graded Leibniz rule, it is enough to replace δ by δ': $\delta' f^p = (-1)^{p-1} \delta f^p$.

Proof. (*i*) This follows from the observation that for any f^p, its coboundary is up to the sign equal to the Gerstenhaber bracket of f^p with the multiplication map $m : A \otimes A \to A$ regarded as an element in $C^2(A, A)$:

$$m \circ f^p(a_0, \ldots, a_p) = m(f^p(a_0, \ldots, a_{p-1}), a_p)$$
$$+ (-1)^{p-1} m(a_0, f^p(a_1, \ldots, a_p))$$
$$= (-1)^{p-1}(a_0 f^p(a_1, \ldots, a_p)$$
$$+ (-1)^{p-1} f^p(a_0, \ldots, a_{p-1}) a_p),$$

$$f^p \circ m(a_0, \ldots, a_p) = \sum_{i=0}^{p-1} (-1)^i f^p(a_0, \ldots, m(a_i, a_{i+1}), \ldots, a_p)$$
$$= \sum_{i=0}^{p-1} (-1)^i f^p(a_0, \ldots, a_i a_{i+1}, \ldots, a_p).$$

So

$$[m, f^p] = (-1)^{p-1} \Big(a_0 f^p(a_1, \ldots, a_p) + (-1)^{p-1} f^p(a_0, \ldots, a_{p-1}) a_p$$
$$- \sum_{i=0}^{p-1} (-1)^i f^p(a_0, \ldots, a_i a_{i+1}, \ldots, a_p) \Big) = (-1)^{p-1} \delta f^p,$$

and $\delta f^p = -[f^p, m]$. Now the Leibniz rule follows from the graded Jacobi identity:

$$\delta([f^p, g^q]) = -[[f^p, g^q], m]$$
$$= (-1)^{p-1}((-1)^{(p-1)(q-1)}[[g^q, m], f^p] + (-1)^{q-1}[[m, f^p], g^q])$$
$$= (-1)^{q(p-1)-1}[\delta g^q, f^p] + (-1)^{q-1}[\delta f^p, g^q]$$
$$= [f^p, \delta g^q] + (-1)^{q-1}[\delta f^p, g^q].$$

(*ii*) Since

$$f^p \cup g^q = (m \circ_0 f^p) \circ_p g^q, \quad g^q \cup f^p = (m \circ_0 g^q) \circ_q f^p,$$

we have (using (8.15))

$$f^p \cup g^q - (-1)^{pq} g^q \cup f^p = (m \circ_1 g^q) \circ_0 f^p - (-1)^{pq}(m \circ_0 g^q) \circ_q f^p.$$

On the other hand
$$\delta(g^q \circ f^p) = [m, g^q \circ f^p] = m \circ (g^q \circ f^p) - (-1)^{p+q}(g^q \circ f^p) \circ m$$
and
$$\delta g^q \circ f^p = (-1)^{q-1}((m \circ g^q) \circ f^p - (-1)^{q-1}(g^q \circ m) \circ f^p)$$
$$g^q \circ \delta f^p = (-1)^{p-1}(g^q \circ (m \circ f^p) - (-1)^{p-1}g^q \circ (f^p \circ m)).$$

So eventually it follows from (8.15) that
$$\delta(g^q \circ f^p) - (-1)^{q-1}\delta g^q \circ f^p - (-1)^{p+q}g^q \circ \delta f^p$$
$$= (m \circ (g^q \circ f^p) - (m \circ g^q) \circ f^p) - (-1)^{p+q}((g^q \circ f^p) \circ m$$
$$- g^q \circ (f^p \circ m)) + (-1)^{q-1}((g^q \circ m) \circ f^p - g^q \circ (m \circ f^p))$$
$$= -(-1)^{q(p-1)}(m \circ_0 g^q) \circ_q f^p - (-1)^{q-1}(m \circ_1 g^q) \circ_0 f^p$$
$$= (-1)^q(f^p \cup g^q - (-1)^{pq}g^q \cup f^p),$$

since all intermediate terms cancel out, similarly to (8.18). □

Summing up the results of Propositions 8.6 and 8.7 we obtain the following proposition:

Corollary 8.8.

(i) *Gerstenhaber bracket induces the structure of **differential graded Lie algebra** (DGL algebra or DGLA for short) on Hochschild complex of any associative algebra A.*

(ii) *Hochschild cohomology of any associative algebra A bears the structure of graded Lie algebra induced by Gerstenhaber bracket on the complex and the graded commutative product, induced from the cup-product on the complex, and the bracket satisfies (graded) Leibniz rule with respect to the cup-product.*

Proof. The only nonevident part of this statement concerns the Leibniz rule. It turns out, that for every three cochains f^p, g^q and h^r in Hochschild complex, there always exists a cochain $H = H(f^p, g^q, h^r)$ such that

$$\delta H(f^p, g^q, h^r) - H(\delta f^p, g^q, h^r)$$
$$- (-1)^p H(f^p, \delta g^q, h^r) - (-1)^{p+q}H(f^p, g^q, \delta h^r)$$
$$= [f^p, g^q \cup h^r] - [f^p, g^q] \cup h^r - (-1)^{q(p-1)}g^q \cup [f^p, h^r].$$

This cochain H can be expressed in terms of the $\circ_i$ operations, similarly to the reasonings which we used to prove Propositions 8.6 and 8.7 (see the original paper by Gerstenhaber [Ge63] for details; one can also find the definition of this cochain in many textbooks on Hochschild cohomology). The original construction of the cochain H was rather cumbersome and we omit it; its existence is closely related with the operads theory, see Appendix B. Now, if f^p, g^q and h^r are closed, it follows that the left-hand side of this formula is exact, so the result follows. □

We are now finally able to formulate the smooth version of Hochschild–Kostant–Rosenberg theorem in full generality:

Theorem 8.9. *For any smooth manifold M, the Hochschild–Kostant–Rosenberg map χ induces an isomorphism in cohomology, so that the wedge-product of polyvectors is identified with the cup-product in cohomology and the Schouten bracket of polyvectors is identified with the Gerstenhaber bracket in cohomology, i.e. with the bracket induced from Gerstenhaber bracket on cochain complex.*

Proof. The only unproved statement in this theorem so far is the identification of Gerstenhaber bracket on cohomology and Schouten bracket on polyvectors: this follows from the uniqueness of the bracket that satisfies the axioms of Schouten bracket, see Theorem 4.8: one only needs to check that Gerstenhaber bracket obeys the same rules, which follows directly from the previous discussion. □

9 Obstructions and deformation theory: Examples

The idea of cohomological obstructions is widespread in modern Mathematics. It naturally appeared in Topology in the study of the mapping extension problem (see for instance Chapter 8 of Spanier's book [Sp66]), that was probably the first instance when the main principles of this approach were clearly articulated. The application of similar constructions to the deformations of algebras and rings was first investigated by Murray Gerstenhaber in a series of papers [Ge64]–[Ge74]. Since that time the scope of deformation theory has tremendously grown and now one can speak about deformation complexes of many different algebraic structures (some idea of the more modernist approach to the deformation can be derived from Appendix A, especially Section A.2.4). In the present section we will give just a brief summary of these constructions and ideas in the simplest form that is most suitable for our purposes, i.e. for the study of the deformation quantisation problem as we formulated it earlier, see Question 1, page 50. An interested reader can consult the aforementioned papers by Gerstanhaber or the seminal survey by Gerstanhaber and Shack [GS88].

9.1 Deformations of associative algebras

One of the main purposes for introducing and studying the Hochschild cohomology in the previous sections is the rôle that it plays in the deformation theory, which we are going to discuss now; however we will not study this theory in full generality here, we will restrict our attention to the context of formal deformations of associative algebras. So let A be an associative algebra over a ground field $\Bbbk$ (usually $\Bbbk = \mathbb{R}$ or $\mathbb{C}$), and let $\hbar$ be a formal variable (here we use notations from the deformation quantisation to underline the analogy of these constructions).

Definition 9.1. *Formal deformation of A is an associative $\hbar$-linear product $\star$ in the space of formal power series $A[[\hbar]]$, such that the map $A[[\hbar]] \to A$, $\hbar \mapsto 0$ is a homomorphism of algebras. In other words, the $\star$-product is equal to the product in A modulo $\hbar$.*

The conditions of this definition amount to the following formula: just like in the case of deformation quantisation the $\star$-product is uniquely determined by the deformation series:

$$a \star b = ab + \sum_{n=1}^{\infty} \hbar^n B_n(a,b), \qquad (9.1)$$

where for $a, b \in A$, ab is their product and B_n are bilinear (over $\Bbbk$) maps $A \otimes A \to A$. As such, the maps B_n are elements in $C^2(A, A)$ and we can try to express the associativity of $\star$-product as equations in Hochschild complex. This is the strategy that we shall pursue now.

9.2 Obstructions

We are going now to study the associativity condition for $\star$-product (9.1),

$$(a \star b) \star c = a \star (b \star c), \quad \forall a, b, c \in A. \qquad (9.2)$$

Since the right-hand side in (9.1) is a formal power series, we can do it in an inductive manner with regard to the powers n of the formal variable $\hbar$. Since the constant term of the series is given by the associative product in A, it gives no condition and we begin with the linear terms.

9.2.1 $n = 1$

Consider the linear parts of expressions on both sides of (9.2): we get

$$B_1(a,b)c + B_1(ab,c) = aB_1(b,c) + B_1(a,bc),$$

or, if we put all the terms to one side, we have $\delta B_1(a,b,c) = 0$, so we conclude: *the formula (9.1) determines an associative product only if $B_1 \in C^2(A, A)$ is a Hochschild cocycle*. Of course, this cocycle can be trivial, i.e. identically equal to 0, but it is natural to assume that it is not. Contrariwise, for any cocycle B_1 we can define $\star_1$-product

$$a \star_1 b = ab + \hbar B_1(a,b),$$

so that this map will satisfy the equation

$$(a \star_1 b) \star_1 c - a \star_1 (b \star_1 c) = o(\hbar), \quad \forall a, b, c \in A,$$

where $o(\hbar)$ denotes the terms of degree 2 and higher in $\hbar$. We shall say that *the cocycle B_1 determines $\star_1$-product*. Let b_1 be its cohomology class.

An important particular case is when $A = C^\infty(M)$ for a smooth manifold M: in this case every cocycle[23] in $C^2(A, A)$ is cohomologous to an image of a bivector $\pi \in \mathcal{T}^2(M)$, i.e. there exists $E \in C^1(A, A)$ such that $B_1 = \chi(\pi) + \delta(E)$. In this case we shall say that *bivector π determines $\star_1$-product*. Observe that the cochain B_1 need not be equal to $\chi(\pi)$, since coboundary δE can be nontrivial.

9.2.2 $n = 2$

We now assume that B_1 is chosen to be a Hochschild cocycle, and we proceed with the next term (quadratic in $\hbar$):

$$B_2(a,b)c + B_2(ab, c) + B_1(B_1(a,b), c)$$
$$= aB_2(b, c) + B_2(a, bc) + B_1(a, B_1(b, c)),$$

or transferring the terms with B_2 on one side and with B_1 on the other side, we get

$$\delta B_2(a, b, c) = B_1(B_1(a,b), c) - B_1(a, B_1(b, c)). \tag{9.3}$$

The expression in the right is evidently equal to $(B_1 \circ B_1)(a, b, c)$, see Equation (8.12). On the other hand, in this case

$$[B_1, B_1] = B_1 \circ B_1 - (-1)^{(2-1)(2-1)} B_1 \circ B_1 = 2 B_1 \circ B_1$$

so we can rewrite (9.3) as

$$\delta B_2 = \frac{1}{2}[B_1, B_1]. \tag{9.3'}$$

Since B_1 is closed, the right-hand side of this formula is also a cocycle due to the properties of Gerstenhaber brackets; the class determined

---

[23] Recall that by default we assume that all Hochschild cochains for $C^\infty(M)$ are differentiable, so we can use the (smooth) Hochschild–Kostant–Rosenberg theorem for it, see Theorem 8.9. It also means that the deformation series (9.1) in this case consists of bidifferential operators, just like in the deformation quantisation Problem 1, Section 4.4.

by this cocycle is equal to $\frac{1}{2}[b_1, b_1] \in HH^3(A)$. Summing up we have the following result: *if the $\star$-product (9.1) is associative, the class $[b_1, b_1]$ is trivial, and contrariwise, if this class is trivial, than for the given B_1 one can find B_2 so that the formula*

$$a \star_2 b = ab + \hbar B_1(a,b) + \hbar^2 B_2(a,b),$$

gives a map for which

$$(a \star_2 b) \star_2 c - a \star_2 (b \star_2 c) = o(\hbar^2), \quad \forall a, b, c \in A,$$

where $o(\hbar^2)$ stands for the terms at $\hbar^3$ and higher degrees of $\hbar$. We shall say that $\star_2$ is associative up to degree 2.

An important particular case is again when $A = C^\infty(M)$. In this case $[b_1, b_1] = [\pi, \pi]$, the Schouten bracket of π with itself, see Theorem 8.9, so we conclude: *the $\star_1$-product determined by a bivector π admits extension to a $\star_2$-product associative up to degree 2, iff $[\pi, \pi] = 0$, i.e. iff the bivector π is Poisson.*

9.2.3 $n > 2$

Let us now assume that for some $n > 2$ we have found $B_1, B_2, \ldots, B_{n-1}$ so that the product

$$a \star_{n-1} b = ab + \hbar B_1(a,b) + \hbar^2 B_2(a,b) + \cdots + \hbar^{n-1} B_{n-1}(a,b) \quad (9.4)$$

is associative up to degree $n-1$, i.e.

$$(a \star_{n-1} b) \star_{n-1} c - a \star_{n-1} (b \star_{n-1} c) = o(\hbar^{n-1}), \quad \forall a, b, c \in A. \quad (9.5)$$

Then our goal is to check, what conditions should hold for $B_1, \ldots, B_{n-1}$ if expression (9.4) is equal to the segment of an associative $\star$-product series (9.1). Expanding the associativity condition for $\star$-product up to degree n we get:

$$\delta B_n(a,b,c) = \sum_{p+q=n} \left(B_p(B_q(a,b), c) - B_p(a, B_q(b,c)) \right).$$

The right-hand side of this expression can again be rewritten as $\frac{1}{2} \sum_{p+q=n} [B_p, B_q](a,b,c)$, so we get the following equation:

$$\delta B_n = \frac{1}{2} \sum_{p+q=n} [B_p, B_q]. \quad (9.6)$$

On the other hand, since $\star_{n-1}$ is $(n-1)$-associative, we see that for all $k = 1, 2, \ldots, n-1$ hold similar equations:

$$\delta B_k = \frac{1}{2} \sum_{p+q=k} [B_p, B_q].$$

Therefore if we apply δ to the right-hand side of (9.6), we get from the graded Leibniz rule, (see Proposition 8.7, part (i))

$$\delta \left(\sum_{p+q=n} [B_p, B_q] \right) = \sum_{p+q=n} ([B_p, \delta B_q] - [\delta B_p, B_q])$$

$$= \frac{1}{2} \sum_{p+q=n} \left(\sum_{r+s=q} [B_p, [B_r, B_s]] - \sum_{r+s=p} [[B_r, B_s], B_q] \right)$$

$$= \frac{1}{2} \sum_{p+q=n} \sum_{r+s=q} [B_p, [B_r, B_s]] + \frac{1}{2} \sum_{p+q=n} \sum_{r+s=p} [B_q, [B_r, B_s]]$$

$$= \sum_{p+q+r=n} [B_p, [B_q, B_r]] = 0.$$

Here we also used the graded Jacobi identity, see Proposition 8.6. So the right-hand side of (9.6) is a cocycle and we conclude that *for a given $\star_{n-1}$-product (9.4) there exists $B_n \in C^2(A, A)$ such that the $\star_n$-product*

$$a \star_n b = ab + \hbar B_1(a,b) + \hbar^2 B_2(a,b) + \cdots + \hbar^n B_n(a,b)$$

is associative up to degree n if and only if the class of $\sum_{p+q=n}[B_p, B_q]$ in $HH^3(A)$ is trivial.

9.3 Uniqueness

Before we consider various applications of this theory, we need to discuss one more question, the uniqueness of the $\star$-product constructed by the method we just described: suppose that for all n we can find a solution B_n of the corresponding equation. Clearly, at every stage there can be many different solutions of this problem. On the other hand, the set of all formal deformations can be separated into classes of equivalent $\star$-products similarly to the relation from Problem 1, see Section 4.4: two products $\star$ and $\star'$ are equivalent if there exists a formal series of elements $T_1, T_2, \cdots \in C^1(A, A)$ such that the map

$$T(a) = a + \hbar T_1(a) + \hbar^2 T_2(a) + \cdots + \hbar^n T_n(a) + \ldots$$

satisfies the condition

$$T(a \star b) = T(a) \star' T(b), \quad \forall a, b \in A. \tag{9.7}$$

As before, we reason by induction: we consider one by one different powers of $\hbar$ on both sides of Equation (9.7). In degree 0 there's no condition whatsoever, so we begin with $n = 1$:

$$T_1(ab) + B_1(a, b) = aT_1(b) + T_1(a)b + B_1'(a, b), \quad \forall a, b \in A \tag{9.8}$$

where we denote by B_k' the coefficients of the series $\star'$. Putting the terms with T on one side and all the other terms on the other side of the equation, we can rewrite (9.8) as

$$\delta T_1 = B_1 - B_1', \tag{9.8'}$$

so that *two $\star$-products can be equivalent only if the corresponding cocycles B_1 and B_1' are cohomologous, i.e. $b_1 = b_1' \in HH^2(A)$*. In particular, *if $A = C^\infty(M)$ for a smooth manifold M, then every formal deformation of A is equivalent to a deformation in which the term at $\hbar$ is equal to $\frac{1}{2}\{a, b\}$, where $\{,\}$ is a Poisson bracket*.

Let now $n > 1$ and suppose we know that $\star$ and $\star'$ are equivalent up to degree $n - 1$, i.e. that there exist operators $T_1, \ldots, T_{n-1}$ such that the map

$$T^{(n-1)}(a) = a + \hbar T_1(a) + \hbar^2 T_2(a) + \cdots + \hbar^{n-1} T_{n-1}(a)$$

satisfies the condition

$$T^{(n-1)}(a \star b) - T^{(n-1)}(a) \star' T^{(n-1)}(b) = o(\hbar^{n-1}), \quad \forall a, b \in A. \tag{9.9}$$

We are now looking for T_n such that $T^{(n-1)}(a) + \hbar^n T_n(a)$ will give the same equality up to degree n. To this end, we conjugate $\star'$ with $T^{(n-1)}$ so that we may now assume that $B_1 = B_1'$, $B_2 = B_2', \ldots$,

$B_{n-1} = B'_{n-1}$. Then we can take $T^{(n)}(a) = a + \hbar^n T_n(a)$; in this case in degree n we obtain the equation:

$$\delta T_n = B_n - B'_n. \tag{9.10}$$

The right-hand side of this equation is clearly closed cochain since by our assumption

$$\delta B_n = \frac{1}{2}\sum_{p+q=n}[B_p, B_q] = \frac{1}{2}\sum_{p+q=n}[B'_p, B'_q] = \delta B'_n$$

and we conclude that *under the assumptions we made, the products $\star$ and $\star'$ are equivalent up to degree n if and only if the class of the difference $B_n - B'_n \in HH^2(A)$ vanishes. In other words, there are as many different nonequivalent continuations of a given $\star_{n-1}$-product to a $\star_n$ product, as many different classes there are in $HH^2(A)$.* This assertion follows directly from the previous one, since for any solution B_n of Equation (9.6) and any cocycle $C \in C^2(A, A)$, their sum $B'_n = B_n + C$ will satisfy the same equation.

9.4 Résumé: Obstructions theory

Summing up the results of Sections 9.2 and 9.3, we can formulate the following theorem

Theorem 9.2.

(i) *The first term B_1 of any deformation series (9.1) is a Hochschild 2-cocycle in $C^2(A, A)$. In particular, if $A = C^\infty(M)$ then B_1 is cohomologous to a bivector field π.*

(ii) *Let $B_1, \ldots, B_{n-1} \in C^2(A, A)$ be such that the series*

$$a \star_{n-1} b = ab + \hbar B_1(a, b) + \hbar^2 B_2(a, b) + \cdots + \hbar^{n-1} B_{n-1}(a, b)$$

gives a product associative up to degree $n - 1$ (in particular, if this is a segment of some $\star$-product series), then one can find $B_n \in C^2(A, A)$ extending this series to a product associative up to degree n if and only if the class of cocycle

$$c_n = \frac{1}{2}\sum_{p+q=n}[B_p, B_q] \in C^3(A, A) \tag{9.11}$$

is trivial in $HH^3(A)$; in this case one can take B_n so that $\delta B_n = c_n$. In particular, if $A = C^\infty(M)$ then the bivector π, cohomologous to B_1 should be Poisson bivector. Below we shall denote by c_n both the cocycle defined in (9.11) and its cohomology class in $HH^3(A)$.

(iii) In the situation of part (ii) different choices of B_n lead to equivalent products iff the class of $B_n - B'_n$ in $HH^2(A)$ is trivial; in other words, there exist as many different nonequivalent extensions of $\star_{n-1}$ to a $\star_n$-product, associative up to degree n as many elements there are in $HH^2(A)$. In particular, if $A = C^\infty(M)$, then any deformation series is equivalent to a series beginning with $a \star b = ab + \frac{\hbar}{2}\{a,b\} + o(\hbar)$.

As one sees, deformation quantisation problem is in this way reduced to the problem of showing that at every stage of the deformation process obstruction $[c_n] \in HH^3_{diff}(C^\infty(M)) \cong \mathcal{T}^3(M)$ can be made trivial by an appropriate choice of the coefficients $B_2, \ldots, B_{n-1}$. Historically the first construction of deformation quantisation for symplectic manifolds by Lecomte and de Wilde was obtained by delicate analysis of these obstructions (we shall sketch its proof in Section 12 below). However in the situation of a generic Poisson manifold this construction leads to too much complications even for $M = \mathbb{R}^n$.

9.5 Example 1: Poincaré–Birkhoff–Witt theorem revisited

We are now going to apply Theorem 9.2 to various situations of deformation quantisation. Our first objective is to obtain another proof of the Poincaré–Birkhoff–Witt theorem, see Section 5.2, Theorem 5.4. To this end we consider deformation of the algebra $S(\mathfrak{g})$ of polynomial functions on $\mathfrak{g}^*$ viewed as a Poisson algebra with respect to Kirillov–Kostant structure. Of course, since we only consider differential cochains on algebras of smooth functions and since Hochschild–Kostant–Rosenberg theorem holds both for polynomials as well as for smooth functions, the series constructed for $S(\mathfrak{g})$ can be made up of bidifferential operators and is applicable to $C^\infty(\mathfrak{g}^*)$; however in our construction the operators will have polynomial coefficients. The main reference for this section is [BG96].

9.5.1 Graded Hochschild cohomology

When $A = S(\mathfrak{g})$ Hochschild cohomology $HH^*(A)$ is equal to the space of polynomial polyvectors on $\mathfrak{g}^*$. This space is infinite-dimensional and we have too little control over the obstructions. However, we can somehow circumvent this difficulty by the following simple observation: every differential operator with polynomial coefficients on $\mathbb{R}^n$ with coordinates $(x_1, \ldots, x_n)$ can be written as a linear combination of monomials of the following type

$$D = \sum_{\alpha,\beta} C_{\alpha,\beta} x_1^{\alpha_1} \ldots x_n^{\alpha_n} \partial_{x_1}^{\beta_1} \ldots \partial_{x_n}^{\beta_n},$$

where $\alpha = (\alpha_1, \ldots, \alpha_n)$, $\beta = (\beta_1, \ldots, \beta_n)$ are multiindices, and $C_{\alpha,\beta} \in \mathbb{R}$ are scalar coefficients; we will often abbreviate $x^\alpha = x_1^{\alpha_1} \ldots x_n^{\alpha_n}$, $\partial_x^\beta = \partial_{x_1}^{\beta_1} \ldots \partial_{x_n}^{\beta_n}$. If we apply this monomial to a generic homogenous polynomial $P(x_1, \ldots, x_n)$ of degree k, the result will be equal to a homogenous polynomial of degree $k + |\alpha| - |\beta|$ where $|\alpha| = \sum_k \alpha_k$ and similarly $|\beta| = \sum_k \beta_k$. We shall say, that $|\alpha| - |\beta|$ is the degree of the monomial $C_{\alpha,\beta} x^\alpha \partial_x^\beta$; so every polynomial differential operator is equal to a sum of homogenous operators with respect to this degree.

More generally, if E is a p-differential operator with polynomial coefficients, then we can represent it as a sum of homogenous polydifferential operators with respect to a similar grading: one says that E is homogenous of degree $k \in \mathbb{Z}$ if for all monomials $a_1, \ldots, a_p$ in variables x_i, one has $\deg E(a_1, \ldots, a_p) = \sum_i \deg a_i + k$.

Now we can make the following observation: let $A = \mathbb{R}[x_1, \ldots, x_n]$ then for every homogenous $E \in C^p(A, A)$ of degree k, $\deg \delta E = k$. This is a direct consequence of the formula for δE: it is expressed in terms of multiplication of the arguments, and this operation preserves the total degree. Hence for $A = \mathbb{R}[x_1, \ldots, x_n]$ the complex $C^{\cdot}(A, A)$[24] breaks up into a direct sum of subcomplexes:

$$C^{\cdot}(A, A) = \bigoplus_{k \in \mathbb{Z}} C^{\cdot}(A, A)(k), \ C^{\cdot}(A, A)(k) = \{E \in C^{\cdot}(A, A) \,|\, \deg E = k\}.$$

---

[24] We should rather consider the complex $C^{\cdot}_{\text{diff}}(A, A)$ of differentiable polynomial cochains; generic Hochschild cochains on $S(\mathfrak{g})$ can be equal to an infinite sum of differentiable ones, but this doesn't change the conclusion about cohomology that we shall make.

Hence the same is true for cohomology: for all $p = 0, 1, \ldots$

$$HH^p(A) = \bigoplus_{k \in \mathbb{Z}} H^p(C^{\cdot}(A,A)(k), \delta).$$

On the other hand, due to the Hochschild–Kostant–Rosenberg theorem, every cohomology class $c \in HH^p(A)$ is represented by a polynomial p-polyvector field, and hence its degree cannot be less than $-p$. Hence we conclude that *every homogenous p-cocycle with degree less than $-p$ is cohomologous to 0*. Also observe that *if E, F are homogenous cochains of degrees k, l in $C^{\cdot}(A, A)$, then their Gerstenhaber bracket $[E, F]$ is homogenous of degree $k + l$*; this follows from the formula of the Gerstenhaber bracket in terms of $\circ_i$ operations, which clearly adds the degrees of the arguments.

9.5.2 Obstructions calculus

We now recall that in terms of linear coordinates on $\mathfrak{g}^*$ Kirillov–Kostant bracket has form

$$\{f, g\} = x_k C_{ij}^k \partial_i f \partial_j g,$$

so this structure is represented by a homogenous 2-cocycle of degree -1 in the sense of the previous section. This cocycle appears at the first stage of our deformation series, i.e. we take $B_1 = \frac{1}{2}\chi(\pi)$; observe that $B_1(a, b) = -B_1(b, a)$ in this case. Since the bivector π is in effect Poisson, its Schouten bracket with itself vanishes, hence the homogenous 3-cocycle of degree -2, given by Gerstenhaber bracket $[B_1, B_1]$ is cohomologous to 0 and we can find $B_2 \in C^2(A, A)(-2)$ such that Equation (9.3') holds, i.e. $\delta B_2 = \frac{1}{2}[B_1, B_1]$. Moreover, one can choose B_2 to be symmetric, i.e. so that $B_2(a, b) = B_2(b, a)$. Indeed let x be any solution of Equation (9.3'), and let $x'(a, b) = x(b, a)$; then since B_1 is skew-symmetric and the algebra A is commutative, we have

$$\begin{aligned}\delta x'(a, b, c) &= ax'(b, c) - x'(ab, c) + x'(a, bc) - x'(a, b)c \\ &= x(c, b)a - x(c, ba) + x(cb, a) - cx(b, a) \\ &= -\delta x(c, b, a) = -B_1(B_1(c, b), a) + B_1(c, B_1(b, a)) \\ &= B_1(B_1(a, b), c) - B_1(a, B_1(b, c)) = \delta x(a, b, c).\end{aligned}$$

Hence we can take $B_2 = \frac{1}{2}(x + x')$, which is clearly symmetric.

Next, at the third stage, we need to find a 2-cochain B_3 such that

$$\delta B_3 = \frac{1}{2}([B_1, B_2] + [B_2, B_1]) = [B_1, B_2].$$

But the anti-symmetrisation of the right-hand side of this equation vanishes identically: indeed, the value of the right-hand side of this formula on elements $a, b, c \in A$ is

$c_3(a, b, c)$
$= B_1(B_2(a,b), c) - B_1(a, B_2(b,c)) + B_2(B_1(a,b), c) - B_2(a, B_1(b,c)).$

Since B_2 is symmetric, all the terms with B_2 inside B_1 will go away in the process of anti-symmetrisation (they will cancel out with similar terms, where the order of the arguments in B_2 is exchanged). The remaining terms contain anti-symmetric cochain B_1 inside, hence their skew-symmetrisation amounts to taking the sum over the cyclic permutations of a, b, c, and so

$SSym(c_3)(a, b, c)$
$= B_2(B_1(a,b), c) - B_2(a, B_1(b,c)) + B_2(B_1(b,c), a)$
$\quad - B_2(b, B_1(c,a)) + B_2(B_1(c,a), b) - B_2(c, B_1(a,b)) = 0,$

because the terms cancel out due to the symmetry of B_2 and skew-symmetry of B_1. It follows (see Remark 8.5) that the cohomology class, represented by c_3 is trivial, and we can find B_3 so that $\delta B_3 = c_3$.[25] The degree of B_3 (in the sense of the previous section) will be equal to $\deg c_3 = -3$. Moreover, B_3 is unique up to an equivalence, because $HH^2(C^{\cdot}(A, A)(-3), \delta) = 0$.

To this moment we only used the general properties of Poisson bivector. From now on, we obtain the result absolutely automatically: for any $n \geq 4$, we have by induction

$$\deg(c_n) = \deg\left(\frac{1}{2} \sum_{p+q=n} [B_p, B_q]\right) = -n < -3,$$

---

[25] The possibility to use similar reasonings based on the symmetry considerations is the main reason for introducing the symmetricity conditions in the definitions of deformation quantisation, see Problem 1, Section 4.4.

and hence the class of $c_n \in HH^3(A)$ is trivial. This means that we can always choose B_n so that $\delta B_n = c_n$ and $\deg B_n = -n$, so the inductive step works. Since we assumed that all the cochains in our construction were differentiable, we conclude that there exists formal deformation of $S(\mathfrak{g})$ which in effect gives a deformation quantisation for $C^\infty(\mathfrak{g}^*)$.

As for the uniqueness, we observe that except for B_2, all the other terms in the deformation series are uniquely defined up to an equivalence by equation $\delta B_n = c_n$ since there are no nonzero cohomology classes in $H^2(C^{\cdot}(A, A)(-n), \delta)$ when $n \geq 3$; and as for the term B_2: it is also unique, if we demand that it is symmetric. This follows from the fact that Hochschild cohomology classes can be represented only by skew-symmetric cochains, see Theorem 8.4.

9.5.3 The Poincaré–Birkhoff–Witt theorem

We can now invert the constructions from Section 5 and derive the Poincaré–Birkhoff–Witt theorem from the existence and uniqueness of deformation of $S(\mathfrak{g})$.

In effect we see that the deformation series determines an associative $\star$-product on $S(\mathfrak{g})[[\hbar]]$, which begins with the terms $f \star g = fg + \frac{\hbar}{2}\{f, g\} + \ldots$ in which all the other terms are given by bi-differential operators of degrees $\deg B_n = -n$, and such that the operator B_2 is symmetric. First, we observe that *for every two polynomials $f, g \in S(\mathfrak{g})$ the formula for $f \star g$ will contain only a finite number of terms*, because all operators with sufficiently large negative degrees will vanish on f, g. This means that instead of $S(\mathfrak{g})[[\hbar]]$ we can consider the $\star$-product on $S(\mathfrak{g})[\hbar]$, the algebra of polynomials in $\hbar$ with coefficients in $S(\mathfrak{g})$.

Next consider the commutators of $X, Y \in \mathfrak{g} \subseteq S(\mathfrak{g})[\hbar]$, where we regard $\mathfrak{g}$ as the space of linear functions on $\mathfrak{g}^*$ with respect to the $\star$-product:

$$X \star Y - Y \star X = (XY + \frac{\hbar}{2}\{X, Y\} + \hbar^2 B_2(X, Y))$$
$$- (YX + \frac{\hbar}{2}\{Y, X\} + \hbar^2 B_2(Y, X))$$
$$= \frac{\hbar}{2}([X, Y] - [Y, X]) = \hbar[X, Y],$$

since all the terms of higher degrees will vanish for linear functions, operator B_2 is symmetric and $\{X, Y\} = [X, Y]$ (the Lie bracket in $\mathfrak{g}$).

Now for every $t \in [0, 1]$ we can consider the associative algebra $\mathcal{B}_t$: as linear space $\mathcal{B}_t$ is isomorphic to $S(\mathfrak{g})$, while the product $*_t$ in $\mathcal{B}_t$ is given by the formula

$$f *_t g = \epsilon_t(f \star g), \text{ where } \epsilon_t : S(\mathfrak{g})[\hbar] \to S(\mathfrak{g}),\ \epsilon_t(\hbar) = t;$$

formally speaking

$$f *_t g = fg + \frac{t}{2}\{f, g\} + \sum_{n=2}^{\infty} t^n B_n(f, g),$$

where the sum is always finite and hence we can substitute for t any real number from the interval $[0, 1]$.

Now in $\mathcal{B}_1$ the commutator relation for linear functions is $X *_1 Y - Y *_1 X = [X, Y]$. By the universal property of $U\mathfrak{g}$, see Definition 5.3, there exists a homomorphism $\varphi_1 : U\mathfrak{g} \to \mathcal{B}_1$, which extends the inclusion of $\mathfrak{g}$ on both sides. On the other hand, since all the terms in the series for $f *_t g$ have degrees smaller than fg, one can prove by induction that $\mathcal{B}_1$ is generated by $\mathfrak{g}$ as an algebra, and hence the map φ_1 is epimorphic.

Let us one more time look at the graded algebra $Gr_{\mathcal{F}}U\mathfrak{g}$, associated with the natural filtration on $U\mathfrak{g}$, see Section 5.3, formula (5.7). Similarly to $U\mathfrak{g}$ one can consider filtration on $\mathcal{B}_1$ induced by the grading in the polynomial algebra $S(\mathfrak{g})$; the map φ_1 clearly preserves these filtrations, so it induces a map: $Gr_{\mathcal{F}}U\mathfrak{g} \to Gr_{\mathcal{F}}\mathcal{B}_1$. But by the very definition, $Gr_{\mathcal{F}}\mathcal{B}_1 = \mathcal{B}_0 \cong S(\mathfrak{g})$, since all the terms in $*_1$ except for the first one are of lower degrees. Composing these maps with the symmetrisation $\sigma : S(\mathfrak{g}) \to U\mathfrak{g}$, which descends to a homomorphism $\sigma_{\mathcal{F}} : S(\mathfrak{g}) \to Gr_{\mathcal{F}}U\mathfrak{g}$, we obtain the following diagram:

$$S(\mathfrak{g}) \xrightarrow{\sigma_{\mathcal{F}}} Gr_{\mathcal{F}}U\mathfrak{g} \xrightarrow{\varphi_{1,\mathcal{F}}} Gr_{\mathcal{F}}\mathcal{B}_1 \xrightarrow{\cong} S(\mathfrak{g}).$$

All the maps in this diagram are epimorphic homomorphisms (since all these algebras are generated by $\mathfrak{g}$), and clearly the composition of these maps is identity on $\mathfrak{g} \subseteq S(\mathfrak{g})$ and hence it is identity on $S(\mathfrak{g})$ and hence they all should be monomorphic as well. We conclude that these maps are isomorphisms of algebras. Hence $\sigma_{\mathcal{F}} : S(\mathfrak{g}) \cong Gr_{\mathcal{F}} U\mathfrak{g}$, and so σ is a linear isomorphism $S(\mathfrak{g}) \cong U\mathfrak{g}$. Also this means that $\varphi_1 : U\mathfrak{g} \to \mathcal{B}_1$ is a canonical isomorphism.

9.6 Example 2: Symplectic manifolds with vanishing $H^3_{dR}(M)$

Obstructions, considered in Sections 9.2–9.4 take values in Hochschild cohomology, which is usually pretty large, so one can not easily prove that they vanish. Later we shall sketch the construction of deformation quantisation in symplectic case, when one can find a way to "kill" these obstructions. However in some simple cases it is enough to modify the construction a little bit so that the obstructions will vanish.

9.6.1 Gerstenhaber brackets and obstructions

Let us consider Equation (9.6): it is clear that there are many different ways to choose B_n so that $\delta B_n = c_n$. In fact for any 2-cocycle $a \in Z^2(A, A)$, $\delta(B_n + a) = \delta B_n$. On the other hand, if we substitute $B'_n = B_n + a$ instead of B_n into the formula for c_{n+1}, we shall get

$$c'_{n+1} = \frac{1}{2} \sum_{p+q=n+1} [B_p, B_q] = \frac{1}{2} \left([B_1, B'_n] + [B'_n, B_1]\right)$$
$$+ \frac{1}{2} \sum_{\substack{1<p,q<n \\ p+q=n+1}} [B_p, B_q]$$
$$= c_{n+1} + [B_1, a]. \tag{9.12}$$

We know that $\delta B_2 = \frac{1}{2}[B_1, B_1]$ and for every $k = 2, \ldots, n-1$ we have $\delta B_{k+1} = c_{k+1}$, so

$$[B_1, B_k] = \delta B_{k+1} - \frac{1}{2} \sum_{\substack{1<p,q<k \\ p+q=k+1}} [B_p, B_q].$$

Also from Jacobi identity we get:

$$[B_1, [B_1, x]] = \frac{1}{2}[[B_1, B_1], x] = [\delta B_2, x]$$

for all cochains $x \in C^{\cdot}(A,A)$. Now we can compute

$$[B_1, c_{n+1}] = \frac{1}{2} \sum_{p+q=n+1} [B_1, [B_p, B_q]]$$

$$= \frac{1}{2} \sum_{p+q=n+1} ([[B_1, B_p], B_q] - [B_p, [B_1, B_q]])$$

$$= \frac{1}{2} \sum_{\substack{p+q=n+1 \\ 1<p,q<n}} ([\delta B_{p+1}, B_q] - [B_p, \delta B_{q+1}])$$

$$+ \frac{1}{2} \sum_{\substack{p+q+r=n+2 \\ 1<p,q,r<n-1}} [B_p, [B_q, B_r]].$$

The term in the last line vanishes due to the Jacobi identity and the previous term is equal to $\delta \left(-\frac{1}{2} \sum_{\substack{p+q=n+2 \\ 1<p,q<n+1}} [B_p, B_q] \right)$.

9.6.2 Generalised Poisson cohomology

The observations in previous subsection lead to the following general construction: let $b \in HH^2(A)$ be a Hochschild cohomology class such that $[b,b] = 0 \in HH^3(A)$, where $[,]$ is the bracket in cohomology induced by the Gerstenhaber bracket; in particular we can take $b = b_1$, the class of the first coefficient of a deformation series. Then on $HH^*(A)$ we have a map:

$$d_b : HH^p(A) \to HH^{p+1}(A), \quad x \mapsto [b, x].$$

Now since the properties of the bracket $[,]$ on $HH^*(A)$ are similar to the properties of Schouten bracket, we conclude, similarly to Proposition 6.2, that $d_b^2 = 0$. Analogously to what we did in Section 6.2.3, we can give the definition:

Definition 9.3. *We shall call the cohomology of the complex $(HH^*(A), d_b)$ **generalised Poisson cohomology** of A with respect to b. We will denote it by $HH_b^*(A)$.*

We are not going to study this cohomology theory in any detail, instead we shall reinterpret the results of Section 9.6.1 in terms of $HH_b^*(A)$. Namely, we have the following proposition:

Proposition 9.4. *For every n-associative deformation of A,*

$$f \star_n g = fg + \hbar B_1(f,g) + \cdots + \hbar^n B_n(f,g),$$

the cohomology class b_1 of the first coefficient B_1 determines a generalised Poisson cohomology complex on $HH^*(A)$. The obstruction class $c_{n+1} \in HH^3(A)$ is closed with respect to the differential d_{b_1} and the class of $[c_{n+1}] \in HH^3_{b_1}(A)$ is trivial iff one can find a cocycle $a \in C^2(A,A)$ such that the $\star'_n$-product

$$f\star'_n g = fg + \hbar B_1(f,g) + \cdots + \hbar^n B'_n(f,g), \quad B'_n(f,g) = B_n(f,g) + a(f,g)$$

is extendable to an associative $\star_{n+1}$-product.

Thus, the classes $[c_{n+1}] \in HH^3_{b_1}(A)$ can be regarded as *the secondary obstruction classes* for the deformation problem.

Proof. The only thing that we have to do is to observe that $[c_{n+1}] = 0 \in HH^3_{b_1}(A)$ iff there exists $\alpha \in HH^2(A)$ such that $d_{b_1}\alpha = c_{n+1}$. If a is any Hochschild cocycle, representing $-\alpha$, then $\delta(B_n + a) = c_n$ and by formula (9.12) $c'_{n+1} = c_{n+1} + [B_1, a]$, so in Hochschild cohomology we have $c'_{n+1} = c_{n+1} - d_{b_1}\alpha = 0$; this happens iff the class of c'_{n+1} in cohomology is trivial, and conclusion follows. □

Remark 9.5. One can use similar constructions to modify the uniqueness results for the deformation series, see Section 9.3; namely, one can show that *the space of nonequivalent deformations of an algebra A with the same B_1 is equal to $\hbar^2 HH^2_{b_1}(A)[[\hbar]]$*. We will not need this construction in our course, so we leave it as an exercise to the reader, see Exercise E.3.11*.

9.6.3 Deformation quantisation: A Neroslavski–Vlassov type result

We conclude this section by the following result, which is based on further reformulation of Proposition 9.4: let $A = C^\infty(M)$ where M is a Poisson manifold and $b_1 = \pi \in \mathcal{T}^2(M)$ the Poisson bivector. Then by the Hochschild–Kostant–Rosenberg Theorem 8.9 $HH^*_{b_1}(A) = H^*_\pi(M)$, the usual Lichnerowicz–Poisson cohomology of M (see Section 6.2.3) and we obtain a series of obstructions in $H^3_\pi(M)$.

In a generic case Poisson cohomology of a manifold is very hard to compute, so this result is of a little practical use. However, if $\pi = \omega^{-1}$, i.e. if it is induced from a symplectic structure, then we can use the isomorphism of Proposition 6.3: $H^*_\pi(M) \cong H^*_{dR}(M)$. Thus, the obstructions we consider take values in de Rham cohomology of M. This cohomology theory is much simpler and can often be calculated by topological methods. In particular, we conclude:

Proposition 9.6. *If M is a symplectic manifold such that $H^3_{dR}(M) = 0$, then there exists a deformation quantisation of M.*

A similar (slightly more general) result, obtained by different methods first appeared in the work of Neroslavski and Vlassov, see [NV81]. An important particular case is when $M = \mathbb{R}^{2n}$ with nonconstant symplectic structure: in this case one cannot use the Moyal product formula since the existence of global Darboux coordinate system is not guaranteed (see Theorem 4.5). But using methods we developed here, we see that the existence of associative $\star$-product is always true because the manifold $\mathbb{R}^{2n}$ is contractible. Moreover, since $H^2_{dR}(M) = 0$ this deformation is unique up to the equivalence relation (see Remark 9.5).

Exercises 3: Hochschild homology and cohomology: Obstructions theory

E.3.1 Compute the Hochschild homology and cohomology of A (with coefficients in A) for the algebras

(a) The ground field $\Bbbk$.

(b*) "Dual numbers" $A = \Bbbk[\varepsilon]/\langle \varepsilon^2 = 0\rangle$. More generally, $A = \Bbbk[\varepsilon]/\langle \varepsilon^k = 0\rangle$, $k \geq 2$. **Hint:** try modifying the construction of Koszul resolution for this case.

(c*) The tensor algebra $T(V) = \bigoplus_{n \geq 0} V^{\otimes n}$ of a vector space V. **Hint:** there exists a small free $T(V)^e$-module resolution of $T(V)$ of length 1, i.e. nonzero modules in it stand only in degrees 0 and 1.

(d*) Weyl algebra
$$W_n(\mathbb{R}) = \mathbb{R}[x^1, \ldots, x^n][\partial_1, \ldots, \partial_n]/\langle [\partial_i, x^j] = \delta_i^j \rangle,$$
i.e. the algebra of polynomial differential operators in n variables. **Hint:** begin with $n = 1$ and observe that $W_{m+n}(\mathbb{R}) \cong W_m(\mathbb{R}) \otimes W_n(\mathbb{R})$.

E.3.2 Prove, that if A is unital, one can compute the Hochschild (co)homology of A using the *normalised* Hochschild complex: $\overline{C}_n(A, M) = M \otimes \overline{A}^{\otimes n}$, where $\overline{A} = A/\Bbbk \cdot 1$ (the differential in $\overline{C}_*(A, M)$ is induced from the usual Hochschild complex differential).

E.3.3 Prove, that if $A = A' \times A''$ (as algebras), then
$$HH_*(A) = HH_*(A') \oplus HH_*(A'').$$

E.3.4 Prove, that the antisymmetrisation map $\sigma : M \otimes \wedge^n A \to M \otimes A^{\otimes n}$ induces a morphism of complexes from the Chevalley complex of $A, [,]$ with coefficients in M (i.e. A, viewed as Lie algebra with respect to the usual commutator) and the Hochschild complex of A with coefficients in M.

E.3.5 Prove that the first Hochschild homology of any algebra A is isomorphic to the bimodule of "universal Kähler differentials", i.e. to the bimodule $\Omega(A)$, spanned by the elements $a\,db$, $a,b \in A$, where $d: A \to \Omega(A)$ is the map, satisfying generalised Leibniz rule:

$$d(ab) = da\,b + a\,db.$$

E.3.6 Observe, that the multiplication of M on either side by the elements of the center of A induces the structure of $Z(A)$-bimodule on the complex $C_*(A,M)$ (and $C^*(A,M)$). Prove that on the level of (co)homology both left- and right-module structures coincide.

E.3.7 Prove that if $B = Mat_n(A)$, then the map $Tr: B \otimes B^{\otimes n} \to A \otimes A^{\otimes n}$, given by the formula

$$Tr(B^0 \otimes B^1 \otimes \cdots \otimes B^n) = \sum_{i_0,i_1,\ldots,i_n} B^0_{i_0 i_1} \otimes B^1_{i_1 i_2} \otimes \cdots \otimes B^n_{i_n i_0}$$

induces a chain map between the corresponding Hochschild complexes. Find a similar map for the Hochschild cochain complexes.[26]

E.3.8 Use results of the Exercise E.3.1 to describe the deformations of dual numbers, tensor algebra and (*)Weyl algebra.

E.3.9* Prove that for any $U\mathfrak{g}$-bimodule M, the following is true: $H_*(U\mathfrak{g}, M) = H_*(\mathfrak{g}, M^{ad})$, where on the right stand the Chevalley–Eilenberg homology and M^{ad} is the $\mathfrak{g}$-complex, obtained from $U\mathfrak{g}$-bimodule by the formula $x \cdot_{ad} m = xm - mx$.

E.3.10* Prove that for any Poisson manifold the term B_2 in deformation series can be chosen as follows:

$$B_2(f,g) = -\frac{1}{8}\pi^{ij}\pi^{kl}\partial_i\partial_k f \partial_j\partial_l g - \frac{1}{12}\pi^{ij}\partial_j\pi^{kl}(\partial_i\partial_k f \partial_l g - \partial_k f \partial_i\partial_l g).$$

---

[26] In effect, one can show that the map Tr induces an isomorphism of (co)homology $HH_*(A) \cong HH_*(B)$ in this case; this statement is a manifestation of the so-called *Morita equivalence* property of Hochschild (co)homology.

Here π^{ij} are the components of the Poisson bivector in coordinates $(x^1, \ldots, x^n)$.

E.3.11*Prove that the space of nonequivalent deformations of an algebra A with the same B_1 is equal to $\hbar^2 HH^2_{b_1}(A)[[\hbar]]$.

10 Deformation quantisation of cotangent bundles

In the previous section, we developed the obstruction theory that allows one answer the question of existence and uniqueness of deformations of an associative algebra so that the Deformation quantisation problem becomes a particular case of this construction: Theorem 9.2, and Proposition 9.4 provide us with a series of obstructions in Hochschild and Poisson cohomology that vanish iff we can construct a deformation series. Unfortunately, these obstructions are usually quite hard to compute. Even in symplectic case, when the Poisson and de Rham cohomology theory coincide it is difficult to identify the obstruction class with anything that can be geometrically described.

For this and also for the historic reason we will begin with a particular case of the deformation quantisation problem (which is by many people regarded as the most important part of this problem): the deformation of symplectic manifolds. In that case the existence of $\star$-product was first proved by Lecomte and de Wilde in [DL83$_2$]; we will discuss their constructions in Section 12. But before that, we will investigate an important particular case of symplectic manifolds: the cotangent bundles of smooth manifolds. The importance of this special case, is due to its relation with differential operators on the base space, see Remark 10.2.

10.1 Graded Hochschild cohomology and obstructions

Let $M = T^*X$ where X is a smooth manifold. Then as we explained in Example 4.2 there exists a canonically defined symplectic structure ω on M, which in local coordinates $(x^1, \ldots, x^n, \xi_1, \ldots, \xi_n)$ on T^*X in which $(x^1, \ldots, x^n)$ are local coordinates in X and $(\xi_1, \ldots, \xi_n)$ are the associated coordinates in cotangent space, has the form

$$\omega = d\xi_1 \wedge dx^1 + \cdots + d\xi_n \wedge dx^n.$$

We are going to construct the deformation quantisation of T^*X with respect to this symplectic structure; the main reference for this section is [DL83₁]. Let

$$C_X^\infty(M) = \left\{ f \in C^\infty(M) \mid f(x^1, \ldots, x^n, \xi_1, \ldots, \xi_n) = \sum_{|\alpha| \le p} f_\alpha(x) \xi^\alpha \right\},$$

where as usual we denote by $\alpha = (\alpha_1, \ldots, \alpha_n)$, $\alpha_k \in \mathbb{N}_0$ a multiindex, $|\alpha| = \sum_i \alpha_i$, and $f_\alpha(x)$ is a C^∞-function in coordinates $(x^1, \ldots, x^n)$ and $\xi^\alpha = \prod_i \xi_i^{\alpha_i}$. We shall say that $f = \sum_\alpha f_\alpha \xi^\alpha$ is a *function, polynomial in vertical direction* or *vertical polynomial*; we say that $\deg f = p$ if there exists $f_\alpha \not\equiv 0$, $|\alpha| = p$.[27] Clearly $\deg fg = \deg f + \deg g$. Since the coordinates change is given by the formula $\xi_{i'} = \frac{\partial x^i}{\partial x^{i'}} \xi_i$ if $x^i = X^i(x^{1'}, \ldots, x^{n'})$ and Jacobi matrix $J\left(\frac{\partial x^i}{\partial x^{i'}}\right)$ is nondegenerate, the space $C_X^\infty(M)$ is well-defined and even the degree p of f is not changed by the change of coordinates.

From now on we will work with functions $f \in C_X^\infty(M)$; every such function is equal to a sum of homogenous in vertical direction polynomials. We are going to prove that there exists a deformation quantisation of $C_X^\infty(M)$. To this end, we consider the complex of differentiable Hochschild cochains $C_{diff}^\cdot(C_X^\infty(M), C_X^\infty(M))$; as in Section 9.5.1, we shall say that a differentiable cochain

$$\varphi \in C_{diff}^k(C_X^\infty(M), C_X^\infty(M))$$

has degree $p \in \mathbb{Z}$ if

$$\deg \varphi(f_1, \ldots, f_k) = p + \sum_{i=1}^k \deg f_i,$$

for all homogenous $f_1, \ldots, f_k \in C_X^\infty(M)$. Informally speaking, $\deg \varphi$ is equal to the vertical degree of its image minus the total number of partial derivations in vertical directions in it. Then as in Section 9.5.1 we see that δ preserves the degree of a cochain (since the product of

---

[27] One can define the degree of homogenous polynomial functions with the help of Euler vector field: it is the vertical field on $M = T^*X$, given by the formula $e(f) = \sum_{i=1}^n \xi_i \frac{\partial f}{\partial \xi_i}$; then $\deg f = p$ (for a homogenous f) iff $e(f) = pf$. This vector field is preserved under the coordinate changes, since $\frac{\partial f}{\partial \xi_{i'}} = \frac{\partial x^{i'}}{\partial x^i} \frac{\partial f}{\partial \xi_i}$.

functions does it) so that the complex $C^{\cdot}_{diff}(C^\infty_X(M), C^\infty_X(M))$ splits into the direct sum of subcomplexes:

$$C^{\cdot}_{diff}(C^\infty_X(M), C^\infty_X(M)) = \bigoplus_{p \in \mathbb{Z}} C^{\cdot}_{diff}(C^\infty_X(M), C^\infty_X(M))(p),$$

$$C^{\cdot}_{diff}(C^\infty_X(M), C^\infty_X(M))(p) = \{\varphi \in C^{\cdot}_{diff}(C^\infty_X(M), C^\infty_X(M)) \mid \deg \varphi = p\}.$$

The same splitting holds for the Hochschild cohomology, turning it into a bigraded space, which we shall refer to as the graded Hochschild cohomology:

$$HH^*(C^\infty_X(M)) = \bigoplus_{p \in \mathbb{Z}} H^*(C^{\cdot}_{diff}(C^\infty_X(M), C^\infty_X(M))(p), \delta).$$

The Gerstenhaber bracket adds up the degrees of cochains. Moreover, since a p-vector field $\varphi \in \mathcal{T}^p(M)$ can have no more than p vertical partial derivatives in it (when all components of the polyvector are vertical), we conclude with the help of the Hochschild-Kostant-Rosenberg theorem that

$$H^k(C^{\cdot}_{diff}(C^\infty_X(M), C^\infty_X(M))(p), \delta) = 0, \text{ when } p < -k.$$

Let now f and g be homogenous vertical polynomials of degrees p and q respectively then the Poisson bracket of two such functions,

$$\{f, g\} = \sum_{i=1}^{n} \left(\frac{\partial f}{\partial x^i} \frac{\partial g}{\partial \xi_i} - \frac{\partial f}{\partial \xi_i} \frac{\partial g}{\partial x^i} \right)$$

is a homogenous polynomial of degree $p + q - 1$. Thus the bidifferential operator $B_1(f, g) = \frac{1}{2}\{f, g\}$ has degree -1 and we can continue reasoning completely in the lines of Section 9.5.2: there exists a unique up to an equivalence symmetric cochain $B_2 \in C^2_{diff}(C^\infty_X(M), C^\infty_X(M))$, $\deg B_2 = -2$ that solves the equation $\delta B_2 = \frac{1}{2}[B_1, B_1]$. In this case, the class of $c_3 = [B_1, B_2]$ in $HH^3(C^\infty_X(M))$ is trivial, since its skew-symmetrisation vanishes identically, and hence there exists $B_3 \in C^2_{diff}(C^\infty_X(M), C^\infty_X(M))$, $\deg B_3 = -3$, such that $\delta B_3 = c_3$; moreover, since $H^2(C^{\cdot}_{diff}(C^\infty_X(M), C^\infty_X(M))(-3), \delta) = 0$, the choice of B_3 is unique up to an equivalence. Finally for all

$n \geq 4$ the obstructions $c_n \in H^3(C^{\cdot}_{diff}(C_X^\infty(M), C_X^\infty(M))(-n), \delta)$ vanish identically, so there exist all the terms B_n of the deformation series and they are uniquely defined up to an equivalence.

Remark 10.1. Observe, that even though the $\star$-product we have just constructed is originally defined only for the algebra of vertical polynomial functions $C_X^\infty(M)$, we can use it to define an associative product in $C^\infty(M)[[\hbar]]$ since differential operators on polynomials and can also act on smooth functions.

Remark 10.2. Similarly to Exercise E.2.1, we can identify the algebra $(C_X^\infty(M)[[\hbar]], \star)$ or rather $(C_X^\infty(M), \star)$ with the algebra of differential operators on X, $\mathscr{DO}(X)$: recall that the linear map $D : C^\infty(X) \to C^\infty(X)$ is a differential operator of order p if in local coordinates $(x^1, \ldots, x^n)$ we have

$$D(F) = \sum_{|\alpha| \leq p} f_\alpha(x) \frac{\partial^{|\alpha|} F}{\partial^\alpha x}.$$

In this case, the principal symbol $\sigma(D) = \sum_{|\alpha|=p} f_\alpha(x) \xi^\alpha$ of the operator gives a well-defined homogenous element in $C_X^\infty(M)$ of degree p; let us denote by $C_X^\infty(M)[p]$ the space of all such elements, $C_X^\infty(M) = \bigoplus_{p=0}^\infty C_X^\infty(M)[p]$. Then the degree of differential operators gives a filtration $\mathcal{F}$ on $\mathscr{DO}(X)$ and the map

$$C_X^\infty(M)[p] \to \mathscr{DO}(X), \quad \sum_{|\alpha|=p} f_\alpha(x) \xi^\alpha \mapsto \sum_{|\alpha|=p} f_\alpha(x) \frac{\partial^{|\alpha|}}{\partial^\alpha x}$$

descends to a well-defined homomorphism

$$C_X^\infty(M) \to Gr_\mathcal{F} \mathscr{DO}(X).$$

Finally we can modify the reasoning from Section 9.5.3 to show that this map is induced by an isomorphism $(C_X^\infty(M), \star) \cong \mathscr{DO}(X)$.

10.2 Cotangent bundles of parallelisable manifolds

The construction from previous section is based on additional structure, grading of vertical polynomial functions. However in a particular case when X is parallelisable, i.e. when the tangent (and hence the cotangent bundle too) of X can be trivialised, one can find a more straightforward proof and hence a more explicit description of the $\star$-product series. Let us briefly sketch this construction, which is due to Cahen and Gutt [CG82], [Gu83]. In particular this will give a description of the deformation quantisation of the Lie group.

10.2.1 The general situation

Let X be a parallelisable manifold, i.e. we have a trivialisation $TX \cong X \times \mathbb{R}^n$, $n = \dim X$; let us fix it by choosing the n-tuple $X_1, \ldots, X_n$ of vector fields on X which generate basis in every fibre $T_x X$ of this vector bundle. In this notation, the commutators of vector fields in X are determined by the "structure functions" $c_{ij}^k(x) = -c_{ji}^k(x)$:

$$[X_i, X_j] = \sum_{k=1}^{n} c_{ij}^k(x) X_k. \tag{10.1}$$

Let $\theta^1, \ldots, \theta^n \in \Omega^1(X)$ be 1-forms, determined by the equation $\theta^i(X_j) = \delta_j^i$ and let $p_1, \ldots, p_n \in C^\infty(T^*X)$ be the functions defined by the equality

$$p_i(\xi) = \xi(X_i). \tag{10.2}$$

Clearly p_j are vertically polynomial functions on T^*X of degree 1, but unlike the functions that we considered in Section 10.1, these functions are globally defined; they depend on the choice of vector fields X_k, but they are independent on local coordinates in X. Moreover, the differential of the forms θ^k is given in terms of the structure functions $c_{ij}^k(x)$: by Cartan's formula

$$d\theta^k(X_i, X_j) = X_i(\theta^k(X_j)) - X_j(\theta^k(X_i)) - \theta^k([X_i, Xj])$$
$$= X_i(\delta_j^k) - X_j(\delta_i^k) - \sum_{r=1}^{n} \theta^k(c_{ij}^r(x) X_r)$$
$$= -c_{ij}^k(x),$$

and so

$$d\theta^k = -\frac{1}{2} \sum_{i,j=1}^{n} c_{ij}^k(x) \theta^i \wedge \theta^j. \tag{10.3}$$

In this notation one can get the following description of the canonical symplectic form on T^*X: let $\pi : T^*X \to X$ be the projection of the cotangent bundle, and let $\pi^*(f)$ be the pull-back of a function $f \in C^\infty(X)$ and similarly $\pi^*(\varphi)$ is the pullback of a differential form $\varphi \in \Omega^*(X)$ to $M = T^*X$. Then the canonical 1-form $\alpha \in \Omega^1(T^*X)$ (see Example 4.2) is given by

$$\alpha = \sum_{i=1}^n p_i \pi^*(\theta^i).$$

In particular this form does not depend on the choice of the fields X_i; then the symplectic form on T^*X is equal to

$$\omega = d\alpha = \sum_{i=1}^n dp_i \wedge \pi^*(\theta^i) - \frac{1}{2} \sum_{i,j,k=1}^n p_i \pi^*(c^i_{jk}(x)) \pi^*(\theta^j) \wedge \pi^*(\theta^k).$$
(10.4)

One can find the formula for the Poisson bracket on M in terms of the structure functions c^k_{ij} too: let $Y_1, \ldots, Y_n$ be the vector fields on T^*X such that $dp_i(Y_k) = 0$, $\pi^*(\theta^i)(Y_k) = \delta^i_k$ for all $i, k = 1, \ldots, n$; similarly let $Z^1, \ldots, Z^n$ be such vector fields on T^*X that $\pi^*(\theta^i)(Z^k) = 0$ and $dp_i(Z^k) = \delta^k_i$ for all $i, k = 1, \ldots, n$: clearly, since $\pi^*(\theta^1), \ldots, \pi^*(\theta^n), dp_1, \ldots, dp_n$ gives a basis in cotangent space at any point of T^*X, the fields Y_i, Z^j are completely determined by these equations; moreover $Y_1, \ldots, Y_n, Z^1, \ldots, Z^n$ give a basis in every tangent space of $TM = T(T^*X)$, and

$$d\pi(Y_i) = X_i, \quad d\pi(Z^j) = 0,$$

i.e. Y^k are "horizontal" and Z^j are "vertical" fields. Commutation relations for these fields are

$$[Y_i, Y_j] = \sum_{k=1}^n \pi^*(c^k_{ij}(x)) Y_k, \quad [Y_i, Z^j] = 0, \quad [Z^i, Z^j] = 0. \quad (10.5)$$

In these terms we can write down the bivector field ω^{-1} as

$$\omega^{-1} = \sum_{i=1}^n Z^i \wedge Y_i - \frac{1}{2} \sum_{i,j,k=1}^n p_i \pi^*(c^i_{jk}(x)) Z^j \wedge Z^k,$$

so

$$\{f,g\} = \sum_{i=1}^{n} \left(Z^i(f) Y_i(g) - Z^i(g) Y_i(f) \right) \\ + \sum_{i,j,k=1}^{n} p_i \pi^*(c^i_{jk}(x)) Z^j(f) Z^k(g). \qquad (10.6)$$

Since $Z^1, \ldots, Z^n$ give a basis in the space of vertical vector fields, we can consider the number of times $Z^1, \ldots, Z^n$ appear in the formulas for polydifferential operators as an analog of the grading from the previous section. However, one should be cautious here: in local coordinates $(x^1, \ldots, x^n, \xi_1, \ldots, \xi_n)$ on T^*X, used above, Z^k does not look like ∂_{ξ_i} for some i, or even as a scalar linear combination of such partial derivations: in generic case Z^k has a nonconstant dependence on the base coordinates, and therefore the formula (10.6) is not homogenous of degree 1 in $Z^1, \ldots, Z^n$; even worse: expressions that represent the fields Y_j in local coordinates might contain operators $\partial_{\xi_1}, \ldots, \partial_{\xi_n}$. That's why one cannot transfer the construction of previous section directly to this situation.

Therefore the original paper by Cahen and Gutt used a more delicate analysis of polydifferential operators, similar to the analysis, which gives the proof of the Hochschild–Kostant–Rosenberg theorem, see Section 8.2.2 to prove the existence of deformation series. Namely, since $Y_1, \ldots, Y_n, Z^1, \ldots, Z^n$ give the basis of vector fields on $M = T^*X$, we can represent every polydifferential operator as a linear combination of compositions of these fields. Then they considered the "multi-degree" of a p-polydifferential operator as the p-tuple of natural numbers (or zeros) $(i_1, \ldots, i_p)$ where i_k is the number of vector fields Y_i, Z^j, applied to the k-th function. Detailed analysis of the multi-degrees of polydifferential operators and their coboundaries allows one to prove the existence of a $\star$-product in $C^\infty(M)$. Moreover, one can show that the coefficients of this product satisfy the symmetry condition (i.e. $B_k(f,g) = (-1)^k(g,f)$) and the following equations hold:

(i) $B_r(\pi^*(f), \pi^*(g)) = 0$ for all $f, g \in C^\infty(X)$, $r \geq 1$;

(ii) $B_r(p_i, p_j) = 0$ for all $i, j = 1, \ldots, n$, $r \geq 2$;

(iii) $B_r(p_i, \pi^*(f)) = 0$ for all $i = 1, \ldots, n$, $f \in C^\infty(X)$, $r \geq 2$;

(iv) $B_r(\{p_i, p_j\}, \pi^*(f)) = 0$ for all $i, j = 1, \ldots, n$, $f \in C^\infty(X)$ and $r \geq 3$;

(v) $B_r(\{p_i, p_j\}, p_k) = 0$ for all $i, j, k = 1, \ldots, n$, $r \geq 3$.

It follows that the $\star$-product, constructed this way, satisfies the following commutation relations:

$$[\pi^*(f), \pi^*(g)] = 0,$$

$$[p_i, p_j] = \{p_i, p_j\} = \sum_{k=1}^{n} p_k \pi^*(c_{ij}^k(x)),$$

for all $i, j = 1, \ldots, n$, $f \in C^\infty(M)$, where $[a, b] = a \star b - b \star a$; in fact the $\star$-product of $\pi^*(f)$ and $\pi^*(g)$ is equal to $\pi^*(fg)$. Similarly we have

$$[p_i, \pi^*(f)] = \{p_i, \pi^*(f)\} = Y_i(\pi^*(f)) = \pi^*(X_i(f)).$$

Also we can compute:

$$[[p_i, p_j], \pi^*(f)] = [\{p_i, p_j\}, \pi^*(f)] = \{\{p_i, p_j\}, \pi^*(f)\}$$
$$= \sum_{k=1}^{n} \{p_k \pi^*(c_{ij}^k(x)), \pi^*(f)\} = \sum_{k=1}^{n} \pi^*(c_{ij}^k(x) X_k(f))$$

and similarly

$$[[p_i, p_j], p_k] = [\{p_i, p_j\}, p_k] = \{\{p_i, p_j\}, p_k\},$$

for all indices i, j, k and $f \in C^\infty(X)$. It is not easy to obtain these relations from the general construction as it is explained in the previous section. Of course the fact that "functions on the base" X commute with each other as differential operators is true, but the behaviour of the functions p_i is not quite evident from the general construction.

10.2.2 Lie groups

An important particular case of parallelisable manifolds are Lie groups. In this situation, $X = G$ and $TG \cong G \times \mathfrak{g}$, where we identify the Lie algebra $\mathfrak{g}$ of G with the space of left- (or right-) invariant vector fields on G or with the tangent space at the unit element $e \in G$.

In this case, one can choose $X_1 \ldots, X_n$ to be the basis in the space of left-invariant fields, so that the commutator relations (10.1) turn into the definition of the Lie bracket (5.1):

$$[X_i, X_j] = C_{ij}^k X_k,$$

i.e. all the structure functions turn into structure constants $c_{ij}^k(x) = C_{ij}^k$. It turns out that in this situation, one can further describe the $\star$-product on $C^\infty(T^*M)$. In order to do it, we introduce the functions $p_1, \ldots, p_n$ and vector fields $Y_i, \ldots, Y_n, Z^1, \ldots, Z^n$ on T^*G as above.

We will now restrict our attention to functions, polynomial in $p_1, \ldots, p_n$: this is a version of the vertically polynomial functions from Section 10.1, although one should be rather careful here: these two constructions are not quite the same since the functions $p_1, \ldots, p_n$ are globally defined on T^*G, unlike the local coordinates $\xi_1, \ldots, \xi_n$. Any such function can be written in a unique way as a linear combinations of functions $Q(p)\pi^*(f)$, where $f \in C^\infty(G)$ and Q is a polynomial. In this case, one can find more or less explicit formulas for the $\star$-product, which extends to the whole algebra $C^\infty(T^*G)$:

Proposition 10.3. *There exists a unique (up to equivalence) $\star$-product on the functions of the form $Q(p)\pi^*(f)$ on T^*G, such that*

(i) *for all $f, g \in C^\infty(G)$*

$$\pi^*(f) \star \pi^*(g) = \pi^*(fg); \tag{10.7}$$

(ii) *for all $f \in C^\infty(G)$ and all polynomials Q we have $Q(p) \star \pi^*(f) = Q(p)\pi^*(f)$ and*

$$\pi^*(f) \star Q(p) = Q(p)\pi^*(f)$$
$$+ \sum_{k=1}^\infty \frac{(-1)^k \hbar^k}{k!} \sum_{i_1,\ldots,i_k=1}^n Z^{i_1} \ldots Z^{i_k}(Q(p))\pi^*(X_{i_1}\ldots X_{i_k}(f)); \tag{10.8}$$

(iii) *for all polynomials Q_1, Q_2 in variables $p_1, \ldots, p_n$*

$$Q(p) \star P(p) = \sigma^{-1}(\sigma(Q(p)) \cdot \sigma(P(p))), \tag{10.9}$$

where $\sigma : \mathbb{R}[p_1, \ldots, p_n] \cong S(\mathfrak{g})[[\hbar]] \cong \widehat{U}_\hbar \mathfrak{g}$ *is the isomorphism of Poincaré-Birkhoff-Witt theorem, and $\cdot$ is the product in the completed universal enveloping algebra $\widehat{U}_\hbar \mathfrak{g}$ see* (5.3') *(compare* (5.10)*);*

(iv) *in particular, if $Q(p) = p_i$ we have*

$$p_i \star Q(p) = \sum_{r=0}^{\infty} \frac{(-1)^r \hbar^r}{r!} B_r \sum_{\substack{1 \leq k, k_1, \ldots, k_r \leq n \\ 1 \leq j_1, \ldots, j_r \leq n}} p_{j_r} C^{j_1}_{k k_1} C^{j_2}_{j_1 k_2} \cdots$$
$$\ldots C^{j_r}_{j_{r-1} k_r} Z^k(p_i) Z^{k_1} \ldots Z^{k_r}(Q(p)), \quad (10.10)$$

where B_r *denote the Bernoulli numbers, compare Lemma* 5.6;

(v) *formulas* (10.7)–(10.10) *determine the $\star$-product on $C^\infty(T^*G)$ in a unique way.*

Proof. The only thing that remains to be proved is the last part of this proposition, i.e. the uniqueness of product, determined by previous data. Indeed, the fact that Equations (10.9) and (10.10) extend to an associative $\star$-product in $S(\mathfrak{g}) = \mathbb{R}[p_1, \ldots, p_n]$ and hence in $C^\infty(\mathfrak{g}^*)$ was proved earlier in Section 5.3.1. Further, we see that

$$Q(p)\pi^*(f) = \pi^*(f) \star Q(p) - \sum_{k=1}^{\infty} \frac{(-1)^k \hbar^k}{k!} \sum_{i_1,\ldots,i_k=1}^{n} Z^{i_1} \ldots Z^{i_k}(Q(p))$$
$$\pi^*(X_{i_1} \ldots X_{i_k}(f)),$$

where all the terms in the series on the right-hand side are of the form $\tilde{Q}(p)\pi^*(\tilde{f})$ and the degree of the polynomial $\tilde{Q}(p)$ is less than the degree of Q. So by induction we get:

$$Q(p)\pi^*(f) = \pi^*(f) \star Q(p) + \sum_{r=1}^{\infty} \pi^*(f_r) \star Q_r(p), \quad (10.11)$$

where $\deg Q_r < \deg Q$ for all r. Hence using (10.11) we see by induction on degrees and using the associativity condition, that the product

$$(Q(p)\pi^*(f)) \star (P(p)\pi^*(g))$$
$$= \left(\pi^*(f) \star Q(p) + \sum_{r=1}^{\infty} \pi^*(f_r) \star Q_r(p)\right) \star (P(p)\pi^*(g))$$
$$= \pi^*(f) \star \big(Q(p) \star (P(p)\pi^*(g))\big)$$
$$+ \sum_{r=1}^{\infty} \pi^*(f_r) \star \big(Q_r(p) \star (P(p)\pi^*(g))\big)$$

is uniquely defined if the first term is uniquely defined. But, similarly to (10.11), we have:

$$P(p)\pi^*(g) = \pi^*(g) \star P(p) + \sum_{s=1}^{\infty} \pi^*(g_s) \star P_s(p), \ \deg P_s < \deg P,$$

so the uniqueness of the product follows by induction from associativity of $\star$-product and formulas (10.7)–(10.9) and (10.11). □

Remark 10.4. One of the important properties of this construction is that the deformation quantisation of the cotangent bundle of a Lie group G contains the universal enveloping algebra of its Lie algebra as a subalgebra; this corresponds to the fact that one can regard $U\mathfrak{g}$ as the subalgebra of left-invariant differential operators on the jets of smooth functions on G at the unit of the group; if we regard the quantisation of T^*G as the algebra of differential operators on G, see Remark 10.2, then the embedding of $U\mathfrak{g}$ is just the embedding of left-invariant part in $\mathscr{DO}(X)$.

11 Lie algebra cohomology: Vey class

Although cotangent bundles are an important example of symplectic manifolds, methods from Section 10 cannot be extended to other situations: additional structures, such as grading, do not survive on a generic symplectic manifold.

On the other hand, the general obstruction theory (see Section 9, in particular Section 9.6.3) in symplectic case gives a series of cohomology obstruction classes in $H^3_{dR}(M)$. Identifying these classes in a generic case is not an easy task: cohomology groups $H^*_{dR}(M)$ depend on the global topology of M, and in general it is not easy to see what particular topological property would guarantee the vanishing of one or another class in cohomology.[28] In this and the next section we shall sketch constructions that eventually allowed Lecomte and de Wilde to prove in [DL83$_2$] the existence of $\star$-products for arbitrary symplectic manifolds, see Theorem 12.1.

11.1 Deformation of Poisson structures

Let us assume that we have an associative $\star$-product on $C^\infty(M)[[\hbar]]$ for a Poisson manifold M (not necessarily symplectic):

$$f \star g = fg + \frac{\hbar}{2}\{f,g\} + \sum_{k=2}^{\infty} \hbar^k B_k(f,g), \ \forall f, g \in C^\infty(M).$$

We shall also suppose that B_k satisfy the symmetry condition, i.e. $B_k(f,g) = (-1)^k B_k(g,f)$. Consider the commutator $[f,g] = f \star g - g \star f$: since $\star$-product is associative, it satisfies the Jacobi identity, and

---

[28] Another possible point of view at this problem is the following: locally every symplectic manifold is isomorphic to $\mathbb{R}^{2n}$ with the standard constant symplectic form at it (see Darboux's theorem, Theorem 4.5). Thus we can introduce Moyal product at an open neighbourhood of every point in M. However, pasting these products together would involve nonlinear coordinate changes from one set of Darboux's coordinates to another. This gives us a cohomology obstructions with values in the Čech cohomology with coefficients in local symplectomorphisms on M. This cohomology theory captures the "local-to-global" obstructions in geometry. But computing it is not a simple problem.

can be written as a formal power series of bidifferential operators:

$$[f,g] = \hbar\{f,g\} + 2\sum_{k=1}^{\infty} \hbar^{2k+1} B_{2k+1}(f,g). \tag{11.1}$$

If we divide both sides of this equation by $\hbar$ and use notation $\lambda = \hbar^2$, $C_k = 2B_{2k+1}(f,g)$, we shall get a bilinear anti-symmetric map $\{,\}_\lambda : C^\infty(M)[[\lambda]] \otimes C^\infty(M)[[\lambda]] \to C^\infty(M)[[\lambda]]$ given by the formula

$$\{f,g\}_\lambda = \{f,g\} + \sum_{p=1}^{\infty} \lambda^p C_p(f,g), \tag{11.2}$$

for some skew-symmetric bidifferential operators C_p. This bracket will satisfy the Jacobi identity, thus *we shall call $\{,\}_\lambda$ formal Lie bracket on $C^\infty(M)[[\lambda]]$, deforming the Poisson bracket $\{,\}$ or simply the deformation of the Poisson structure* on M. Informally, one can say now that half of the deformation quantisation problem can be phrased as the problem of finding a deformation of the Poisson bracket.

It is not hard to develop an obstruction theory for the deformation of Poisson structures, completely analogous to the theory from Section 9, see for instance [Vey75]: consider the Jacobi identity for $\{,\}_\lambda$ and compare the terms at different powers of λ. We begin with the linear terms:

$$\{C_1(f,g), h\} + C_1(\{f,g\}, h) + \{C_1(g,h), f\}$$
$$+ C_1(\{g,h\}, f) + \{C_1(h,f), g\} + C_1(\{h,f\}, g) = 0.$$

Using the skew-symmetry of $\{,\}$ and C_1, and renaming f, g, h as f_1, f_2, f_3 we can rewrite this formula as follows:

$$\{f_1, C_1(f_2, f_3)\} - \{f_2, C_1(f_1, f_3)\} + \{f_3, C_1(f_1, f_2)\}$$
$$- C_1(\{f_1, f_2\}, f_3) + C_1(\{f_1, f_3\}, f_2) \; C_1(\{f_2, f_3\}, f_1) = 0. \tag{11.3}$$

Comparing this formula with the constructions from Section 6.2.4, we conclude that C_1 *is a cocycle in the Chevalley–Eilenberg complex* $C^\cdot(C^\infty(M), C^\infty(M))$ *of the algebra $C^\infty(M)$ viewed as the Lie algebra with respect to the Poisson bracket $\{,\}$ and with coefficients in itself.*

In the same way, if we consider elements at λ^2 we get the equation:

$$\delta C_2(f,g,h) = C_1(C_1(f,g),h) + C_1(C_1(g,h),f) + C_1(C_1(h,f),g). \quad (11.4)$$

The right-hand side of this equation is equal (up to the factor $\frac{1}{2}$) to the "skew-symmetric version of Gerstenhaber bracket", see Exercise E.4.2, we shall denote it by $\frac{1}{2}[C_1, C_1]$. Thus we see that *the class of $\frac{1}{2}[C_1, C_1]$ in the Lie algebra cohomology $H^3(C^\infty(M), \{,\})$ is the obstruction to the second step of deformation process.*

Reasoning further as in Section 9 (see Section 9.2.3), we obtain equations

$$\delta C_n(f,g,h) = \sum_{p+q=n} (C_p(C_q(f,g),h)$$
$$+ C_p(C_q(g,h),f) + C_p(C_q(h,f),g)), \quad (11.5)$$

and again *the class of*

$$\sum_{p+q=n} (C_p(C_q(f,g),h) + C_p(C_q(g,h),f) + C_p(C_q(h,f),g))$$

$$= \frac{1}{2} \sum_{p+q=n} [C_p, C_q]$$

in the Lie algebra cohomology $H^3(C^\infty(M), \{,\})$ is the obstruction to the n-th step of deformation process for Poisson bracket. Similarly to Section 9.3, there is a classification result, that identifies the set of equivalence classes of such deformations at every step with $H^2(C^\infty(M), \{,\})$. Thus in order to describe the obstructions of the deformations of a Poisson structure we need to know the Chevalley–Eilenberg cohomology of the Poisson algebra $C^\infty(M)$ in degrees 2 and 3. It turns out that one can describe these cohomolgy spaces in a universal way, independent (or almost independent) of topology of M. In the following sections, we shall briefly outline the way this result can be (in effect, was) obtained.

11.2 Lie algebras related with a manifold and their cohomology

Our purpose is to describe the elements in cohomology of the Lie algebra $\mathfrak{P}(M) = (C^\infty(M), \{,\})$. This is an infinite-dimensional

Lie algebra associated with any symplectic manifold (and even with any Poisson manifold). In order to describe classes in its cohomology $H^*(\mathfrak{P}(M), \mathfrak{P}(M))$, we need to study cohomology of other infinite-dimensional Lie algebras. This subject has been extensively studied since the pioneering works by Gelfand and Fuks in 1970s, see [GF69], [GF70$_1$], [GF70$_2$], or Fuks' book [Fu86]. We cannot give a comprehensive introduction into this theory here due to its complexity, so we shall now only sketch the major ideas.

11.2.1 Algebras of formal vector fields on $\mathbb{R}^{2n}$

First of all, let $\widehat{S}_{2n}$ be the algebra of formal power series in $2n$ variables, i.e. $\widehat{S}_{2n} = \mathbb{R}[[p_1, \ldots, p_n, q^1, \ldots, q^n]]$; we can regard it as the space of Taylor series[29] of smooth functions on $\mathbb{R}^{2n}$ at the origin. The standard symplectic form $\omega = \sum_i dp_i \wedge dq^i$ on $\mathbb{R}^{2n}$ induces a Poisson bracket on $\widehat{S}_{2n}$; in effect, it is not necessary to use canonical Darboux's coordinates in the definition, since every nondegenerate skew-symmetric matrix can be transformed to the canonic form by a linear transformation, therefore we shall often denote the coordinates in $\mathbb{R}^{2n}$ simply by $(x^1, \ldots, x^{2n})$ so that $\omega = \omega_{ij} dx^i \wedge dx^j$. We shall denote by $\mathfrak{P}_{2n}$ (or simply $\mathfrak{P}$ for short) the Lie algebra $(\widehat{S}_{2n}, \{,\})$.

Similarly, we consider the Lie algebra of formal symplectic vector fields on $\mathbb{R}^{2n}$; we shall denote it by $\mathfrak{S}_{2n}$ or just $\mathfrak{S}$ for short. In explicit terms

$$\mathfrak{S} = \left\{ \xi = \sum_{i=1}^n \left(\xi^i \frac{\partial}{\partial p_i} + \xi^{n+i} \frac{\partial}{\partial q^i} \right), \xi^i \in \widehat{S}_{2n} \mid \mathcal{L}_\xi \omega = 0 \right\}.$$

Here $\mathcal{L}_\xi$ denotes the formal Lie derivative of ω along the formal vector field ξ and the Lie algebra structure is given by the usual commutators of vector fields:

$$[\xi, \eta]^i = \xi^k \frac{\partial \eta^i}{\partial x^k} - \eta^k \frac{\partial \xi^i}{\partial x^k}, \quad i = 1, \ldots, 2n, \quad (x^1, \ldots, x^{2n}) = (p_1, \ldots, q^n). \tag{11.6}$$

---

[29] Reader familiar with the notion of jets of functions will undoubtedly recognise this concept here and in the discussion that follows below.

Of course every formal symplectic field on $\mathbb{R}^{2n}$ is in fact Hamiltonian; we shall denote the formal Hamilton field, associated with a formal function $f \in \widehat{S}_{2n}$ by X_f, see Definition 4.3. Also let $\mathfrak{s}_{2n} = \mathfrak{s}$ be the subalgebra of $\mathfrak{G}$ of fields, vanishing at the origin:

$$\mathfrak{s} = \{\xi \in \mathfrak{G} \mid \xi(0) = 0\};$$

formula (11.6) implies that the commutator of two elements in $\mathfrak{s}$ is again in $\mathfrak{s}$. We let the algebra $\mathfrak{G}$ act on $\widehat{S}_{2n}$ by derivations.

11.2.2 Cohomology of formal vector fields

Computing the Chevalley–Eilenberg cohomology of the Lie algebras of vector fields is a rather difficult problem. Although the definition is quite elementary, the technics used in computations goes much further from what we can explain in our short notes. In particular, few important computations are based on the properties of spectral sequences, which is completely beyond the scope of this course. We hope that the reader will forgive us the absence of explanations; we restrict ourselves to just giving reference to the books that explain this method. Some basics of this theory can be found in the books by Weibel [Wei94] and Gelfand and Manin [GM03]; also a nice introduction into the theory gauged for the applications in differential geometry is given in the book [BT82]. A comprehensive introduction to this theory with many important examples is given in [MCl01]. The results of this section are taken mostly from the seminal Vey's paper [Vey75], which was a direct predecessor of many further works on Deformation Quantisation.

The first statement that we will need is given by the following proposition:

Proposition 11.1. *There is a natural isomorphism in cohomology $H^*(\mathfrak{s}, \mathbb{R}) = H^*(\mathfrak{G}, \widehat{S}_{2n})$, where we regard $\mathbb{R}$ as a trivial $\mathfrak{s}$-module.*

Proof (sketch). First of all we notice that $\mathfrak{G} \cong \mathfrak{s} \oplus \mathfrak{t}_{2n}$ as vector space, where $\mathfrak{t}_{2n}$ denotes the commutative Lie subalgebra in $\mathfrak{G}$ that consists of vector fields with constant coefficients ξ^i. Now Poincaré-Birkhoff-Witt theorem extended to the case of formal vector fields shows that

$U\mathfrak{G} \cong U\mathfrak{s} \otimes U\mathfrak{t}_{2n} = U\mathfrak{s} \otimes S(\mathfrak{t}_{2n})$ as left $U\mathfrak{s}$-modules,[30] so

$$\widehat{S}_{2n} \cong \widehat{S}(\mathfrak{t}_{2n}^*) = \mathrm{Hom}_{U\mathfrak{s}}(U\mathfrak{G}, \mathbb{R})$$

as $U\mathfrak{G}$-modules. On the other hand

$$H^*(\mathfrak{G}, \widehat{S}_{2n}) = Ext^*_{U\mathfrak{G}}(\mathbb{R}, \widehat{S}_{2n}) \text{ and } H^*(\mathfrak{s}, \mathbb{R}) = Ext^*_{U\mathfrak{s}}(\mathbb{R}, \mathbb{R})$$

so the statement follows by dualisation:

$$\mathrm{Hom}_R(A, \mathrm{Hom}_L(R, C)) \cong \mathrm{Hom}_L(A, C)$$

if the ring $R \cong L \otimes B$ as L-module and A is free R-module. □

Now let $a, b \in \widehat{S}_{2n}$ and let $\xi = X_a$, $\eta = X_b$ be their Hamilton vector fields; we assume that $X_a, X_b \in \mathfrak{s}$. Let us define the cochain $\beta \in C^2(\mathfrak{s}, \mathbb{R})$ by the formula:

$$\beta : \mathfrak{s} \wedge \mathfrak{s} \to \mathbb{R},$$
$$\beta(\xi, \eta) = \pi^3(a, b)(0) = (\pi^{i_1 j_1} \pi^{i_2 j_2} \pi^{i_3 j_3} \partial_{i_1} \partial_{i_2} \partial_{i_3} a \partial_{j_1} \partial_{j_2} \partial_{j_3} b)(0), \quad (11.7)$$

for all ξ, η of this form. Here $\{a, b\} = \pi^{ij} \partial_i a \partial_j b$, i.e. $(\pi^{ij}) = \omega^{-1}$.

Proposition 11.2. *The formula (11.7) determines a well-defined 2-cochain β, which is in effect a cocycle, i.e. it satisfies the equation $\delta(\beta) = 0$.*

Proof. The formal functions a, b are determined by condition $\xi = X_a$, $\eta = X_b$ up to constant terms, so there's no ambiguity in the definition of $\beta(\xi, \eta)$ i.e. its values do not depend on the choice of representing functions; moreover it is anti-symmetric, since $\pi^{ij} = -\pi^{ji}$.

In order to see the cocycle relation, observe that since π is a constant bivector, we can use it to define the Moyal product on $\widehat{S}_{2n}$, see Section 3.2.1. Then, $\frac{1}{48}\beta(\xi, \eta) = B_3(a, b)(0)$. As we know the third term of any $\star$-product is up to the factor $\frac{1}{2}$ equal to the second term C_2 of the deformation of the Poisson bracket (see Section 11.1)

---

[30]The description of Lie algebra cohomology as the Ext functor is completely analogous to the similar description of Hochschild cohomology. We leave filling in the details of this statement to the reader, see Exercise E.4.4.

which satisfies the condition (11.5): $\delta C_2 = \frac{1}{2}[C_1, C_1]$. On the other hand, see Definition 4.3 for all $f \in \widehat{S}_{2n}$ we have

$$\{a, f\}(0) = -X_a(f)(0) = -\xi(f)(0) = 0$$

since $\xi(0) = 0$. So

$$\delta C_2(a, b, c)(0) = \frac{1}{2}(\{\{a, b\}, c\} + \{\{b, c\}, a\} + \{\{c, a\}, b\})(0) = 0$$

for all formal functions $a, b, c \in \widehat{S}_{2n}$ such that $X_a, X_b, X_c \in \mathfrak{s}$. Similarly

$$\{a, C_2(b, c)\}(0) = \{b, C_2(a, c)\}(0) = \{c, C_2(a, b)\}(0) = 0,$$

and hence, since $[X_a, X_b] = X_{\{a,b\}}$, we have

$$\delta\beta(\xi, \eta, \zeta) = \frac{1}{24}\delta C_2(a, b, c)(0) = 0.$$

$\square$

It turns out that the following fact is true (see [Vey75]):

Theorem 11.3 (Vey). *Cohomology group $H^2(\mathfrak{s}, \mathbb{R})$ is generated by the class of β as a linear space, and $H^1(\mathfrak{s}, \mathbb{R}) = 0$. If $n \geq 2$, then $\dim H^3(\mathfrak{s}, \mathbb{R}) = 1$, $\dim H^4(\mathfrak{s}, \mathbb{R}) \geq 2$.*

We are not going to prove this hard theorem here, it is based on a detailed analysis of degrees of polydifferential operators and the study of relative cohomology of $\mathfrak{s}$ with respect to the subalgebra of linear symplectic transformations $\mathfrak{sp}$. The reader can look it up in the original paper of Vey [Vey75]. We will denote by $\tilde{\beta} \in H^2(\mathfrak{S}, \widehat{S}_{2n})$ the class that corresponds to β under the isomorphism of Proposition 11.1. One can call $\tilde{\beta}$ the *local Vey class*, due to the role it plays in the definition of the latter, see Section 11.2.4 below.

11.2.3 Cohomology of formal symplectic space

We are using the notation from Section 11.2.2. Let μ denote the map $\mu : a \mapsto X_a$, where $a \in \widehat{S}_{2n}$ and X_a denotes the Hamilton vector field of a; as one sees from Jacobi identity for the Poisson bracket and the definition of X_a, μ is a surjective homomorphism of Lie algebras from the algebra $(\widehat{S}_{2n}, \{,\})$, denoted by $\mathfrak{P}_{2n} = \mathfrak{P}$, to $\mathfrak{S}$ with kernel equal to the space of constant functions, giving the following extension of Lie algebras:

$$0 \to \mathfrak{t}_1 \to \mathfrak{P} \xrightarrow{\mu} \mathfrak{S} \to 0, \tag{11.8}$$

where $\mathfrak{t}_1 = \mathbb{R}$ with trivial Lie bracket. Using this extension and the spectral sequence associated with it, one can describe the Lie algebra cohomology of $\mathfrak{P}$ with coefficients in $\mathfrak{P} = \widehat{S}_{2n}$ starting with the cohomologies of $\mathfrak{S}$ and of $\mathfrak{t}_1$, which leads to the following proposition:[31]

Proposition 11.4. *There exists a natural isomorphism of graded spaces* $H^*(\mathfrak{P}, \widehat{S}_{2n}) \cong H^*(\mathfrak{S}, \widehat{S}_{2n}) \hat{\otimes} \Lambda[\alpha]$. *Here* $\Lambda[\alpha]$ *is the exterior algebra with one generator* α *in degree 1 and* $\hat{\otimes}$ *denotes the graded tensor product.*

One can find the cocycle in $C^1(\mathfrak{P}, \widehat{S}_{2n})$ that generates the class corresponding to α in cohomology: let $e = \sum_{i=1}^{2n} x^i \frac{\partial}{\partial x^i}$, be the *Euler vector field* on $\mathbb{R}^{2n}$, so that for every homogenous polynomial $P(x) \in \mathbb{R}[x^1, \ldots, x^{2n}]$, $e(P) = \deg P \cdot P(x)$. Then α is represented by the following cochain

$$\alpha(u) = e(u) - 2u, \quad u \in \widehat{S}_{2n}.$$

Indeed, since the Poisson bracket in $\widehat{S}_{2n}$ drops the degree of polynomials by 2, we have for all homogenous $u, v \in \widehat{S}_{2n}$ of degrees p and q respectively

$$\delta\alpha(u,v) = \{u, \alpha(v)\} - \{v, \alpha(u)\} - \alpha(\{u,v\})$$
$$= (q-2)\{u,v\} - (p-2)\{v,u\} - (p+q-4)\{u,v\} = 0,$$

so α is a cocycle. Further the map $\mu : \mathfrak{P} \to \mathfrak{S}$ induces pull-back map μ^* in cohomology complexes from $C^{\cdot}(\mathfrak{S}, \widehat{S}_{2n})$ to $C^{\cdot}(\mathfrak{P}, \widehat{S}_{2n})$:

$$\mu^*(\varphi)(u_1, \ldots, u_p) = \varphi(X_{u_1}, \ldots, X_{u_p})$$

---

[31] We omit the proof, but reader familiar with the spectral sequences will easily fill in the missing arguments; just notice that Chevalley–Eilenberg complex of $\mathfrak{t}_1$ has zero differential.

for all $u_1, \ldots, u_p \in \widehat{S}_{2n}$ and all $\varphi \in C^p(\mathfrak{S}, \widehat{S}_{2n})$. Also the product in $\widehat{S}_{2n}$ induces a wedge-product in Lie algebra cohomology with coefficients in $\widehat{S}_{2n}$, see Exercise E.4.3. Using all these maps, Theorem 11.3 and identifications we made above, we conclude that if $n \geq 2$

$$H^1(\mathfrak{P}, \widehat{S}_{2n}) = \mathbb{R} \text{ with generator } \alpha,$$
$$H^2(\mathfrak{P}, \widehat{S}_{2n}) = \mathbb{R} \text{ with generator } \mu^*(\tilde{\beta}), \tag{11.9}$$
$$H^3(\mathfrak{P}, \widehat{S}_{2n}) = \mathbb{R}^2 \text{ with one of the generators } \mu^*(\tilde{\beta}) \wedge \alpha.$$

11.2.4 Cohomology of $\mathfrak{P}(M)$

We finally can say something about the structure of the cohomology space $H^*(\mathfrak{P}(M), C^\infty(M))$ for a symplectic manifold M in low dimensions. To this end, we consider the exact sequence of sheaves[32] of Lie algebras, analogous to (11.8):

$$0 \to \mathscr{T}_1 \to \mathscr{P}(M) \xrightarrow{\mu} \mathscr{S}(M) \to 0, \tag{11.10}$$

where $\mathscr{S}(M)$ is the sheaf of local smooth symplectic vector fields on M, $\mathscr{P}(M)$ is the sheaf of smooth functions on M with Lie algebra structure given by the Poisson bracket and $\mathscr{T}_1$ is the constant sheaf of commutative 1-dimensional Lie algebras on M.

Since these sheaves bear the structure of Lie algebras, we can associate with them sheaves of Chevalley complexes with coefficients in $C^\infty(M)$ and then compute the cohomology of the corresponding double complexes; but the sheaves $\mathscr{S}(M)$, $\mathscr{P}(M)$ are flabby, so the spectral sequences that compute this cohomology beginning with the sheaf cohomology (one can say that these spectral sequences go "in the row direction"), collapse at second stage, and we see that the corresponding cohomology spaces are equal to $H^*(\mathfrak{S}(M), C^\infty(M))$, where $\mathfrak{S}(M)$ is the algebra of symplectic vector fields on M, and $H^*(\mathfrak{P}(M), C^\infty(M)) = H^*(\mathfrak{P}(M), \mathfrak{P}(M))$ respectively.

Now we can consider the other spectral sequences (this time "in column direction") associated with the double complexes of sheaves

---

[32] The reader, not familiar with sheaf theory and sheaf cohomology can skip this section, or just regard sheaves as functors, which associate with open sets in M spaces of functions or vector fields on them, and sheaf cohomology as given by a Čech-type cohomology construction.

of Chevalley complexes and look at the corresponding E_∞ sheet. We denote this spectral sequences by $'E_r^{p,q}$ and $''E_r^{p,q}$ respectively. Then since the formal Lie algebras $\mathfrak{S}$, $\mathfrak{P}$ are stalks of the sheaves we consider here, one can compute the second sheets of these sequences as cohomology of these algebras; so we get:

$$'E_2^{p,q} = H^p(\mathfrak{S}, \widehat{S}_{2n}) \hat{\otimes} H_{dR}^q(M) \Rightarrow H^*(\mathfrak{S}(M), C^\infty(M))$$
$$''E_2^{p,q} = H^p(\mathfrak{P}, \widehat{S}_{2n}) \hat{\otimes} H_{dR}^q(M) \Rightarrow H^*(\mathfrak{P}(M), C^\infty(M)).$$
(11.11)

Of course, using Proposition 11.1 we can replace $H^p(\mathfrak{S}, \widehat{S}_{2n})$ with $H^p(\mathfrak{s}, \mathbb{R})$. Further analysis, similar to the proof of Theorem 11.3 then shows that *the cohomology class of $\mu^*(\tilde\beta)$ "survives" to give nontrivial classes in $H^*(\mathfrak{P}(M), C^\infty(M))$*, denoted by S_Γ^3; as for the element $\mu^*(\tilde\beta) \wedge \alpha$, one may say that it gives rise to a series of classes, parametrised by the first de Rham cohomology of M, see Remark 11.11 below. Thus in particular the cocycle, corresponding to $\mu^*(\tilde\beta)$ can be used as the first step in the process of deformation of the Poisson structure. Below we shall give a description of this class in terms of symplectic connections on manifolds, but to this end we will need the following digression into the Differential Geometry.

11.3 Symplectic connections and cohomology

Before we can proceed with the study of cohomology theory of the Lie algebras of vector fields, we need to talk a little about the definitions and properties of the affine connections adapted to symplectic case. We hope that the reader is familiar with the standard facts from the theory of affine connections, in particular with Riemannian or Levi-Civita connections. The proper reference here is any standard course on Differential Geometry, for instance [KN63] or [He78].

11.3.1 Existence of symplectic connections

For every manifold M we shall identify the affine connections on TM (and hence on T^*M) with the corresponding covariant derivatives ∇ on vector fields (and differential forms); i.e. we shall say that connection ∇ is a bilinear (over scalars) map, that for every two

vector fields X, Y determines a third field $\nabla_X Y$, such that
$$\nabla_{fX} Y = f \nabla_X Y, \quad \nabla_X(fY) = f \nabla_X Y + X(f) Y,$$
for all smooth functions $f \in C^\infty(M)$.

Definition 11.5. *We say, that a connection ∇ on a symplectic manifold (M, ω) is **symplectic** if it is symmetric (or torsion-free) and $\nabla(\omega) = 0$, where ω is the symplectic form; equivalently ∇ is symplectic, if for all vector fields X, Y, Z on M the following equations hold:*
$$\nabla_X Y - \nabla_Y X = [X, Y]$$
i.e. **the torsion tensor of** ∇
$$\Omega(X, Y) = \nabla_X Y - \nabla_Y X - [X, Y]$$
vanishes and for any three vector fields X, Y, Z we have
$$Z(\omega(X, Y)) = \omega(\nabla_Z X, Y) + \omega(X, \nabla_Z Y).$$

Symplectic connections are an important tool in many branches of modern Geometry. A survey of results related with symplectic connections may be found in [BCGRS06]. In particular they play a crucial role in the deformation quantisation of symplectic manifolds, therefore we are going to prove the existence of such connections in the general case:

Proposition 11.6. *On every symplectic manifold, there exist symplectic connections.*

Proof. It is well-known that there exist symmetric connections on all smooth manifolds (e.g. the Levi-Civita connection). Let ∇^0 be any such connection; then the map
$$F(X, Y, Z) = (\nabla^0_X \omega)(Y, Z) = X(\omega(Y, Z)) - \omega(\nabla^0_X Y, Z) - \omega(Y, \nabla^0_X Z)$$
is $C^\infty(M)$-linear in all variables X, Y, Z, i.e. $F(fX, Y, Z) = F(X, fY, Z) = F(X, Y, fZ) = fF(X, Y, Z)$ for all vector fields X, Y, Z and all $f \in C^\infty(M)$. Since ω is nondegenerate, there exists a $C^\infty(M)$-bilinear map N of on vector fields such that
$$F(X, Y, Z) = \omega(N(X, Y), Z).$$

Then, since $\omega(Y, Z) = -\omega(Z, Y)$, we have

$$\omega(N(X,Y), Z) = F(X, Y, Z) = -F(X, Z, Y) = -\omega(N(X, Z), Y). \tag{11.12}$$

Also, since $d\omega = 0$ and $d\omega$ is equal to the skew-symmetrisation of $\nabla^0 \omega$ for any torsion-free connection ∇^0, we have

$$d\omega(X, Y, Z) = \frac{1}{3}\left((\nabla^0_X \omega)(Y, Z) + (\nabla^0_Y \omega)(Z, X) + (\nabla^0_Z \omega)(X, Y)\right) = 0,$$

and hence

$$\omega(N(X, Y), Z) + \omega(N(Y, Z), X) + \omega(N(Z, X), Y) = 0. \tag{11.13}$$

We now put: $\nabla_X Y = \nabla^0_X Y + \frac{1}{3}(N(X, Y) + N(Y, X))$. It is clear that this map satisfies the conditions that determine covariant derivations, and hence it only remains to show that $\nabla \omega = 0$; we compute:

$$\begin{aligned}(\nabla_X \omega)(Y, Z) &= (\nabla^0_X \omega)(Y, Z) - \frac{1}{3}(\omega(N(X, Y), Z) + \omega(N(Y, X), Z) \\ &\quad + \omega(Y, N(X, Z)) + \omega(Y, N(Z, X))) \\ &= \frac{2}{3}\omega(N(X, Y), Z) - \frac{1}{3}(\omega(N(Y, X), Z) \\ &\quad + \omega(Y, N(X, Z)) + \omega(Y, N(Z, X))) \\ &= \frac{1}{3}(\omega(N(X, Y), Z) + \omega(N(Y, Z), X) + \omega(N(Z, X), Y)) \\ &\quad + \frac{1}{3}(\omega(N(X, Y), Z) + \omega(N(X, Z), Y)) = 0,\end{aligned}$$

where we used Equations (11.12) and (11.13). □

Remark 11.7. Unlike torsion-free Riemannian connections (Levi-Civita connections), which are uniquely determined by a Riemann metric, symplectic connections on a symplectic manifold are never unique: in effect, if we add to ∇ any bilinear map S on TM with values in TM, which is symmetric (i.e. $S(X, Y) = S(Y, X)$) and commutes with ω (i.e. $\omega(S(X, Y), Z) = \omega(S(X, Z), Y)$), we see that $\nabla'_X Y = \nabla_X Y + S(X, Y)$ is torsion-free and

$$(\nabla'_X \omega)(Y, Z) = -\omega(S(X, Y), Z) - \omega(Y, S(X, Z)) = 0.$$

Remark 11.8. It is sometimes convenient to have a coordinate expression of symplectic connection that we have just constructed; we will use it below, when we speak about Fedosov quantisation, see Section 13.3. In order to give this expression, let us recall that any connection ∇ is determined by the set of *Christoffel symbols* Γ^k_{ij}:

$$(\nabla_X Y)^k = X^i \frac{\partial Y^k}{\partial x^i} + \Gamma^k_{ij} X^i Y^j$$

where $(x^1,\ldots,x^n)$ are the local coordinates and $X^i, Y^j, i,j = 1,\ldots,n$ are the components of vector fields X, Y in these coordinates; in this notation ∇ is symmetric iff $\Gamma^k_{ij} = \Gamma^k_{ji}$ for all i, j, k. Let now $\Gamma_{ijk} = \omega_{ip}\Gamma^p_{jk}$ (the sum over p is assumed); of course Γ^k_{ij} is determined by Γ_{ijk} in a unique way. The condition on Γ^k_{ij} which means that ∇ is symplectic now looks as follows:

$$\frac{\partial \omega_{ij}}{\partial x^k} = \Gamma_{ijk} - \Gamma_{jik}.$$

We can change Γ^k_{ij} by adding a tensor $\Delta\Gamma^k_{ij}$ satisfying the condition $\Delta\Gamma^k_{ij} = \Delta\Gamma^k_{ji}$ to it; let $\Delta\Gamma_{ijk} = \omega_{ip}\Delta\Gamma^p_{jk}$. Then the condition that the new connection is symplectic is equivalent to

$$(\nabla_k \omega)_{ij} = \Delta\Gamma_{ijk} - \Delta\Gamma_{jik}. \qquad (11.14)$$

So taking

$$\Delta\Gamma_{ijk} = \frac{1}{3}\left((\nabla_k \omega)_{ij} + (\nabla_j \omega)_{ik}\right),$$

where

$$(\nabla_a \omega)_{bc} = \frac{\partial \omega_{bc}}{\partial x^a} - \Gamma^p_{ab}\omega_{pc} - \Gamma^p_{ac}\omega_{bp}$$

will give a symplectic connection. The nonuniqueness of the choice now is readily seen: if $S_{ijk} = S_{jik}$ then $\Delta\Gamma_{ijk} + S_{ijk}$ satisfies Equation (11.14) as long as $\Delta\Gamma_{ijk}$ does.

11.3.2 Vey's class S^3_Γ

Now one can give a description of the classes in $H^2(\mathfrak{P}(M), C^\infty(M))$, that correspond to $\mu^*(\tilde{\beta})$ (and also in some degree to the classes in

$H^3(\mathfrak{P}(M), C^\infty(M))$ that correspond to $\mu^*(\tilde\beta) \wedge \alpha)$ in the cohomology of formal Lie algebras that we considered in Section 11.2.2.

Let us denote by $\mathfrak{V}(M)$ the Lie algebra of vector fields on M with respect to the commutator of vector fields as the Lie bracket. Let $X \in \mathfrak{V}(M)$ be a vector field and ∇ a symmetric connection on M; consider the Lie derivation of ∇: it is the map that sends every pair of fields Y, Z on M to the field $\mathcal{L}_V(\nabla)(Y, Z)$, given by

$$\mathcal{L}_X(\nabla)(Y, Z) = [X, \nabla_Y Z] - \nabla_{[X,Y]} Z - \nabla_Y([X, Z]). \quad (11.15)$$

We let the reader as an exercise (Exercise E.4.7) show that this map is in effect bilinear over $C^\infty(M)$ as function of Y, Z.

Now let ∇ be symplectic; consider the map ϑ:

$$(U; A, B, C) \in \mathfrak{S}(M) \times (\mathfrak{V}(M))^{\times 3} \stackrel{\vartheta}{\mapsto} \omega((\mathcal{L}_U \nabla)(A, B), C) \in C^\infty(M).$$

As far as dependence on A, B, C is concerned, ϑ is linear over $C^\infty(M)$ and symmetric: since $\nabla_A B - \nabla_B A = [A, B]$, we have

$$\mathcal{L}_U(\nabla)(A, B) - \mathcal{L}_U(\nabla)(B, A) = [U, [A, B]] - [[U, A], B] - [A, [U, B]] = 0,$$

and the symmetry $\vartheta(U; A, C, B) = \vartheta(U; A, B, C)$ follows from the fact that both $\nabla_A \omega = 0$ and $\mathcal{L}_U \omega = 0$, since $U \in \mathfrak{S}(M)$ and ∇ is symplectic. Freezing the argument U, we may regard $\vartheta(U)$ as a 3-linear symmetric function on TM and equivalently as a section of $S^3 T^* M$ (the space of symmetric 3-linear forms on TM). Then the formula

$$\pi^3(\nu^3, \rho^3) = \pi(\nu, \rho)^3, \quad \nu, \rho \in T^* M, \quad (11.16)$$

(where as before $\pi = \omega^{-1}$, if ω is the symplectic form) determines an anti-symmetric pairing on sections of $S^3 T^* M$ with values in $C^\infty(M)$, so we may put

$$\tilde\beta_\nabla(U, V) = \pi^3(\vartheta(U), \vartheta(V)) \in C^2(\mathfrak{S}(M), C^\infty(M)),$$

and eventually we set

$$\sigma_\Gamma^3 = \mu^*(\tilde\beta_\nabla), \quad (11.17)$$

where $\mu : \mathfrak{P}(M) \to \mathfrak{S}(M)$ is the homomorphism that sends every function $u \in \mathfrak{P}(M)$ to its Hamilton field X_u, similar to the extension (11.8). Then a direct computation shows that σ_Γ^3 *is a cocycle, representing the class* S_Γ^3 (see [Vey75]).

11.3.3 Alternative description of S^3_Γ

Another way to describe a cocycle, representing S^3_Γ (see de Wilde and Lecomte [DL85]) is the following.

Consider again the Lie derivative $\mathcal{L}_X \nabla(Y, Z)$ of a symmetric connection ∇. Fixing X, Y we can regard $\theta^\nabla_X(Y) = \mathcal{L}_X(\nabla)(Y, \cdot)$ as an endomorphism of TM, and θ^∇_X as an element of $\Omega^1(M, \mathrm{End}(TM))$, the space of 1-forms on M with values in endomorphisms of TM. Since endomorphisms of TM is a Lie algebra with respect to the point-wise commutator, one can define the wedge-product of $\mathrm{End}(TM)$-valued differential forms (just like one defines the wedge-product of tensorial forms in the case of principal bundles); e.g. for any $U, V \in \mathfrak{V}(M)$

$$(\theta^\nabla_U \wedge \theta^\nabla_V)(Y, Z) = \frac{1}{2}([\theta^\nabla_U(Y), \theta^\nabla_V(Z)] - [\theta^\nabla_U(Z), \theta^\nabla_V(Y)]).$$

Then $\theta^\nabla_U \wedge \theta^\nabla_V \in \Omega^2(M, \mathrm{End}(TM))$, and we can take point-wise trace of this form, thus obtaining for all U, V the usual 2-form $Tr(\theta^\nabla_U \wedge \theta^\nabla_V) \in \Omega^2(M)$. We define $\tilde{\Phi}_\nabla$ as the skew-symmetrisation of the map $(U, V) \mapsto Tr(\theta^\nabla_U \wedge \theta^\nabla_V)$:

$$\Phi_\nabla(U, V) = SSym(Tr(\theta^\nabla_U \wedge \theta^\nabla_V)) = \frac{1}{2}(Tr(\theta^\nabla_U \wedge \theta^\nabla_V) - Tr(\theta^\nabla_V \wedge \theta^\nabla_U)).$$

Then Φ_∇ is an element in the Chevalley–Eilenberg complex of $\mathfrak{V}(M)$ with coefficients in $\Omega^2(M)$. We claim that

Proposition 11.9. *The cochain Φ_∇ is a closed; thus we obtain a cocycle in $C^2(\mathfrak{V}(M), \Omega^2(M))$.*

The proof of this proposition follows directly from the definitions and the standard property of Lie derivatives:

$$\mathcal{L}_X(\mathcal{L}_Y(\nabla)) - \mathcal{L}_Y(\mathcal{L}_X(\nabla)) = \mathcal{L}_{[X,Y]}(\nabla). \tag{11.18}$$

We leave it as an exercise to the reader.

Now the homomorphism $\mu : \mathfrak{P}(M) \to \mathfrak{S}(M) \subseteq \mathfrak{V}(M)$ allows us to pull back the cocycle Φ_∇ to a cocycle $\mu^*(\Phi_\nabla) \in C^2(\mathfrak{P}(M), \Omega^2(M))$. If we go further then the composition of μ^* with pairing of differential 2-forms with bivector $\pi = \omega^{-1}$ gives a cochain

map μ' from $C^{\cdot}(\mathfrak{V}(M), \Omega^2(M))$ to $C^{\cdot}(\mathfrak{P}(M), C^\infty(M))$, given by the formula
$$\mu'(\varphi) = \langle \mu^*(\varphi), \pi \rangle. \tag{11.19}$$

We put
$$\tilde{\sigma}_\Gamma^3(U, V) = \mu'(\Phi_\nabla). \tag{11.20}$$

Then *if the connection ∇ is symplectic, the cocycle $\tilde{\sigma}_\Gamma^3 = \mu'(\Phi_\nabla)$ coincides with σ_Γ^3 above (Section 11.3.2), which represents the class S_Γ^3.*

Remark 11.10. Using either of the constructions in Section 11.3.2 or 11.3.3, we can obtain the local expression for the cochain σ_Γ^3, representing the Vey class S_Γ^3. Fix the coordinate chart $(x^1, \ldots, x^n)$ on M and let Γ_{ij}^k be the Christoffel symbols of a symplectic connection ∇ and ω_{ij}, $\pi^{ij} = (\omega^{-1})^{ij}$ be the coefficients of symplectic form and its inverse in these coordinates; then

$$\sigma_\Gamma^3(u, v) = \pi^{i_1 j_1} \pi^{i_2 j_2} \pi^{i_3 j_3} \mathcal{L}_{X_u}\left(\omega_{i_1 k} \Gamma_{i_2 i_3}^k\right) \mathcal{L}_{X_v}\left(\omega_{j_1 l} \Gamma_{j_2 j_3}^l\right), \tag{11.21}$$

where X_u, X_v are the Hamilton fields associated with $u, v \in \mathfrak{P}(M)$.

Remark 11.11. The classes, that correspond to $\mu^*(\tilde{\beta}) \wedge \alpha$ can now be described as follows: let $X \in \mathfrak{S}(M)$, consider the 1-form $\alpha_X = i_X \omega \in \Omega^1(M)$ (i.e. i_X is the convolution of differential forms with the vector field X). Then $\mu^*(\alpha_X)(u) = \alpha_X(X_u)$, $u \in \mathfrak{P}(M)$ is a 1-cocycle on $\mathfrak{P}(M)$ and if $X = X_v$ is Hamiltonian, then this cocycle is in effect a coboundary (we leave to the reader as an easy but tedious exercise checking this claim). Then speaking a bit loosely, one may say that the class $\mu^*(\tilde{\beta}) \wedge \alpha$ in $H^3(\mathfrak{P}, \widehat{S}_{2n})$ gives rise to the classes $S_\Gamma^3 \wedge \alpha_{[X]}$, where for $X \in \mathfrak{S}(M)$ we denote by $[X]$ its class in Lichnerowicz-Poisson cohomology $H_\pi^1(M) \cong H_{dR}^1(M)$ (we used the identification of Proposition 6.3).

12 Lecomte and de Wilde's theorem: Quantisation of symplectic manifolds

Historically the first construction of the deformation series for symplectic manifolds was due to Marc De Wilde and Pierre Lecomte [DL83$_2$]; they proved

Theorem 12.1. *For every symplectic manifold M, there exist associative $\star$-products on $C^\infty(M)[[\hbar]]$, given by deformation series*

$$f \star g = fg + \frac{1}{2}\hbar\{f,g\} + \sum_{k \geq 2} \hbar^k B_k(f,g). \qquad (12.1)$$

These $\star$-products are associative and the constant functions $\mathbb{R}[[\hbar]]$ are in the center of the corresponding algebra; coefficients of these series satisfy the condition $B_k(f,g) = (-1)^k B_k(g,f)$, but in general 1 is not a unit of the product. The $\star$-products for which the function 1 is not unit are sometimes called weak $\star$-products.[33]

In this lecture we will sketch the original proof of this theorem, although we will skip many details: the evident reason for this is that the constructions that De Wilde and Lecomte used in their proof were very technical, exposing them in full detail will take very much efforts and time, and the second reason is that shortly after this result was published, Fedosov in [Fed86] (published in Russian) suggested a much simpler geometric construction, which we shall explain in the following sections.

Let us recall that in previous section we connected the deformation quantisation problem with the similar deformation problem for Poisson structures: find a formal power series (in λ)

$$\{f,g\}_\lambda = \{f,g\} + \sum_k \lambda^k C_k(f,g),$$

---

[33]With a little more work one can prove that the $\star$-product can always be chosen so that 1 will be the unit, although we shall not give details of this statement here. The second condition (concerning the symmetricity of coefficients) is important for the proof, but not for the conclusion.

of skew-symmetric bidifferential operators $C_k(f,g) = -C_k(g,f)$ of strictly positive bidegrees, such that $\{,\}_\lambda$ satisfies the Jacobi identity. We further reduced this problem to the study of obstructions that belong to the third differentiable cohomology group of the Poisson Lie algebra $\mathfrak{P}(M)$ (i.e. of the algebra of smooth functions on M regarded as a Lie algebra with respect to the Poisson bracket $\{,\}$; the term "differentiable" signifies that all cochains in the Chevalley complex should be given by polydifferential operators on M) and sketched a way that allows one compute the cohomology groups $H^p(\mathfrak{P}(M), C^\infty(M))$, $p = 2, 3$; in particular we constructed classes $S^3_\Gamma \in H^2(\mathfrak{P}(M), C^\infty(M))$ and $S^3_\Gamma \wedge \alpha_{[X]} \in H^3(\mathfrak{P}(M), C^\infty(M))$, $[X] \in H^1_\pi(M)$.

These observations and further analysis of the spectral sequences (11.11) allowed Vey to show that $H^3(\mathfrak{P}(M), C^\infty(M)) = 0$, if $H^1_{dR}(M) = H^3_{dR}(M) = 0$ and $p_1(M) \neq 0$, where $p_1(M)$ is the first *Pontryagin class of* TM (we shall describe its construction below, see Section 12.1.2), hence in this case all obstructions for the deformation of $\{,\}$ vanish identically. With a bit of further efforts (partly based on the fact that locally we always have the deformation, given by Moyal formula in Darboux's coordinates) Vey proved that there always exists a nontrivial deformation of the Poisson bracket on M, provided $H^3_{dR}(M) = 0$.[34]

In their original proof of Theorem 12.1 Lecomte and de Wilde improved the methods of Vey and showed that the deformation of Poisson structure exists for all symplectic manifolds; then they used this result to construct the deformation series (12.1) for the $\star$-product. We are going to follow the steps of their proof, so we begin with the description of the Lie cohomology of the algebra $\mathfrak{P}(M)$ in dimensions $2, 3$.

Warning! In this lecture we will deal both with Hochschild cohomology complex and Chevalley–Eilenberg complex on the algebra of functions $C^\infty(M)$ where we view it as the Lie algebra with respect to the Poisson bracket (Poisson Lie algebra) in the second case.

---

[34]Observe that in this case Proposition 9.6 is applicable, so there's no obstructions for the original deformation problem in this case either; this observation is even simpler than Vey's result.

In order to distinguish between these two complexes, we will use the following notation: the symbol δ will be reserved for the Chevalley complex differential, and $\bar\delta$ will denote Hochschild coboundary.

12.1 Cohomology of Poisson Lie algebra $\mathfrak{P}(M)$ in low dimensions

We begin by constructing cohomology classes of $\mathfrak{P}(M)$; it turns out that these classes are in some sense the only nontrivial elements in the cohomology groups of this Poisson Lie algebra in dimensions 2 and 3.

12.1.1 Differential forms and differentiable cohomology $H^*(\mathfrak{P}(M), C^\infty(M))$

First of all we are going to construct more elements in these cohomology groups. Let $\mathfrak{V}(M)$ be the Lie algebra of vector fields on M; for any $\varphi \in \Omega^p(M)$ we may regard φ as a skew-symmetric multi-linear map $\varphi : \mathfrak{V}(M)^{\times p} \to C^\infty(M)$. Then it follows from *Cartan's formula*

$$d\varphi(X_0,\ldots,X_p) = \sum_{k=0}^{p}(-1)^k X_p(\varphi(X_0,\ldots,\widehat{X_k},\ldots,X_p))$$
$$+ \sum_{0\leq i<j\leq p}(-1)^{i+j}\varphi([X_i,X_j],X_0,\ldots$$
$$\ldots,\widehat{X_i},\ldots,\widehat{X_j},\ldots,X_p),$$

that de Rham differential for the differential forms turns into the Chevalley differential in $C^\cdot(\mathfrak{V}(M), C^\infty(M))$, if we regard φ as a Chevalley cochain. The homomorphism

$$\mu : \mathfrak{P}(M) \to \mathfrak{S}(M) \subseteq \mathfrak{V}(M), u \mapsto X_u$$

now induces the cochain morphism $\mu^* : C^\cdot(\mathfrak{V}(M), C^\infty(M)) \to C^p(\mathfrak{P}(M), C^\infty(M))$; it follows immediately that for any $\varphi \in \Omega^*(M)$ we have $\delta\mu^*(\varphi) = \mu^*(d\varphi)$. In this way we obtain a map that sends the de Rham cohomology of M into $H^*(\mathfrak{P}(M), C^\infty(M))$; in abstract terms this map is induced from the second spectral sequence in (11.11) (see Exercise E.4.5). We shall denote by μ^* both the morphism from $\Omega^*(M)$ to $C^{\cdot}(\mathfrak{P}(M), C^\infty(M))$ and the morphism $C^{\cdot}(\mathfrak{V}(M), C^\infty(M)) \to C^{\cdot}(\mathfrak{P}(M), C^\infty(M))$, induced by homomorphism μ.

12.1.2 Chern–Weil forms and cohomology of vector fields on manifolds

It turns out that under certain topological conditions one can add one more nontrivial cocycle in dimension 3. To this end, let us consider a symmetric connection ∇ on M. As one knows, the square of the operator ∇ can be regarded as a differential 2-form on M with values in endomorphisms of the vector fields on M, called *curvature of* ∇: the value of this form on the fields X, Y takes a third field Z on M to $R(X, Y; Z)$, given by the formula

$$R(X,Y;Z) = \nabla_X \nabla_Y Z - \nabla_Y \nabla_X Z - \nabla_{[X,Y]} Z. \qquad (12.2)$$

In local coordinates the map ∇ is determined by Christoffel symbols Γ_{ij}^k, and one can find an expression for R in terms of these symbols as well. In shorthand notation, let $A_j^k = \Gamma_{ij}^k dx^i$ be the locally defined differential 1-forms. We can regard $A = (A_j^k)$ as the local 1-form on M with values in the endomorphism bundle of TM; A is called *local gauge potential of connection* ∇. Then $R(X,Y;Z) = F(X,Y)(Z)$ where $F \in \Omega^2(M, \mathrm{End}(TM))$ is the endomorphism-valued 2-form on M, given by:

$$F = dA + A \wedge A. \qquad (12.2')$$

The form F is globally defined; it is called *the curvature form of* ∇. It follows that F satisfies the following Bianchi identity

$$dF = F \wedge A - A \wedge F. \qquad (12.3)$$

The concept of curvature plays an important role in modern Mathematics and is used in various situations. Here we recall the classical result due to Chern and Weil (see for instance [KN63]): it follows from the Bianchi identity (12.3) that the even differential forms

$$ch_{2n} = Tr(\underbrace{F \wedge \cdots \wedge F}_{n \text{ times}})$$

are closed. One further can show that the classes in de Rham cohomology, determined by these forms are independent on the choice of ∇. The classes corresponding to odd n turn out to be trivial on oriented manifolds, and the remaining classes are expressible in terms of the Pontryagin classes of TM; in particular $ch_4 = Tr(F \wedge F)$ up to some constant factor represents the first Pontryagin class $p_1(TM)$ (one can regard the construction we discuss here as a definition of this Pontryagin class).

Based on the formula (12.2′) one can find a canonical way to choose a local differential form (i.e. a system of forms defined on coordinate charts of our manifold and depending in a nontrivial way on these charts) cs_3 such that $d(cs_3) = ch_4$ on a fixed coordinate chart:

$$cs_3 = Tr\left(F \wedge A - \frac{1}{3} A \wedge A \wedge A\right).$$

Indeed, it follows from (12.2′), that $dA = F - A \wedge A$, and hence

$$dF = dA \wedge A - A \wedge dA = F \wedge A - A \wedge F.$$

So, we compute

$$d(cs_3) = Tr(F \wedge A \wedge A - A \wedge F \wedge A + F \wedge F - F \wedge A \wedge A)$$
$$- \frac{1}{3} Tr(F \wedge A \wedge A - A \wedge F \wedge A + A \wedge A \wedge F) = Tr(F \wedge F),$$

since $Tr(A \wedge B) = (-1)^{pq} Tr(B \wedge A)$ for all matrix-valued p-form A and q-form B. However, this expression is only local and cannot be "glued" into a globally defined differential 3-form on M, so in general $p_1(TM) \neq 0$ (in effect, one can show that cs_3 determines a well-defined form on the principal frame bundle $P(M)$ of TM but not on M in the general situation).

Let us now take a different point of view on the covariant derivative ∇: freezing the variable Y in expression $\nabla_X Y$ we get an endomorphism of tangent bundle $A^\nabla(Y)(\cdot) = \nabla.Y$. Let $cs_1^\nabla(Y) = Tr(A^\nabla(Y)) \in C^1(\mathfrak{V}(M), C^\infty(M))$, then

$$\delta(cs_1^\nabla)(X,Y) = Tr(\nabla_X(A^\nabla(Y))) - Tr(\nabla_Y(A^\nabla(X))) - Tr(A^\nabla([X,Y]))$$
$$= Tr(\nabla_X \nabla.Y - \nabla_{\nabla_X(\cdot)} Y) - Tr(\nabla_Y \nabla.X - \nabla_{\nabla_Y(\cdot)} X) - Tr(\nabla.([X,Y]))$$
$$= Tr(\nabla_X \nabla.Y - \nabla_{[X,\cdot] - \nabla.X} Y) - Tr(\nabla_Y \nabla.X - \nabla_{[Y,\cdot] - \nabla.Y} X)$$
$$\qquad - Tr(\nabla.\nabla_X Y - \nabla.\nabla_Y X)$$
$$= Tr(\nabla_X \nabla.Y - \nabla.\nabla_X Y - \nabla_{[X,\cdot]} Y) - Tr(\nabla_Y \nabla.X - \nabla.\nabla_Y X - \nabla_{[Y,\cdot]} X)$$
$$\qquad - Tr(\nabla_{\nabla.X} Y - \nabla_{\nabla.Y} X)$$
$$= Tr(R(X,\cdot;Y) - R(Y,\cdot;X)) = Tr(F)(X,Y),$$

here we used the symmetricity of the connection, the identity $\mathcal{L}_X(Tr(A)) = Tr(\nabla_X(A))$ for all $End(TM)$-valued fields A on M, the equality

$$Tr(A^\nabla(X) \circ A^\nabla(Y) - A^\nabla(Y) \circ A^\nabla(X)) = 0$$

and the Bianchi identity for R:

$$R(X,Y;Z) + R(Y,Z;X) + R(Z,X;Y) = 0.$$

In other words we showed that the form ch_2 represents a closed cochain in $C^2(\mathfrak{V}(M), C^\infty(M))$, although the element cs_1^∇ is not induced from a differential form.

Totally similar construction now can be applied to ch_2: consider the 3-cochain $cs_3^\nabla \in C^3(\mathfrak{V}(M), C^\infty(M))$,

$$cs_3^\nabla(X,Y,Z)$$
$$= SSym\left(Tr\left(F(X,Y) \circ A^\nabla(Z) - \frac{1}{3} A^\nabla(X) \circ A^\nabla(Y) \circ A^\nabla(Z)\right)\right),$$
(12.4)

then computations similar to the previous ones (we leave them as an exercise to the reader) show that

$$\delta cs_3^\nabla = Tr(F \wedge F)$$

in $C^4(\mathfrak{V}(M), C^\infty(M))$. If $p_1(TM) = 0$, then one can find a differential 3-form α, such that $d\alpha = Tr(F \wedge F)$. Then in $C^{\cdot}(\mathfrak{V}(M), C^\infty(M))$

we have $\delta(cs_3^\nabla - \alpha) = 0$. Eventually we put $T_\theta = cs_3^\nabla - \alpha \in C^3(\mathfrak{V}(M), C^\infty(M))$, then $\delta T_\theta = 0$, so $\mu^*(T_\theta) \in C^3(\mathfrak{P}(M), C^\infty(M))$ is a differentiable 3-cocycle, moreover, the cohomology class of T_θ does not depend on the choice of ∇.[35] Using spectral sequences now it is possible to show that this cocycle is not closed.

12.1.3 De Wilde, Gutt and Lecomte's theorem

It turns out that if we add to the Vey class and other classes that we constructed in the previous section, the classes described in Sections 12.1.1 and 12.1.2, we will exhaust the differentiable cohomology groups $H^2(\mathfrak{P}(M), C^\infty(M))$, $H^3(\mathfrak{P}(M), C^\infty(M))$ in all situations, for all symplectic manifolds. More accurately, one can prove the following theorem

Theorem 12.2. *Let M be a symplectic manifold, and let $\mathfrak{P}(M)$ denote the Lie algebra of smooth functions on M with respect to the Poisson bracket. Then the following is true.*

(i) Let C_2 be a differentiable 2-cocycle of $\mathfrak{P}(M)$ with values in $C^\infty(M)$, then

$$C_2 = rS_\Gamma^3 + \mu^*(\varphi) + \delta E, \tag{12.5}$$

where $r \in R$ is a real number and $\varphi \in \Omega^2(M)$ is a closed form; it follows that the differentiable cohomology of $\mathfrak{P}(M)$ in degree 2 is equal to $H^2(\mathfrak{P}(M), C^\infty(M)) = \mathbb{R} \oplus H_{dR}^2(M)$.

(ii) Let C_3 be a differentiable 3-cocycle of $\mathfrak{P}(M)$ with values in $C^\infty(M)$; suppose that $p_1(TM) = 0$, then

$$C_3 = S_\Gamma^3 \wedge \alpha_{[X]} + \mu^*(rT_\theta + \psi) + \delta E, \tag{12.6}$$

where $r \in R$ is a real number, $[X]$ is a class in $H_\pi^1(M) = H_{dR}^1(M)$ and $\psi \in \Omega^3(M)$ is a closed form; it follows that the differentiable cohomology of $\mathfrak{P}(M)$ in degree 3 is equal to $H^3(\mathfrak{P}(M), C^\infty(M)) = \mathbb{R} \oplus H_{dR}^1(M) \oplus H_{dR}^3(M)$. If $p_1(TM) \neq 0$, then the same statement holds with $r = 0$, so that $H^3(\mathfrak{P}(M), C^\infty(M)) = H_{dR}^1(M) \oplus H_{dR}^3(M)$ in that case.

---

[35]This class depends on the choice of α, up to the addition of a class in de Rham cohomology; however, the choice of α does not change the final result, i.e. Theorem 12.2.

12.2 Deformations of the Poisson bracket

12.2.1 Exact symplectic structures and Lie derivatives

Let us first assume that the symplectic form ω is exact, i.e. that there exists a 1-form α on M such that $d\alpha = \omega$. This condition is equivalent to the following: *there exists a vector field ξ on M, for which $\mathcal{L}_\xi \omega = \omega$* (we leave the proof of this statement as an exercise to the reader).

In this case Lie derivative with respect to ξ can be applied to the geometric objects involved in our discussion, including the elements of the Chevalley complex $C^{\cdot}(\mathfrak{P}(M), C^\infty(M))$. For instance, for any two functions $f, g \in \mathfrak{P}(M)$, we will get

$$\begin{aligned}\xi(\{f,g\}) &= \xi(\pi(df, dg)) \\ &= \mathcal{L}_\xi(\pi)(df, dg) + \pi(\mathcal{L}_\xi(df), dg) + \pi(df, \mathcal{L}_\xi(dg)) \\ &= \{f,g\}_\xi + \{\xi(f), g\} + \{f, \xi(g)\},\end{aligned}$$

where $\{f,g\}_\xi = \mathcal{L}_\xi(\pi)(df, dg)$ and $\pi = \omega^{-1}$, so $\mathcal{L}_\xi(\pi) = -\pi$. Now if $u \in C^\infty(M)$, and X_u is the Hamiltonian field associated with u, then for any $f \in C^\infty(M)$

$$\begin{aligned}\mathcal{L}_\xi(X_u)(f) &= \xi(X_u(f)) - X_u(\xi(f)) = \xi(\{f,u\}) - \{\xi(f), u\} \\ &= \{f, \xi(u)\} - \{f, u\} = X_{\xi(u)}(f) - X_u(f),\end{aligned}$$

and so

$$\mathcal{L}_\xi(X_u) = X_{\xi(u)} - X_u. \tag{12.7}$$

Using analogous methods, one proves the following identities, satisfied by the operation $\mathcal{L}_\xi$ on $C^{\cdot}(\mathfrak{P}(M), C^\infty(M))$:

$$\mathcal{L}_\xi \circ \delta = \delta \circ \mathcal{L}_\xi - \delta, \tag{12.8}$$

$$\mathcal{L}_\xi \circ \mu^* = \mu^* \circ \mathcal{L}_\xi - N \circ \mu^*, \tag{12.9}$$

$$\mathcal{L}_\xi \circ \mu' = \mu' \circ \mathcal{L}_\xi - (N+1) \circ \mu'. \tag{12.10}$$

Here μ^* is the cochain map
$$\mu^* : C^{\cdot}(\mathfrak{V}(M), C^{\infty}(M)) \to C^{\cdot}(\mathfrak{P}(M), C^{\infty}(M)),$$
induced by the Lie algebra morphism $\mu : u \mapsto X_u$, μ' is the cochain map
$$\mu' : C^{\cdot}(\mathfrak{V}(M), \Omega^2(M)) \to C^{\cdot}(\mathfrak{P}(M), C^{\infty}(M)),$$
given by the pairing of $\mu^*(\varphi)$ with the bivector $\pi = \omega^{-1}$, see formula (11.19) and N is the "degree" operator on cochain complex, i.e. $N(\varphi) = p\varphi$ if φ is a p-cochain. We leave the proof of these equalities as an exercise to the reader.

12.2.2 Cohomological properties of Lie derivatives

It follows from the formulas (12.7)–(12.10) and Theorem 12.2 that *the operators $\mathcal{L}_\xi + k$ induce self-isomorphisms on $H^2(\mathfrak{P}(M), C^{\infty}(M))$ if $k \ne 2, 3$ and self-isomorphisms of $H^3(\mathfrak{P}(M), C^{\infty}(M))$, when $k \ne 3, 4$* (this follows from the fact that the elements listed in Theorem 12.2 are mapped by $\mathcal{L}_\xi$ to something clearly understandable from the formulas (12.9), (12.10), see Exercise E.4.9).

Using this observation we conclude that the following proposition holds:

Proposition 12.3. *Let M be an exact symplectic manifold and ξ the field, satisfying the equation $\mathcal{L}_\xi \omega = \omega$; let $A \in C^3(\mathfrak{P}(M), C^{\infty}(M))$ be a differentiable cocycle and $B \in C^2(\mathfrak{P}(M), C^{\infty}(M))$ be a differentiable cochain such that*

$$(\mathcal{L}_\xi + k)A = \delta B, \quad k \ne 3, 4. \tag{12.11}$$

Then there exists $C \in C^2(\mathfrak{P}(M), C^{\infty}(M))$ such that $A = \delta C$ and the cocycle $(\mathcal{L}_\xi + k - 1)C - B$ is exact. Moreover, this cochain C is uniquely determined up to a coboundary.

Proof. It follows from equality (12.11) that the cocycle $(\mathcal{L}_\xi + k)A$ represents trivial cohomology class, and hence it follows from the observation we made above that so does A. Let C' be such that $A = \delta C'$, then

$$\delta((\mathcal{L}_\xi + k - 1)C' - B) = (\mathcal{L}_\xi + k)\delta C' - \delta B = 0,$$

so $B' = (\mathcal{L}_\xi + k - 1)C' - B$ is a 2-cocycle. But the map $\mathcal{L}_\xi + k - 1$ is an isomorphism in cohomology $H^2(\mathfrak{P}(M), C^\infty(M))$, so there exists a cocycle C'' such that $(\mathcal{L}_\xi + k - 1)C'' = B' + \delta E$, hence, taking $C = C' - C''$ we have $\delta C = A$, $(\mathcal{L}_\xi + k - 1)C - B = \delta E$. If $\tilde{C}$ is another cochain with this property, then $C - \tilde{C}$ is a cocycle and $(\mathcal{L}_\xi + k - 1)(C - \tilde{C})$ is exact, so by the same observation, $C - \tilde{C} = \delta \tilde{E}$. □

Remark 12.4. In effect, Lecomte and de Wilde showed that one can choose *a homotopy retraction* of $\mu^* : C^2(\mathfrak{V}(M), C^\infty(M)) \to C^2(\mathfrak{P}(M), C^\infty(M))$, i.e. a map

$$\tau : C^2(\mathfrak{P}(M), C^\infty(M)) \to C^2(\mathfrak{V}(M), C^\infty(M)),$$

for which

$$\mu^* \circ \tau = \mathbb{1}, \text{ and } \tau \circ \mu^* \equiv \mathbb{1} \mod \delta C^1(\mathfrak{V}(M), C^\infty(M)).$$

This choice facilitates the construction a little.

12.2.3 Deformation of the Poisson algebras for exact symplectic forms

As one knows, locally every closed differential form is in effect exact, so we can obtain the existence of local $\star$-products on symplectic manifolds from the properties of exact symplectic structures. To this end, in the lines of the construction of [DL85], let us first prove the existence and uniqueness of Poisson algebra deformations for exact symplectic structures. To this end, we consider the map $\Theta : C^\infty(M)[[\lambda]] \to C^\infty(M)[[\lambda]]$, given by the formula

$$\Theta\left(\sum_{k=0}^\infty \lambda^k f_k\right) = \sum_{k=0}^\infty (2k - 1)\lambda^k f_k. \qquad (12.12)$$

Put

$$\mathcal{D} = \sum_{k=0}^\infty \lambda^k D_k + \Theta,$$

where D_k are differential operators on M; it is an operator on the space $C^\infty(M)[[\lambda]]$. Following Lecomte and de Wilde we will say, that this operator $\mathcal{D}$ is a *derivation of type* Θ of a formal deformation $\{,\}_\lambda$, if for any $f, g \in C^\infty(M)[[\lambda]]$ we have

$$\mathcal{D}(\{f,g\}_\lambda) = \{\mathcal{D}(f),g\}_\lambda + \{f,\mathcal{D}(g)\}_\lambda.$$

The following statement was proved by Lecomte and de Wilde (we collect several their propositions into one here):

Theorem 12.5.

(i) *If the formal deformation $\{,\}_\lambda$ of a Poisson structure on a symplectic manifold (M,ω) has a derivation $\mathcal{D}$ of type Θ, then ω is exact, $D_0 = \mathcal{L}_\xi$, where ξ is the vector field, for which $\mathcal{L}_\xi \omega = \omega$ and the first cochain C_1 in the deformation series is of type $rS_\Gamma^3 + \delta E$ for some $E \in C^1(\mathfrak{P}(M), C^\infty(M))$.*

(ii) *Let (M,ω) be an exact symplectic manifold, and ξ be a field for which $\mathcal{L}_\xi \omega = \omega$. Then for each cocycle $C_1 \in C^2(\mathfrak{P}(M), C^\infty(M))$ of the form $C_1 = rS_\Gamma^3 + \delta E$ for some $E \in C^1(\mathfrak{P}(M), C^\infty(M))$, there exists a formal deformation $\{,\}_\lambda = \{,\} + \lambda C_1 + \sum_{k>1} \lambda^k C_k$ of the Poisson structure on M which admits at least one derivation $\mathcal{D}$ of type Θ.*

(iii) *Two formal deformations $\{,\}_\lambda$ and $\{,\}'_\lambda$, which admit derivations $\mathcal{D}, \mathcal{D}'$ of type Θ are equivalent[36] iff the classes of C_1 and C'_1 in $H^2(\mathfrak{P}(M), C^\infty(M))$ coincide.*

The *proof* of part (i) is trivial and the proof of part (ii) is based on the repetitive use of Proposition 12.3. Namely, we construct simultaneously $\{,\}_\lambda$ and the derivation $\mathcal{D}$: using the assumption that they both are known up to degree p we conclude that Equation (12.11) holds for the obstruction class with $k = 2p + 4$. It follows at once that the solutions C_{p+1} and D_{p+1} of the obstruction problems (see for instance Equation (11.5)) exists; we skip the details and leave filling them in as an exercise to the reader.

---

[36] Recall that this means that there exists a formal power series of differential operators beginning with identity, such that conjugation by it gives isomorphism between $\{,\}_\lambda, \mathcal{D}$ and $\{,\}'_\lambda, \mathcal{D}'$.

Part (iii) of the theorem is also proved by induction in very much the same way. However, we need to make one important observation here. Let us first consider two vector fields ξ, ξ' for which $\mathcal{L}_\xi \omega = \mathcal{L}_{\xi'}\omega = \omega$: we assume that ξ is the field used to define $\mathcal{D}$ and ξ' plays the same role for $\mathcal{D}'$ (see part (i)). Then the difference $\eta_0 = \xi' - \xi$ of these fields is an infinitesimal symplectomorphism or symplectic field, i.e. $\mathcal{L}_{\eta_0}\omega = 0$. On the other hand on symplectic manifolds any symplectic field is locally equal to Hamilton field of a function f_{η_0}. Such functions are uniquely determined up to constants and hence the map $\mathcal{D}_\eta(g) = \{f_{\eta_0}, g\}_\lambda$ is a well-defined global derivation of the deformed Poisson structure $\{,\}_\lambda$. Conjugating $\{,\}_\lambda$ and $\mathcal{D}$ with $\exp(\mathcal{D}_{\eta_0})$ we conclude that from now on we can assume that $\xi = \xi'$.

The same observations will appear at further stages of the construction: using Proposition 12.3, we can assume that the difference $\beta = (D'_k - D_k) - (\mathcal{L}_\xi + 2k)E$ at stage k (where E is a cochain $E \in C^1(\mathfrak{P}(M), C^\infty(M))$ such that $\delta E = C'_k - C_k$), is closed. However 1-cocycles in $C^1(\mathfrak{P}(M), C^\infty(M))$ are easily seen to come from symplectic fields on M (see for instance (11.9) and the spectral sequences (11.11)). So we again have a symplectic field η_k and the differentiation $\mathcal{D}_{\eta_k}$, so that the conjugating by $\exp(\mathcal{D}_{\eta_k})$ we "kill" the difference β. As before, we leave the details to the reader. $\square$

Remark 12.6. The use of contracting homotopy τ from Remark 12.4 allows one to uniformly find the solutions of the obstruction problems we considered.

12.2.4 Deformation of Poisson structures on general symplectic manifolds

Let us finally sketch the way in which one can obtain the deformation of Poisson structure in the case of an arbitrary (not necessarily exact) symplectic manifold. To this end, we recall again that *according to the Poincaré lemma, every closed form on a manifold is locally exact*. In particular, the symplectic form ω is locally exact, and hence we can apply the theory from the previous subsection and construct deformation of the Poisson bracket together with a type Θ derivation thereof.[37] It now remains to describe the "glueing" procedure that pastes together these deformations on different charts; some details can be found in [GR99] and [Gu11].

So we suppose that on every (small) open set we have chosen a deformation of the Poisson structure, as we have explained earlier. Then on intersections of open charts, where ω is exact we will have two deformations $\{,\}_\lambda$, $\{,\}'_\lambda$ of the Poisson structure on the sets U and U' respectively, together with their derivations $\mathcal{D}$, $\mathcal{D}'$ of type Θ; these derivations (and hence the deformations $\{,\}_\lambda$, $\{,\}'_\lambda$) depend on the choice of the vector fields ξ, ξ' such that $\mathcal{L}_\xi \omega = \omega = \mathcal{L}_{\xi'} \omega$ on U and U', respectively. According to part (iii) of Theorem 12.5, these deformations are equivalent iff the classes of C_1 and C'_1 coincide; on the other hand we know that the deformations exist iff $C_1 = rS_\Gamma^3 + \delta E$, $C'_1 = r'S_\Gamma^3 + \delta E'$ (here S_Γ^3 denotes the restriction of the Vey class S_Γ^3 on M to the corresponding open subset; clearly it coincides with the Vey class defined for the subset itself), hence for the equivalence of our deformations it is necessary and sufficient that $r = r'$.

Let us assume that the coefficient r at the Vey class in the formula $C_1 = rS_\Gamma^3 + \delta E$ is given once and for all, and let us now fix the equivalence isomorphism $T_{UU'}$ of $\{,\}_\lambda$ and $\{,\}'_\lambda$ on the intersection $U \cap U'$. As we saw from the discussion in the end of the (sketched) proof of Theorem 12.5, (see also Remark 12.6) this map depends on the choices of functions u_k, $k = 0, 1, 2, \ldots$ such that $X_{u_k} = \eta_k$, where $\eta_0 = \xi' - \xi$; all the other η_k are the symplectic vector fields, which appear in the same way in the process of constructing the equivalence isomorphism. On triple intersections $U \cap U' \cap U''$ of open charts, on which ω is exact, we shall have three maps: $T_{UU'}$, $T_{U'U''}$ and $T_{UU''}$. Since the choice of u_k is unique up to the addition of constants, which does not change the morphism, we have the cocycle relation $T_{UU''} = T_{UU'}T_{U'U''}$. Hence, using a partition of unity and the contraction map τ (see Remark 12.6), we can globalise the deformation of Poisson bracket on M without a problem; some details can be found in the

---

[37]In effect, locally due to the Darboux theorem and Moyal product formula one always has the Moyal $\star$-products and hence the deformations of Poisson structures, but these constructions are phrased in terms of the Darboux coordinates, which makes controlling their behaviour on the intersections of the charts rather complicated.

paper [GR99]. Observe, that this does not work for the derivations of type Θ: contracting homotopy τ cannot be applied there, so we cannot make sure that the solutions to obstruction problems would match on the intersections.

Remark 12.7. When the coefficient r is fixed, one can prove that the process by which we construct the isomorphisms $T_{UU'}$ produces a series of cohomology classes in $H^2(M, \mathbb{R}) = H^2_{dR}(M)$. Namely, when we compare the functions u_k, u'_k, u''_k, that determine the operators $T_{UU'}$, $T_{U'U''}$ and $T_{UU''}$, respectively, we see that $f_k(UU'U'') = u_k - u''_k + u'_k$ is a constant on the triple intersection $U \cap U' \cap U''$, so the collection $f_k = \{f_k(UU'U'')\}$ is a cochain in the Čech complex[38] of M, corresponding to the cover of M by open sets, on which ω is exact. It is easy to see that the Čech coboundary of f_k vanishes, so it produces a class in the second Čech cohomology group of M with real coefficients (by the well-known de Rham theorem, proved, for instance, via the spectral sequence arguments, it is isomorphic to de Rham cohomology).

Let $[f_0]$, $[f_1]$, $\cdots \in H^2(M, \mathbb{R})$ be the cohomology classes, corresponding to Čech cocycles f_k. It turns out that they determine the class of the deformation $\{,\}_\lambda$ in a unique way up to an equivalence: let $\widetilde{\{,\}}_\lambda$ be another deformation, corresponding to the same coefficient r, let $[\tilde{f}_k] \in H^2(M, \mathbb{R})$ be the corresponding classes in Čech cohomology. Assuming without loss of generality that both deformations are defined for the same covering of M by open sets, where ω is exact, we see that the deformations $\{,\}_\lambda$ and $\widetilde{\{,\}}_\lambda$ are equivalent iff we can find local formal equivalence isomorphisms S_U, generated by a local collection of functions v_k on U (this is similar to the method by which isomorphisms $T_{UU'}$ are generated by the functions u_k). Then on the intersections $U \cap U'$ we have $S_{U'}^{-1} \widetilde{T}_{UU'} S_U = T_{UU'}$. This means that $\tilde{u}_k - u_k = v_k - v'_k$ on this intersection. But $c_k = v_k - v'_k$ is a locally constant function on $U \cap U'$, so on triple intersection we shall have

$$\tilde{f}_k - f_k = \delta c_k.$$

---

[38] A reader unfamiliar with Čech cohomology construction, can skip the remaining part of this section: the classes $[f_k]$ correspond to the classes that will naturally appear in Fedosov's construction. An introduction into Čech cohomology construction, including the de Rham theorem can be found in many textbooks.

In other words *two deformations of the Poisson structure on M for the same coefficient r are equivalent iff the classes $[f_k] = [\tilde{f}_k] \in H^2(M, \mathbb{R})$ for all k*; thus we can identify the space of deformations of the Poisson structure on M with the space $\mathbb{R} \cdot S_\Gamma^3 \oplus H_{dR}^2(M)[[\lambda]]$.

12.3 Deformations of the Poisson structure and ⋆-products: Proof of Theorem 12.1

We saw in Section 11.1 that every ⋆-product gives rise to a deformation of the Poisson structure. Now that we know that for any symplectic manifold there always exist deformations of the Poisson structure, we can try to reverse this observation and restore the ⋆-product from a deformation $\{,\}_\lambda$.

To this end we consider a formal deformation of the Poisson bracket $\{,\}$,

$$\{f, g\}_\lambda = \{f, g\} + \sum_{k=1}^\infty \lambda^k C_k(f, g),$$

where C_k are skew-symmetric and operators with positive bidegrees and $\{,\}_\lambda$ satisfies the Jacobi identity. We are looking for the ⋆-product (12.1) that would satisfy the following equation (up to an equivalence relation):

$$\{f, g\}_{\hbar^2} = \frac{1}{\hbar}(f \star g - g \star f).$$

We shall assume that the coefficients B_k of the ⋆-product are known up to degree $2n-1$, and that they satisfy the condition

$$B_{2k}(f, g) = B_{2k}(g, f),\ B_{2k+1}(f, g) = -B_{2k+1}(g, f),\ k \leq n-1.$$

We also assume that all these operators have strictly positive bidegrees and $C_k = 2B_{2k+1}$ for this set of indices. Then we are going to find B_{2n} such that

$$f \star_{2n+1} g = fg + \frac{\hbar}{2}\{f, g\} + \sum_{k=1}^{2n} \hbar^k B_k(f, g) + \frac{\hbar^{2n+1}}{2} C_n(f, g)$$

is a ⋆-product, associative up to degree $2n+1$. At the first step we see that the choice of B_2 is prescribed by the choice of Poisson structure (see Exercise E.3.9*) and the choice of $C_1 = \frac{1}{2}B_3$ is a little restricted by the choice of $B_1 = \{,\}$ and B_2; namely the coefficient at S_Γ^3 in the formula (12.5) must be equal to $\frac{1}{6}$.

Further, we know from the obstructions theory (see Theorem 9.2 and formula (9.6)) that the operator B_{2p} is determined by the following equation (where $\bar{\delta}$ is the Hochschild cohomology coboundary operator)

$$\bar{\delta}B_{2n}(a,b,c) = \sum_{p+q=2n} (B_p(B_q(a,b),c) - B_p(a,B_q(b,c))). \quad (12.13)$$

Anti-symmetrisation $\tilde{c}_{2n}$ of the right-hand side of this formula will preserve the Hochschild cohomology class of this cocycle (see Remark 8.5); at the same time it "kills" all the terms with even terms B_{2k} on the right-hand side since these operators are symmetric. What remains on the right-hand side can be identified with the difference of the left and the right sides of Equation (11.5), that determines the obstruction for the deformation of Poisson structure. Since $C_1, \ldots, C_{n-1}$ constitute a deformation of the Poisson structure up to degree $n-1$, we conclude that $\tilde{c}_{2n} = 0$ so that B_{2n}, solving the Equation (12.13) exists. Moreover, since the symmetrisation preserves Hochschild differential (see Section 9.5.2), we can assume that B_{2n} is symmetric bidifferential operator.

Let us now consider the obstruction equation at the next step:

$$\bar{\delta}B_{2n+1}(a,b,c) = \sum_{p+q=2n+1} (B_p(B_q(a,b),c) - B_p(a,B_q(b,c))). \quad (12.14)$$

The parity of p and q on the right is opposite, hence we see, just like we did in Section 9.5.2 that the skew-symmetrisation of the right-hand side of this equation vanishes. Hence according to Remark 8.5 we conclude that the Hochschild cohomology class of the right-hand side of Equation (12.14) vanishes and there exists an anti-symmetric B_{2n+1}, which we seek. We are going to modify the construction a little so that we can take $2B_{2n+1} = C_n$. This will conclude the inductive step.

Since the formula

$$f \star'_{2n+1} g = fg + \frac{\hbar}{2}\{f,g\} + \sum_{k=1}^{2n+1} \hbar^k B_k(f,g)$$

gives a $\star$-product associative up to degree $2n+1$, the difference

$$c_{2n+2} = \bar{\delta} B_{2n+1} - \sum_{p+q=2n+2} [B_p, B_q]$$

represents the obstruction for the next step of the deformation process, see Equation (9.6) (recall that $\bar{\delta}$ stands for Hochschild differential there), so $\bar{\delta} c_{2n+2} = 0$. By the Hochschild–Kostant–Rosenberg theorem, c_{2n+2} is equal to the image of a polyvector field α under the HKR-homomorphism. Passing to commutator with respect to the product $\star_{2n+1}$ we see that $2B_{2n+1}$ can be regarded as a cochain in $C^2(\mathfrak{P}(M), C^\infty(M))$, which satisfies the equation

$$2\delta B_{2n+1}(f,g,h) = \sum_{p+q=n} \bigl(C_p(C_q(f,g),h) + C_p(C_q(g,h),f)$$
$$+ C_p(C_q(h,f),g)\bigr) + \mu^* \tilde{\alpha},$$

where the right-hand side comes from Equation (11.5) (here δ stands for the Chevalley differential) and $\tilde{\alpha}$ is the image of α under the isomorphism $\Lambda^2 TM \cong \Omega^2(M)$ induced by ω. But C_n is a solution for the problem of deformation of the Poisson structure in degree n. Hence $\delta(2B_{2n+1} - C_n) = \mu^* \tilde{\alpha}$, so the difference $2B_{2n+1} - C_n$ is described by the formula (12.5):

$$C_n = 2B_{2n+1} + r S_\Gamma^3 + \mu^*(\varphi) + \delta E,$$

for some $r \in \mathbb{R}$, $E \in C^1(\mathfrak{P}(M), C^\infty(M))$ and $\varphi \in \Omega^2(M)$. Now it only remains to modify B_{2n} to $B'_{2n} = B_{2n} + \frac{1}{2} r B_2 - \bar{\delta} E$ and change $B_{2n-2}(f,g)$ to $B'_{2n-2}(f,g) = B_{2n-2}(f,g) - \frac{1}{2} r fg$, then

$$f \star_{2n+1} g = fg + \frac{\hbar}{2}\{f,g\} + \sum_{k=1}^{2n-3} \hbar^k B_k(f,g) + \hbar^{2n-2} B'_{2n-2}(f,g)$$
$$+ \hbar^{2n-1} B_{2n-1}(f,g) + \hbar^{2n} B'_{2n}(f,g) + \hbar^{2n+1} C_n(f,g),$$

will be a $\star$-product, associative up to the order $2n+1$ and the inductive step is complete.

Remark 12.8. Let us make a couple of remarks about the results of this section. First, we have in fact proved that any deformation of

Poisson structure on M with $C_1 = \frac{1}{6}S_\Gamma^3$ in Chevalley cohomology is equal to the commutator of a weak $\star$-product; more accurately, with some extra work one can show that equivalent deformations of the Poisson structure induce equivalent $\star$-products, so we have a bijective correspondence between the equivalence classes of weak $\star$-products and deformations $\{,\}_\lambda$ of the Poisson bracket with the class of C_1 equal to $\frac{1}{6}S_\Gamma^3$. Taking into consideration the results of Remark 12.7 we conclude that *there is a bijection between the set of equivalence classes of weak $\star$-products on M and the space $H_{dR}^2(M)[[\lambda]]$.*

Second, instead of weak $\star$-product with some extra work one can obtain the $\star$-product for which 1 is a unit; to this end one can modify the construction of the deformation of the Poisson bracket so as to prove that any deformation up to degree n of $\{,\}$ can be extended to a formal deformation of arbitrary degree. Alternatively, one can use the constructions, similar to what we discussed in Section 12.2.4 in order to describe the way one can "sew" the Moyal quantisations on Darboux charts into a global deformation.

Finally, the construction of the class S_Γ^3 and the proof of Theorem 12.2 substantially depended on the existence of symplectic connection; in particular, if this connection can be chosen so that it has some additional properties, then the class S_Γ^3 will be transferred into the context of cohomology that "respects" such properties, and the whole deformation procedure will respect them. For instance, if there's a group G acting on M in a way that preserves the symplectic form ω, then the existence of G-invariant symplectic connection will provide the possibility to find a G-invariant deformation of $\{,\}$ and hence a G-invariant $\star$-product.

Exercises 4: Obstructions and deformation quantisation of symplectic manifolds

E.4.1 Follow the lines of Remark 10.2 and prove that the deformation quantisation of cotangent bundle T^*X is isomorphic to the algebra of differential operators on X.

E.4.2 Prove that the skew-symmetrisation of Gerstenhaber bracket:

$$[\varphi,\psi](X_1,\ldots,X_{p+q-1}) = \sum_{s\in Sh(q,p-1)} (-1)^s \varphi(\psi(X_{s(1)},\ldots$$
$$\ldots,X_{s(q)}),X_{s(q+1)},\ldots,X_{s(p+q-1)})$$
$$-(-1)^{(p-1)(q-1)} \sum_{s\in Sh(p,q-1)} (-1)^s \psi(\varphi(X_{s(1)},\ldots$$
$$\ldots,X_{s(p)}),X_{s(p+1)},\ldots,X_{s(p+q-1)}),$$

where $\varphi \in C^p(\mathfrak{g},\mathfrak{g})$, $\psi \in C^p(\mathfrak{g},\mathfrak{g})$, $X_1,\ldots,X_{p+q-1} \in \mathfrak{g}$ and $Sh(m,n)$ is the set of all (m,n)-shuffles, determines a graded Lie bracket on the Chevalley-Eilenberg complex $C^{\cdot}(\mathfrak{g},\mathfrak{g})$ of a Lie algebra with coefficients in itself, similar to the usual Gerstenhaber bracket.

E.4.3 Prove that if the Lie-module $\mathfrak{M}$ over a Lie algebra $\mathfrak{g}$ has the structure of a commutative algebra, such that $\mathfrak{g}$ acts on $\mathfrak{M}$ by derivations, then the formula

$$\varphi \wedge \psi(X_1,\ldots,X_{p+q})$$
$$= \sum_{s\in Sh(p,q)} (-1)^s \varphi(X_{s(1)},\ldots,X_{s(p)})\psi(X_{s(p+1)},\ldots,X_{s(p+q)})$$

where $\varphi \in C^p(\mathfrak{g},\mathfrak{g})$, $\psi \in C^p(\mathfrak{g},\mathfrak{g})$, $X_1,\ldots,X_{p+q} \in \mathfrak{g}$ and $Sh(m,n)$ is the set of all (m,n)-shuffles, determines a graded commutative product on the Chevalley-Eilenberg complex $C^{\cdot}(\mathfrak{g},\mathfrak{g})$ of a Lie algebra with coefficients in itself, similar to the $\cup$-product in Hochschild complex.

E.4.4 Prove that the cohomology of a Lie algebra can be computed as the derived extension functor, $H^*(\mathfrak{g},\mathfrak{M}) = Ext^*_{U\mathfrak{g}}(\mathbb{R},\mathfrak{M})$.

Hint: use the identification of modules over Lie algebras and (left-) modules over its universal enveloping algebra and choose a suitable resolution of $\mathbb{R}$.

E.4.5 Prove that the cochain map $\mu^* : \Omega^*(M) \to C^{\cdot}(\mathfrak{P}(M), C^{\infty}(M))$, induced by the homomorphism $u \mapsto X_u$ (here X_u is the Hamilton field of u) on the level of cohomology coincides with the coboundary morphism of the spectral sequence $''E_n^{p,q}$ of (11.11).

In the following two exercises we sketch the main steps in the construction of cohomology classes in the cohomology of Poisson algebras.

E.4.6 Construction of Vey class:

(a) Prove that the map

$$\mathcal{L}_X(\nabla)(Y,Z) = [X, \nabla_Y Z] - \nabla_{[X,Y]} Z - \nabla_Y([X,Z]),$$

where X, Y, Z are vector fields on M, is $C^{\infty}(M)$-linear in the arguments Y and Z. Thus, $\theta_X^{\nabla}(Y) = \mathcal{L}_X(\nabla)(Y, \cdot)$ is a differentiable 1-cochain on $\mathfrak{V}(M)$ with values in $\text{End}(TM)$.

(b) Check that $\Phi_{\nabla}(X,Y) = SSym(Tr(\theta_X^{\nabla} \wedge \theta_Y^{\nabla}))$ is a cocycle in $C^2(\mathfrak{V}(M), \Omega^2(M))$.

(c) Prove that if ∇ is symplectic, then $\sigma_{\Gamma}^3 = \mu' \Phi_{\nabla}$, where σ_{Γ}^3 is cochain, given by the formula (11.17) (in particular, this proves that σ_{Γ}^3 is a cocycle). **Hint:** prove that both cocycles are represented in local coordinates by formula (11.21).

(d*) Prove that the class of σ_{Γ}^3 is not trivial on a generic symplectic manifold.

E.4.7 Construction of "Poisson–Chern–Simons class"

(a) Let ∇ be a torsion-free connection on M; consider for each $X \in Vect(M)$ the linear map

$$A^{\nabla}(X) : TM \to TM, \ A^{\nabla}(X)(Y) = \nabla_Y X.$$

Also let
$$F : \wedge^2 Vect(M) \to End(TM)$$
be the curvature map of ∇. Prove that the cochain
$$cs_3^\nabla \in C^3_{CE}(Vect(M), C^\infty(M)),$$
given by
$$cs_3^\nabla(X, Y, Z)$$
$$= SSym\left(Tr\left(F(X,Y) \circ A^\nabla(Z) - \frac{1}{3} A^\nabla(X) \circ A^\nabla(Y) \circ A^\nabla(Z)\right)\right),$$
satisfies the equation $\delta cs_3^\nabla = Tr(F \wedge F)$.

(b) Suppose, that the first Pontrjagin class on M is trivial, and α is a differential 3-form on M, such that $d\alpha = ch_4$ (we use notation from Section 12.1.2). Let $T_\theta = cs_3^\nabla - \mu^* \alpha$ then $\delta(T_\theta) = 0$. Prove that the class of this cocycle does not depend on the choice of ∇.

(c*) Prove that the class of T_θ in $H^3(\mathfrak{P}(M), C^\infty(M))$ in this case is nontrivial.

(d*) Find the cochains $cs_{2n-1}^\nabla \in C^{2n-1}(\mathfrak{P}(M), C^\infty(M))$ similar to cs_3^∇ for the classes ch_{2n}, i.e. such that
$$\delta cs_{2n-1}^\nabla = ch_{2n}.$$

E.4.8 Prove that the existence of the field ξ on a symplectic manifold (M, ω) such that $\mathcal{L}_\xi \omega = \omega$ is equivalent to the exactness of the symplectic form ω. Moreover, if ξ, ξ' are two fields with this property, then $\xi - \xi'$ is a Hamiltonian field of some function. As an example, find a vector field ξ with this property, when $M = T^*X$, the cotangent bundle of a smooth manifold.

E.4.9 Preliminary constructions for the quantisation of exact symplectic manifolds:

(a) Let M, ω be an exact symplectic manifold, let ξ be the vector field for which $\omega = \mathcal{L}_\xi \omega$. We use the Lie derivative on polydifferential operators to extend the action of ξ to the differentiable Chevalley complex $C^{\cdot}(\mathfrak{P}(M), \Omega^q(M))$, $q = 0, 1, 2, \ldots$. Let δ denote the Chevalley–Eilenberg differential and X_u the Hamiltonian vector field, associated function $u \in C^\infty(M)$. Prove the following identities (c.f. formulas (12.8)–(12.10)):

(i) $\mathcal{L}_\xi \delta = \delta \mathcal{L}_\xi - \delta$;
(ii) $\mathcal{L}_\xi \mu^*(C) = \mu^* \mathcal{L}_\xi C - p\mu^* C$, $C \in C^p(\mathfrak{V}(M), \Omega^k(M))$;
(iii)
$$\mathcal{L}_\xi \mu'(\Phi) = \mu' \mathcal{L}_\xi \Phi - (p+1)\mu' \Phi, \ \Phi \in C^p(\mathfrak{V}(M), \Omega^2(M)),$$

where μ^* and μ' are the maps, induced by representation $\mu : \mathfrak{P}(M) \to \mathfrak{V}(M)$, $u \mapsto X_u$ and the pairing of $\Omega^2(M)$ with the bivector $\pi = \omega^{-1}$.

(b) Prove that for all ξ the map $\mathcal{L}_\xi$ commutes with the differential on the Chevalley complex of $\mathfrak{V}(M)$ with coefficients in $\Omega^*(M)$. Also show that $\mathcal{L}_\xi + p$ (resp. $\mathcal{L}_\xi + (p+1)$) sends cocycles of the form $\mu^* C$ (resp. $\mu' \Phi$) to coboundaries (notation as above).

(c) Prove that for all real $k \neq 2, 3$ (resp. $k \neq 3, 4$) the map $\mathcal{L}_\xi + k$ induces on $H^2(\mathfrak{P}(M), C^\infty(M))$ (resp. on $H^3(\mathfrak{P}(M), C^\infty(M))$) a bijection.

13 Fedosov quantisation: Abelian connections

In this section, we begin the exposition of Fedosov's simple and powerful method to construct the $\star$-products on symplectic manifolds. First formulated in a short article [Fed86] and later explained in greater details in the survey [Fed91], both published in Russian, it has not been widely known till the publication of the paper [Fed94] and the book [Fed96]. Since that time these ideas have become an indispensable part of the deformation theory and adjacent areas of Mathematics and keep reemerging in different contexts till now.

13.1 The main idea: An overview

Lecomte's and de Wilde's quantisation construction has many advantages. First of all, it is based on the natural idea of obstructions, which makes it close to the Physic's perturbative approach to quantisation. Another important feature of this result is that being based on a scrupulous study of cohomology classes, that appear in the process of deformation it brings forth to light the relations between various cohomology theories and cohomology classes involved in the process; for instance, Vey's class that first appeared in the works on deformation of Poisson structures later found numerous applications in Geometry.

However, in general this approach is pretty hard and the formulas that it gives are not easily converted into explicit expressions; in effect, we have only a rather limited control over the solutions of the deformation problem apart from showing that these solutions always exist. So the global geometric meaning of these classes are often hard to grasp, even if the local expressions are known (in fact locally we will always have just Moyal product up to an equivalence). In effect global structure in Lecomte and de Wilde's construction appeared through a pretty messed up process of globalisation which depends on the choice of a partition of unity. All this makes the $\star$-product obtained in this way hard to apply in other Mathematical theories.

Thus it is not surprising that when Fedosov's works [Fed86], [Fed94] where he proposed in particular another deformation quantisation procedure, were published, they attracted much interest. His construction was based on a totally different idea: instead of solving step by step the obstruction problem for the coefficients of the deformation series, he suggested finding an $\hbar$-linear embedding Q of the space $C^\infty(M)[[\hbar]]$ into a suitable noncommutative associative algebra $W_\hbar(M)$, so that the image of this embedding gives a subalgebra in $W_\hbar(M)$. Then the formula

$$f \star g = Q^{-1}(Q(f) \cdot Q(g)), \; f, g \in C^\infty(M)$$

where $\cdot$ is the multiplication in $W_\hbar(M)$ will give an associative $\star$-product in $C^\infty(M)[[\hbar]]$: in effect associativity in this context is evident and we only need to check that the coefficients of this operation are given by bidifferential operators and that the first two terms of this map are $fg + \frac{\hbar}{2}\{f, g\}$ (one may also check if the constant function 1 plays the role of neutral element with respect to $\star$: this condition is not quite trivial, since the map Q is not a homomorphism of algebras).

Recall, that we have already met with a similar idea earlier, when we discussed the "Poincaré-Birkhoff-Witt quantisation" of the functions on $\mathfrak{g}^*$, see Section 5.3, Proposition 5.5; in that case the map Q was the symmetrisation isomorphism $\sigma : S(\mathfrak{g})[[\hbar]] \to \widehat{U}_\hbar \mathfrak{g}$, see (5.8'), (5.9), where $\widehat{U}_\hbar \mathfrak{g}$ is the completed universal enveloping algebra, see Equation (5.3').

In the situation of Proposition 5.5 there was no need to specify the subalgebra of $\widehat{U}_\hbar \mathfrak{g}$ that lay in the image of σ: in that case it was equal to the whole algebra $\widehat{U}_\hbar \mathfrak{g}$. It is not so in the case that we consider here: since the space of functions is too large, it would be a bit too much to expect that the algebra $C^\infty(M)$ can be identified with some naturally constructed noncommutative algebra. Thus we need to find a suitable way to produce subalgebras in arbitrary associative algebras. The idea that was used by Fedosov is that for any differentiation D of an algebra $W_\hbar(M)$, i.e. a linear map

$$D : W_\hbar(M) \to W_\hbar(M), \; D(a \cdot b) = D(a) \cdot b + a \cdot D(b)$$

the null subspace

$$W_D(M) = \{a \in W_\hbar(M) \mid D(a) = 0\}$$

is closed under the multiplication:

$$D(a \cdot b) = D(a) \cdot b + a \cdot D(b) = 0, \text{ if } a, b \in W_D(M).$$

It is this null-subspace of a derivative of the algebra $W_\hbar(M)$ that will appear as the image of Q.

The algebra $W_\hbar(M)$ that Fedosov uses in his work appears rather naturally as the globalisation of the Weyl algebras (see Exercise E.2.1) to the situation when the symplectic form depends on a point in M. The main achievement due to Fedosov was the construction of the differential operator D on it, or rather on the extension of $W_\hbar(M)$ by differential forms on M. This operator can be regarded as a choice of a special connection on M; connections of this type have since then been called *Fedosov connections*. Such a connection D is represented by a formal infinite series of differential operators, obtained by an iterative process; moreover, one can obtain the solutions of the corresponding iterative equations with the help of a linear operator δ^*, similar to the Hodge dual of de Rham differential. All these features make Fedosov's construction and its geometric properties very intuitive and hence quite convenient for applications and generalisations, some of which we will discuss in Section 15.

13.2 The Weyl algebras bundle

13.2.1 Fibrewise Weyl algebras

Let (M, ω) be a symplectic manifold (in particular, M is even-dimensional: $\dim M = 2n$). Restriction of the form ω to the tangent space of M at a point $x \in M$ determines a nondegenerate anti-symmetric bilinear form on T_xM, which in local coordinates $(x^1, \ldots, x^{2n})$ on M can be identified with the matrix-valued function $\omega(x) = (\omega_{ij}(x))$, where $\omega_{ij} = -\omega_{ji}$. Recall (c.f. Exercise E.2.1), that the complex Weyl algebra, associated with such a matrix is

$$W_\hbar^\mathbb{C}(T_xM, \omega(x)) = \mathbb{C}[\hbar]\langle y^1, \ldots, y^{2n}\rangle / ([y^i, y^j] = -\hbar\sqrt{-1}\pi^{ij}(x)), \tag{13.1}$$

where $(y^1, \ldots, y^{2n})$ are formal variables, corresponding to the coordinate functions on the tangent vector space T_xM induced by the local coordinates $(x^1, \ldots, x^{2n})$ on M, $\pi(x) = \omega^{-1}(x)$ and $\langle \cdot \rangle$ denotes the free associative algebra, generated by whatever is inside the angular brackets. Here we stick to Fedosov's notation, which is based on a more Physics-inspired commutation relation: $[P_i, Q^j] = -\sqrt{-1}\hbar \delta_i^j$. Therefore we must use complex coefficients in the definition of Weyl algebra. For this reason and in order to facilitate the notation we shall denote the algebra $C^\infty(M, \mathbb{C})$ of complex-valued functions on M simply by $C^\infty(M)$, and similarly for the differential forms. Of course most part of the contents of our discussion of Fedosov quantisation survives over real numbers.

This construction gives us a family of associative unital algebras parametrised by the points in the coordinate chart. If $(x^{1'}, \ldots, x^{2n'})$ is another local coordinate system, we have (summation over repeating upper and lower indices is assumed):

$$y^{i'} = y^i \frac{\partial x^{i'}}{\partial x^i}, \quad \pi^{i'j'} = \pi^{ij} \frac{\partial x^{i'}}{\partial x^i} \frac{\partial x^{j'}}{\partial x^j},$$

and so

$$[y^{i'}, y^{j'}] = \left[y^i \frac{\partial x^{i'}}{\partial x^i}, y^j \frac{\partial x^{j'}}{\partial x^j} \right] = [y^i, y^j] \frac{\partial x^{i'}}{\partial x^i} \frac{\partial x^{j'}}{\partial x^j}$$

$$= \frac{\hbar}{\sqrt{-1}} \pi^{ij} \frac{\partial x^{i'}}{\partial x^i} \frac{\partial x^{j'}}{\partial x^j} = \frac{\hbar}{\sqrt{-1}} \pi^{i'j'}.$$

Thus this construction does not depend on the choice of coordinates in M and can be globalised to a family of algebras $W_\hbar^{\mathbb{C}}(M, \omega)$, parametrised by the points in M. Speaking in abstract terms $W_\hbar^{\mathbb{C}}(M, \omega)$ can be regarded as a fibre bundle $W_\hbar^{\mathbb{C}}(M, \omega) \to M$: to this end we topologise $W_\hbar^{\mathbb{C}}(M, \omega)$ so that locally we have homeomorphisms

$$W_\hbar^{\mathbb{C}}(M, \omega)|_U = \bigcup_{x \in U} W_\hbar^{\mathbb{C}}(T_x M, \omega(x)) \cong U \times W_\hbar^{\mathbb{C}}(\mathbb{R}^{2n}, \omega_0);$$

here $U \subset M$ is a coordinate chart and ω_0 is the canonical symplectic form on $\mathbb{R}^{2n}$ (for instance we can assume that the coordinates $(x^1, \ldots, x^{2n})$ are canonical). We will refer to this bundle $W_\hbar^{\mathbb{C}}(M, \omega)$ as the *Weyl algebras bundle*.

If one is familiar with the basic ideas of the fibre bundles theory, one can describe the fibre bundle $W_\hbar^{\mathbb{C}}(M,\omega)$ as follows: let $P_{Sp_{2n}}(TM)$ be the principal bundle of symplectic frames in TM, i.e. the principal bundle with structure group $Sp_{2n}(\mathbb{R})$ such that $TM \cong P_{Sp_{2n}}(TM) \times_{Sp_{2n}} \mathbb{R}^{2n}$. The symplectic group $Sp_{2n}(\mathbb{R})$ acts on the Weyl algebra $W_\hbar^{\mathbb{C}}(\mathbb{R}^{2n}, \omega_0)$ by automorphisms, see Exercise E.5.2. We then put

$$W_\hbar^{\mathbb{C}}(M,\omega) = P_{Sp_{2n}}(TM) \times_{Sp_{2n}} W_\hbar^{\mathbb{C}}(\mathbb{R}^{2n}, \omega_0).$$

Remark 13.1. The same construction can be applied to arbitrary *symplectic vector bundle* $E \to M$, i.e. to a vector bundle over M with a fixed fibre-wise symplectic structure on it. Most part of the following discussions applies in a word-for-word manner to the generic symplectic vector bundles, but we will not use this observation here.

13.2.2 Algebras $W_\hbar(M)$, $\widehat{W}_\hbar(M)$ and $\widehat{W}_\hbar^*(M)$

Fibres of the bundle $W_\hbar^{\mathbb{C}}(M,\omega)$ being associative algebras in such a way that the changes of bundle coordinates "respect" the algebra structure, we can introduce the product into the space of sections of this bundle. This remark brings forward the following definition

Definition 13.2. *Let* $W_\hbar(M) = \Gamma^\infty(W_\hbar^{\mathbb{C}}(M,\omega))$ *denote the space of smooth global sections of the bundle* $W_\hbar^{\mathbb{C}}(M,\omega)$. *Then* $W_\hbar(M)$ *is an algebra with respect to the fibre-wise multiplication in* $W_\hbar^{\mathbb{C}}(M,\omega)$.

It is this algebra that we would like to use as mentioned in Section 13.1. Locally elements in $W_\hbar(M)$ are given by expressions

$$a = a(x,y,\hbar) = \sum_{k \geq 0, l \geq 0} \sum_{1 \leq i_1, i_2, \ldots, i_l \leq 2n} \hbar^k a_{k,l,i_1,\ldots,i_l}(x) y^{i_1} \ldots y^{i_l}, \tag{13.2}$$

where the elements $y^1, \ldots, y^{2n}$ satisfy the relations from (13.1): observe that *the expression in the formula* (13.2) *is not unique, since we do not assume that the indices* $i_1, \ldots, i_l$ *are ordered*; $a_{k,l,i_1,\ldots,i_l}(x)$ are complex-valued functions, only a finite number of which are different from zero. Observe that since the center of usual Weyl algebra is easily found to be spanned by the constants, we see that the center of $W_\hbar(M)$ is equal to $C^\infty(M)[\hbar]$.

However later we will need to work with elements of the form (13.2), described by infinite sums of nonzero monomials in variables $y^1, \ldots, y^{2n}$. To this end, we should pass to suitable completions of the algebras $W_\hbar^{\mathbb{C}}(T_xM, \omega(x)) \cong W_\hbar^{\mathbb{C}}(\mathbb{R}^{2n}, \omega_0)$ for all $x \in M$. Roughly speaking, we need to complete $W_\hbar^{\mathbb{C}}(\mathbb{R}^{2n}, \omega_0)$ with respect to the filtration by the powers of monomials (in $y^1, \ldots, y^{2n}$); in simple words this means that we should allow infinite sums of elements $a_{i_1,\ldots,i_p} y^{i_1} \ldots y^{i_p}$ with growing "powers" with respect to the variables $y^1, \ldots, y^{2n}$. The problem is that changing the order of the variables y^i one can obtain monomials with lower degrees of y^i. So in order to provide unambiguous notion of convergence for such sums we will incorporate variable $\hbar$ into the play; to this end we will introduce the degrees of variables:

Definition 13.3. *Let* $\deg y^i = 1$ *and* $\deg \hbar = 2$; *then the commutation relation of Weyl algebras preserves the total degree of the elements. Completions of the algebras with respect to the total degree of their elements will be denoted by* $\widehat{}$ *in this and the next sections. Observe that the product in Weyl algebra preserves the sum of total degrees of the factors, hence the algebra structures on completed spaces are well-defined.*

For applications it is also useful to allow negative powers of $\hbar$, so we modify the construction a bit more. Eventually at every point $x \in M$ we will consider the algebra

$$\widehat{W}_\hbar^{\mathbb{C}}(T_xM, \omega(x)) = \widehat{\mathbb{C}}[\hbar^{-1}, \hbar]] \langle y^1, \ldots, y^{2n} \rangle / ([y^i, y^j] = \hbar \pi^{ij}(x)), \tag{13.1'}$$

(recall that $\widehat{}$ denotes the completion). We will denote by $\widehat{W}_\hbar^{\mathbb{C}}(M, \omega)$ the fibre bundle with fibres $\widehat{W}_\hbar^{\mathbb{C}}(T_xM, \omega(x))$ and eventually we denote by $\widehat{W}_\hbar(M)$ the algebra of smooth sections of this bundle: $\widehat{W}_\hbar(M) = \Gamma^\infty(\widehat{W}_\hbar^{\mathbb{C}}(M, \omega))$. Elements of the algebra $\widehat{W}_\hbar(M)$ in local coordinates are given by *infinite sums* of the form

$$a = a(x, y, \hbar) = \sum_{k \geq N, l \geq 0} \sum_{1 \leq i_1, i_2, \ldots, i_l \leq 2n} \hbar^k a_{k,l,i_1,\ldots,i_l}(x) y^{i_1} \ldots y^{i_l}, \tag{13.2'}$$

for some $N \in \mathbb{Z}$, provided the total degree $2k + l$ goes to infinity. The center of this algebra is equal to $C^\infty(M, \mathbb{C})[\hbar^{-1}, \hbar]]$.

In addition to the algebra $\widehat{W}_\hbar(M)$ of sections in $\widehat{W}_\hbar^{\mathbb{C}}(M,\omega)$ we shall also consider the differential forms with values in $\widehat{W}_\hbar^{\mathbb{C}}(M,\omega)$, i.e. the sections of the tensor product bundles $\widehat{W}_\hbar^{\mathbb{C}}(M,\omega) \otimes \Omega^*(M)$; we will denote these forms by $\widehat{W}_\hbar^p(M)$, $p = 0, 1, \ldots$ (of course $\widehat{W}_\hbar^0(M) = \widehat{W}_\hbar(M)$). These sections also form a graded noncommutative algebra (with grading induced from $\Omega^*(M)$), denoted by $\widehat{W}_\hbar^*(M)$. In local coordinates a generic element in $\widehat{W}_\hbar^p(M)$ can be written in the form:

$$\alpha = \alpha(x, y, \hbar)$$
$$= \sum_{k \geq N, l \geq 0} \sum_{1 \leq i_1, \ldots, i_l \leq 2n} \sum_{1 \leq j_1 < \cdots < j_p \leq 2n} \hbar^k a_{k,l,I,J}(x) y^{i_1} \ldots$$
$$\ldots y^{i_l} dx^{j_1} \wedge \cdots \wedge dx^{j_p}. \quad (13.3)$$

Here $I = (i_1, \ldots, i_l)$, $J = (j_1, \ldots, j_p)$ are multi-indices and $a_{k,l,I,J}(x)$ are complex-valued functions. Center of this algebra (in graded sense) is spanned by $\Omega^*(M; \mathbb{C})[\hbar^{-1}, \hbar]]$, the complex-valued forms on M with added formal variable $\hbar$.

13.2.3 Symmetrisation

As we already mentioned, expressions in the formulas (13.2), (13.2′) (and similarly (13.3)) are not uniquely determined by an element in the corresponding algebras, since the commutation relations let one obtain monomials of lower degrees in y^i by permuting the variables. In order to get rid of this ambiguity, one can do the following: let

$$\sigma : \mathbb{C}[y^1, \ldots, y^{2n}] \to C\langle y^1, \ldots, y^{2n} \rangle$$

be the symmetrisation map:

$$\sigma(y^{i_1} \ldots y^{i_p}) = \frac{1}{p!} \sum_{s \in S_p} y^{i_{s(1)}} \ldots y^{i_{s(p)}}.$$

Here $\mathbb{C}\langle y^1, \ldots, y^{2n} \rangle$ denotes the "algebra of noncommutative polynomials" in $y^1, \ldots, y^{2n}$ (clearly, it is just the tensor algebra of $\mathbb{R}^{2n}$). We shall denote by the same symbol σ the map, obtained by composition of this symmetrisation with the natural projection of $\mathbb{C}\langle y^1,\ldots,y^{2n}\rangle$ onto the Weyl algebra. In this case (see Exercise E.2.1), σ descends to a linear isomorphism; in particular *every element* $a = a(x, y, \hbar) \in \widehat{W}_\hbar(M)$ *can be uniquely expressed in the form*

$$a = a(x, y, \hbar) = \sum_{k \geq N, q \geq 0} \sum_{1 \leq i_1 \leq \cdots \leq i_q \leq 2n} \hbar^k a_{k,q,i_1,\ldots,i_q}(x) y^{i_1} \circ \ldots \circ y^{i_q}, \quad (13.2'')$$

where $y^{i_1} \circ \cdots \circ y^{i_l} = \sigma(y^{i_1} \ldots y^{i_l})$. Similarly, every $\alpha \in \widehat{W}_\hbar^p(M)$ is uniquely representable as

$$\alpha = \alpha(x, y, \hbar)$$
$$= \sum_{k \geq N, q \geq 0} \sum_{1 \leq i_1 \leq \cdots \leq i_q \leq 2n} \sum_{1 \leq j_1 < \cdots < j_p \leq 2n} \hbar^k a_{k,q,I,J}(x) y^{i_1} \circ \ldots$$
$$\cdots \circ y^{i_q} dx^{j_1} \wedge \cdots \wedge dx^{j_p}. \quad (13.3')$$

Observe that the operation $y \circ a$ is well-defined and associative, although it does not preserve the product in Weyl algebra: in effect, it is just the product in the symmetric algebra, transferred to the Weyl algebra by linear isomorphism. Also observe that the formula (13.3') determines a unique representation of $\alpha \in \widehat{W}_\hbar^*(M)$ as the sum of its bihomogeneous components:

$$\alpha = \sum_{p,q \geq 0} \alpha_{p,q}, \quad (13.4)$$

where

$$\alpha_{p,q} = \sum_{1 \leq i_1 \leq \cdots \leq i_q \leq 2n} \sum_{1 \leq j_1 < \cdots < j_p \leq 2n} a_{I,J}(x, \hbar) y^{i_1} \circ \cdots \circ y^{i_q} dx^{j_1} \wedge \cdots \wedge dx^{j_p}. \quad (13.5)$$

13.2.4 Alternative constructions

There are other possible points of view on the main constructions of this section:

(*i*) One can describe the bundle $\widehat{W}_\hbar^{\mathbb{C}}(M,\omega)$ as the quotient bundle of the bundle of tensor algebras

$$T^{*\otimes}M[\hbar^{-1},\hbar]] = \left(\bigoplus_{k \geq 0} (T^*M)^{\otimes k}\right)[\hbar^{-1},\hbar]]$$

by the subbundle of relations $R^\otimes M \subseteq T^{*\otimes}M[\hbar^{-1}, \hbar]]$. The subbundle $R^\otimes M$ is the bundle of ideals in tensor algebras, generated by the subbundle

$$R_1^\otimes M = \left\{ a \otimes b - b \otimes a - \frac{\hbar}{\sqrt{-1}}\pi(a,b) \mid a, b \in T^*M \right\}.$$

(*ii*) In a similar way, one can regard the algebra $\widehat{W}_\hbar(M)$ as the algebra generated by the $C^\infty(M)$-module of differential 1-forms $\Omega^1(M)$ factorised over the relations

$$\alpha \cdot \beta - \beta \cdot \alpha = \frac{\hbar}{\sqrt{-1}}\pi(\alpha, \beta), \ \alpha, \beta \in \Omega^1(M).$$

It is not difficult to see that all these constructions lead to the same result.

13.3 Fedosov connections

13.3.1 Symplectic connections and the bundle $\widehat{W}_\hbar^{\mathbb{C}}(M, \omega)$

Let ∇ be any symmetric connection on TM; we shall denote by the same symbol its unique natural extension to a connection on the tensor algebra bundle $T^\otimes M$, which satisfies the condition

$$\nabla_X(S \otimes T) = (\nabla_X S) \otimes T + S \otimes (\nabla_X T)$$

for all vector fields $X \in \mathfrak{V}(M)$ and all tensor fields S, T on M. If the connection ∇ is symplectic (see Section 11.3.1), i.e. $\nabla \omega = 0$, then the relations that generate $\widehat{W}_\hbar^{\mathbb{C}}(M, \omega)$ will be preserved by ∇, and hence *every symplectic connection ∇ on M determines a unique connection on $\widehat{W}_\hbar^{\mathbb{C}}(M, \omega)$*, such that

$$\nabla_X(a \cdot b) = (\nabla_X a) \cdot b + a \cdot (\nabla_X b) \tag{13.6}$$

for all sections $a, b \in \widehat{W}_\hbar(M)$. In dual notation one can regard ∇ as a map $\nabla : \widehat{W}_\hbar(M) \to \widehat{W}_\hbar^1(M)$ and extend it in a unique way to maps $\nabla : \widehat{W}_\hbar^p(M) \to \widehat{W}_\hbar^{p+1}(M)$. In this case, the Leibniz rule (13.6) is changed to the graded Leibniz rule:

$$\nabla(\alpha \cdot \beta) = (\nabla \alpha) \cdot \beta + (-1)^p \alpha \cdot (\nabla \beta), \ \forall \alpha \in \widehat{W}_\hbar^p(M), \ \beta \in \widehat{W}_\hbar^q(M). \tag{13.7}$$

The main idea of Fedosov was to modify ∇ so that the space of covariant flat sections

$$\widehat{W}_\nabla(M) = \{a \in \widehat{W}_\hbar(M) \mid \nabla a = 0\} \subseteq \widehat{W}_\hbar(M)$$

becomes linearly isomorphic to $C^\infty(M)[\hbar^{-1}, \hbar]]$. Flat sections form a subalgebra, so this isomorphism will give a $\star$-product; on the other hand the equation that determines flat sections is linear, and hence, if the connection ∇ is flat, one can hope to use the methods of homological algebra to treat its solutions.

Unfortunately, symplectic connection is usually nonflat, so one might not hope to use ∇ directly, however the way its square acts on $\widehat{W}_\hbar(M)$ is quite easily understandable. Indeed, since ∇ acts on $\widehat{W}_\hbar(M)$ by derivations, its square is also a derivation of this algebra, so by the Leibniz rule (13.7) it is enough to consider only generators of this algebra. Let now $(x^1, \ldots, x^{2n})$ be a local coordinate system on M, and let $y^l \in \Omega^1(M)$ be a generator of the algebra $\widehat{W}_\hbar(M)$ in these coordinates (see part (ii) in Section 13.2.4). Then, since the coordinate vector fields commute, we have

$$\nabla^2_{ij}(y^l) = (\nabla_i \nabla_j - \nabla_j \nabla_i)(y^l) = F_{ij}(y^l) = F^l_{ij;k} y^k, \tag{13.8}$$

where $F = (F^k_{ij;l} dx^i \wedge dx^j)$ is the matrix of curvature form of ∇ in the coordinates $(x^1, \ldots, x^{2n})$. Let $F_{ij;kl} = F^m_{ij;k} \omega_{ml}$ and consider the formal sum $\Omega_{ij}(y) = \frac{\sqrt{-1}}{2\hbar} F_{ij;lk} y^l y^k$. Then the element $\Omega(y) = \Omega_{ij}(y) dx^i \wedge dx^j \in \widehat{W}^2_\hbar(M)$ does not depend on the choice of coordinates (Exercise E.5.4) and since $F_{ij,kl} = F_{ij;lk}$ (Exercise E.5.3), we get

$$[\Omega_{ij}(y), y^l] = \frac{\sqrt{-1}}{2\hbar} F_{ij;pq}[y^p y^q, y^l] = \frac{1}{2\hbar} F_{ij;pq}(y^p[y^q, y^l] + [y^p, y^l] y^q)$$

$$= \frac{1}{2}(F^m_{ij;p} \omega_{mq} \pi^{ql} y^p + F^m_{ij;q} \omega_{mp} \pi^{pl} y^q)$$

$$= \frac{1}{2}(F^m_{ij;p} \delta^l_m y^p + F^m_{ij;q} \delta^l_m y^q) = F^l_{ij;k} y^k.$$

So the right-hand side of Equation (13.8) can be rewritten as follows:

$$\nabla^2(y^l) = [\Omega(y), y^l],$$

and hence a similar formula

$$\nabla(\alpha) = [\Omega(y), \alpha], \tag{13.9}$$

holds for any element $\alpha \in \widehat{W}_\hbar^p(M)$ and for all $p = 0, 1, \ldots, 2n$.

In effect, locally the action of ∇ can also be written as a commutator in the algebra $\widehat{W}_\hbar(M)$. Namely, fix the coordinates $(x^1, \ldots, x^{2n})$; then the action of ∇ on an element $a = a_k(x)y^k$ is given by the formula

$$\nabla_i(a) = \frac{\partial a_k}{\partial x^i}y^k + \Gamma_{ij}^k a_k y^j, \text{ so that } \nabla_i(y^k) = \Gamma_{ij}^k y^j,$$

where Γ_{ij}^k are the Christoffel symbols of ∇. Let us put $\Gamma_i(y) = \frac{\sqrt{-1}}{2\hbar}\Gamma_{ijk}y^j y^k$, where $\Gamma_{ijk} = \Gamma_{ij}^l \omega_{lk}$, then by a similar computation we have

$$\nabla_i(a) = \frac{\partial a}{\partial x^i} + [\Gamma_i(y), a],$$

and more generally: if $\Gamma(y) = \Gamma_i(y)dx^i$, then for any $\alpha \in \widehat{W}_\hbar^p(M)$, we have

$$\nabla(\alpha) = d\alpha + [\Gamma(y), \alpha], \tag{13.10}$$

where $[\Gamma(y), \alpha]$ is *graded commutator*:

$$[a, b] = a \cdot b - (-1)^{pq} b \cdot a, \ a \in \widehat{W}_\hbar^p(M), \ b \in \widehat{W}_\hbar^q(M).$$

One should keep in mind however that the $\widehat{W}_\hbar^\mathbb{C}(M, \omega)$-valued 1-form $\Gamma(y)$ is only locally defined, since Christoffel symbols are changed in a more complex way when the coordinates change.

13.3.2 Fedosov's theorem

It is our goal now to find a connection D on $\widehat{W}_\hbar^\mathbb{C}(M, \omega)$ such that for any $\alpha \in \widehat{W}_\hbar^p(M)$ the image of α under the square of D vanishes. Of course it would be naïve to hope that the curvature of D as an endomorphism-valued differential form is identically zero. However if we take into consideration formula (13.9), we come up with the following idea: *if $D^2(\alpha)$ can also be expressed in the form of a commutator of some 2-form Ω_D on M with values in $\widehat{W}_\hbar^\mathbb{C}(M, \omega)$, then*

$D^2(\alpha) \equiv 0$ iff Ω_D is central with respect to the graded product in $\widehat{W}_\hbar^*(M)$. In this case the curvature of D is not zero, but its action on sections vanish! This idea was introduced by Fedosov, and it eventually lead to the solution of the deformation quantisation problem in this situation.

The formula (13.10) justifies the following idea: let $\Delta\Gamma \in \widehat{W}_\hbar^1(M)$ denote a 1-form on M with values in $\widehat{W}_\hbar^\mathbb{C}(M,\omega)$; let ∇ be a symplectic connection (which we extend to $\widehat{W}_\hbar^*(M)$); consider the map

$$D\alpha = \nabla\alpha + [\Delta\Gamma, \alpha]. \tag{13.11}$$

Then

$$D^2(\alpha) = \nabla^2\alpha + \nabla([\Delta\Gamma, \alpha]) + [\Delta\Gamma, \nabla\alpha] + [\Delta\Gamma, [\Delta\Gamma, \alpha]]$$

$$= \nabla^2\alpha + [\nabla\Delta\Gamma, \alpha] + \frac{1}{2}[[\Delta\Gamma, \Delta\Gamma], \alpha]$$

$$= [\Omega(y) + \nabla\Delta\Gamma + \frac{1}{2}[\Delta\Gamma, \Delta\Gamma], \alpha] = [\tilde{\Omega}(y), \alpha],$$

where

$$\tilde{\Omega}(y) = \Omega(y) + \nabla\Delta\Gamma + \frac{1}{2}[\Delta\Gamma, \Delta\Gamma]. \tag{13.12}$$

Here we used the graded Leibniz rule:

$$\nabla([\Delta\Gamma, \alpha]) + [\Delta\Gamma, \nabla\alpha] = [\nabla\Delta\Gamma, \alpha] - [\Delta\Gamma, \nabla\alpha] + [\Delta\Gamma, \nabla\alpha] = [\nabla\Delta\Gamma, \alpha]$$

and the graded Jacobi identity (c.f. Proposition 6.2).

Thus the problem now is reduced to the question of finding such correction term $\Delta\Gamma$ that $\tilde{\Omega}(y)$ is in the center of $\widehat{W}_\hbar^2(M)$. This center is easy to describe: as we have earlier explained it is equal to $\Omega^2(M)[\hbar^{-1}, \hbar]]$. So we have to find such correction term $\Delta\Gamma$ that $\tilde{\Omega}(y)$ is in $\Omega^2(M)[\hbar^{-1}, \hbar]]$. Connections with this property were called *Abelian* by Fedosov; nowadays they are often called *Fedosov connections*. It turns out that Fedosov connections always exist; moreover, there are plenty of them. In effect, one can prove the following theorem

Theorem 13.4 (Fedosov). *Let φ be an arbitrary element in $\Omega^2(M)[[\hbar]]$ closed with respect to the de Rham differential, then one can find an element $\Delta\Gamma \in \widehat{W}_\hbar^1(M)$ such that the curvature $\tilde{\Omega}(y)$ of Fedosov connection $D = \nabla + [\Delta\Gamma, \cdot]$ is equal to $-\frac{\sqrt{-1}}{\hbar}\omega + \varphi$.*

Remark 13.5. The form $\tilde{\Omega}(y) \in \Omega^2[\hbar^{-1}, \hbar]]$ should be closed because of the Bianchi identity: on one hand we should have $D(\tilde{\Omega}(y)) = 0$, and on the other, since $\tilde{\Omega}(y)$ is in the center of Weyl algebra bundle, $D(\tilde{\Omega}(y)) = d\tilde{\Omega}(y)$.

13.3.3 Proof of Fedosov's theorem: Change of variables

Let $(x^1, \ldots, x^{2n})$ be a local coordinate chart, so that $y^1, \ldots, y^{2n}$ are the corresponding local generators of the Weyl algebra bundle; put $\omega(y) = \frac{\sqrt{-1}}{\hbar} \omega_{ij} y^i dx^j$. This expression does not depend on the choice of coordinates, so $\omega(y) \in \widehat{W}_\hbar^1(M)$ is a global differential 1-form with values in Weyl algebras. Then for every y^k we have

$$[\omega(y), y^k] = \frac{\sqrt{-1}}{\hbar} \omega_{ij}[y^i, y^k] dx^j = \omega_{ij} \pi^{ik} dx^j = -\delta_i^k dx^i = -dx^k.$$

After Fedosov, we will regard the graded commutator of $\alpha \in \widehat{W}_\hbar^p(M)$ with $-\omega(y)$ as an operator $\delta : \widehat{W}_\hbar^p(M) \to \widehat{W}_\hbar^{p+1}(M)$; this is a graded differentiation of this algebra of degree 1; locally it is completely determined by the formula $\delta(y^k) = -[\omega(y), y^k] = dx^k$. Observe that in particular $\delta^2 = 0$: this is so because $[\omega(y), dx^k] = 0$ for all k; also one has the following identity:

$$[\omega(y), \omega(y)] = -\delta(\omega(y)) = -\frac{\sqrt{-1}}{\hbar} \omega_{ij} dx^i \wedge dx^j = -\frac{\sqrt{-1}}{\hbar} \omega$$

(see Exercise E.5.6 for details). We are going to look for the correction term $\Delta\Gamma$ in the following form:

$$\Delta\Gamma = \omega(y) + \zeta, \quad \zeta \in \widehat{W}_\hbar^1(M).$$

In this case the equation $\tilde{\Omega}(y) = -\frac{\sqrt{-1}}{\hbar}\omega + \varphi$ turns into the following formula (c.f. Equation (13.12)):

$$-\frac{\sqrt{-1}}{\hbar}\omega + \varphi = \Omega(y) + \nabla\omega(y) - \frac{\sqrt{-1}}{\hbar}\omega - \delta\zeta + \nabla\zeta + \frac{1}{2}[\zeta, \zeta]. \quad (13.13)$$

Below we shall denote the right-hand side of this formula by Ω_ζ: this is the curvature of the operator D, corresponding to ζ. After reducing the similar terms we get the following equation

$$\delta\zeta = \Omega_0 + \nabla\zeta + \frac{1}{2}[\zeta, \zeta], \quad (13.13')$$

Fedosov quantisation: Abelian connections   191

where $\Omega_0 = \Omega(y) + \nabla \omega(y) - \varphi$. It is this equation that we shall consider below.

13.3.4 Proof of Fedosov's theorem: Operators δ and δ^*

In order to solve Equation (13.13′), let us consider the map δ more attentively. Choose local coordinates $(x^1, \ldots, x^{2n})$ in M so that the corresponding generators of Weyl algebra are $(y^1, \ldots, y^{2n})$. One can write the operator δ somewhat loosely as

$$\delta = dx^i \wedge \frac{\partial}{\partial y^i}; \qquad (13.14)$$

observe that although partial derivatives with respect to y^i are ill-defined in general, if y^i does not commute with y^j, this definition of the operator δ is not ambiguous, since differentiation kills the right-hand side of the commutator relations. In this form, one can find an almost inverse operator for δ. Namely, put

$$\delta^* = y^i \circ \frac{\partial}{\partial dx^i}, \qquad (13.15)$$

where $y^i \circ$ denotes the commutative product of variables, see Section 13.2.3. More accurately, let $X^1, \ldots, X^{2n}$ be the local vector fields, which give dual basis for $dx^1, \ldots, dx^{2n}$ (i.e. $X^i = \frac{\partial}{\partial x^i}$); then

$$\delta^* = y^i \circ \iota_{X^i}. \qquad (13.15′)$$

Observe again that *in formulas* (13.15), (13.15′) *we do not fix the order of multiplication by* y^i, instead we take the commutative multiplication $\circ$ in variables $y^1, \ldots, y^{2n}$: this operation is well-defined in spite of the noncommutativity of the Weyl algebra. It is easy to see that the operator δ^* does not depend on the choice of coordinates and that for any

$$a = a_{I,J} y^I dx^J \in \widehat{W}_\hbar^p(M),$$

where $I = (i_1, \ldots, i_q)$, $1 \le i_k \le 2n$ for $k = 1, \ldots, q$ and $y^I = y^{i_1} \circ \cdots \circ y^{i_q}$; similarly $J = (1 \le j_1 < \cdots < j_p \le 2n)$ and $dx^J = dx^{j_1} \wedge \cdots \wedge dx^{j_p}$, we have

$$(\delta^*)^2(a_{I,J} y^I dx^J) = \delta^* \left(\sum_{l=1}^{p} (-1)^l y^{j_l} \circ y^I dx^{J \setminus \{j_l\}} \right)$$

$$= \sum_{l=1}^{p} \sum_{j_m \in J \setminus \{j_l\}} (-1)^{l+m'} y^{j_m} \circ (y^{j_l} \circ y^I) dx^{J \setminus \{j_l, j_m\}} = 0,$$

since $\circ$ is a commutative operation and $m' = m$ if $m < l$ and $m' = m - 1$ otherwise. Similarly

$$\delta^* \delta(a_{I,J} y^I dx^J) = \delta^* \left(a_{I,J} \sum_{k=1}^{q} y^{I \setminus \{i_k\}} dx^{i_k} \wedge dx^J \right)$$

$$= a_{I,J} \left(\sum_{k=1}^{q} \left(y^{i_k} \circ y^{I \setminus \{i_k\}} dx^J + \sum_{l=1}^{q} (-1)^{l+1} y^{j_l} \circ y^{I \setminus \{i_k\}} dx^{i_k} \wedge dx^{J \setminus \{j_l\}} \right) \right)$$

$$= p a_{I,J} y^I dx^J + \sum_{l=1}^{q} (-1)^{l+1} y^{j_l} \circ y^{I \setminus \{i_k\}} dx^{i_k} \wedge dx^{J \setminus \{j_l\}}.$$

Here we used the observation that the top degree part of the product in Weyl algebra satisfies the equation $y^{i_k} \circ y^{I \setminus \{i_k\}} = y^I$. Similarly

$$\delta \delta^*(a_{I,J} y^I dx^J) = \delta \left(a_{I,J} \sum_{l=1}^{p} (-1)^l y^{j_l} \circ y^I dx^{J \setminus \{j_l\}} \right)$$

$$= a_{I,J} \left(\sum_{l=1}^{p} \left(y^I dx^J + \sum_{l=1}^{q} (-1)^l y^{j_l} \circ y^{I \setminus \{i_k\}} dx^{i_k} \wedge dx^{J \setminus \{j_l\}} \right) \right)$$

$$= q a_{I,J} y^I dx^J + \sum_{l=1}^{q} (-1)^l y^{j_l} \circ y^{I \setminus \{i_k\}} dx^{i_k} \wedge dx^{J \setminus \{j_l\}}.$$

Summing up these equalities, we get the formula for the "Laplace operator":

$$(\delta \delta^* + \delta^* \delta)(a_{p,q}) = (p+q) a_{p,q}, \tag{13.16}$$

for all $a_{p,q} \in \widehat{W}_\hbar^p(M)$ homogenous in variables $y^1, \ldots, y^{2n}$ with degree q (see the decomposition (13.4)); also we have $\delta^2 = (\delta^*)^2 = 0$. Let us put for all bi-homogenous $a_{p,q}$ (see formula (13.5))

$$\delta^{-1}(a_{p,q}) = \begin{cases} \frac{1}{p+q}\delta^* a_{p,q}, & p+q \neq 0 \\ 0, & p = q = 0. \end{cases} \quad (13.17)$$

Then $(\delta^{-1})^2 = 0$ and

$$(\delta\delta^{-1} + \delta^{-1}\delta)(a_{p,q}) = \begin{cases} a_{p,q}, & p+q \neq 0 \\ 0, & p = q = 0; \end{cases} \quad (13.18)$$

in particular, for any $a \in \widehat{W}_\hbar^*(M)$ we will have

$$a = (\delta\delta^{-1} + \delta^{-1}\delta)(a) + a_{00}, \quad (13.19)$$

where a_{00} is the homogenous part of a of bi-degree $(0,0)$ in decomposition (13.4). We will also need the following property of δ^{-1} (which is the consequence of the definition of δ^*): *operator δ^{-1} increases the total degree of the elements in $\widehat{W}_\hbar^*(M)$, see Definition 13.3.*

13.3.5 Proof of Fedosov's theorem: Iterative process

Now we can return to the Fedosov's Equation (13.13′). First of all, we shall make more assumptions about ζ: suppose that $\delta^{-1}\zeta = 0$ and $\zeta_{00} = 0$. In this case, the decomposition (13.19) reduces to $\delta^{-1}\delta\zeta = \zeta$, so if we again abbreviate $\delta^{-1}\Omega_0 = \zeta_0$ then Fedosov's equation turns into the following problem, which is a corollary of the original one: find ζ, satisfying the conditions

$$\zeta = \zeta_0 + \delta^{-1}\left(\nabla\zeta + \frac{1}{2}[\zeta,\zeta]\right), \quad \delta^{-1}\zeta = 0, \quad \zeta_{00} = 0. \quad (13.20)$$

Fedosov suggested iterative procedure to solve this problem: at the initial step we put:

$$\zeta_1 = \zeta_0 + \delta^{-1}\left(\nabla\zeta_0 + \frac{1}{2}[\zeta_0,\zeta_0]\right), \quad (13.20_0)$$

then we find

$$\zeta_2 = \zeta_0 + \delta^{-1}\left(\nabla\zeta_1 + \frac{1}{2}[\zeta_1,\zeta_1]\right), \tag{13.20$_1$}$$

$$\zeta_3 = \zeta_0 + \delta^{-1}\left(\nabla\zeta_2 + \frac{1}{2}[\zeta_2,\zeta_2]\right), \tag{13.20$_2$}$$

and so on. So that at the n-th stage we will get the equation:

$$\zeta_{n+1} = \zeta_0 + \delta^{-1}\left(\nabla\zeta_n + \frac{1}{2}[\zeta_n,\zeta_n]\right). \tag{13.20$_n$}$$

We are going to show that the sequence ζ_n does actually converge to a solution of the Problem (13.20). To this end we subtract the Equation (13.20$_{n-1}$) from (13.20$_n$) and get:

$$\zeta_{n+1} - \zeta_n = \delta^{-1}\left(\nabla(\zeta_n - \zeta_{n-1}) + \frac{1}{2}[\zeta_n + \zeta_{n-1}, \zeta_n - \zeta_{n-1}]\right),$$

since

$$[\zeta_n + \zeta_{n-1}, \zeta_n - \zeta_{n-1}] = [\zeta_n,\zeta_n] - [\zeta_{n-1},\zeta_{n-1}] - [\zeta_n,\zeta_{n-1}] + [\zeta_{n-1},\zeta_n]$$

and $[\zeta_n,\zeta_{n-1}] = [\zeta_{n-1},\zeta_n]$ as both ζ_n and ζ_{n-1} are from $\widehat{W}_\hbar^1(M)$. Then, since δ^{-1} increases the total degree of the elements in Weyl algebras, while the other operations do not change it (or also increase it), we see that the total degree of $\zeta_{n+1} - \zeta_n$ is strictly greater than that of $\zeta_n - \zeta_{n-1}$. Thus it grows indefinitely and by Definition 13.3, there exists

$$\zeta \in \widehat{W}_\hbar^1(M), \quad \zeta = \lim_{n\to\infty} \zeta_n.$$

We will show that this element ζ solves the Problem (13.20). First, passing to the limit in Equation (13.20$_n$) we get

$$\zeta = \lim_{n\to\infty} \zeta_{n+1} = \lim_{n\to\infty}\left(\zeta_0 + \delta^{-1}\left(\nabla\zeta_n + \frac{1}{2}[\zeta_n,\zeta_n]\right)\right)$$

$$= \zeta_0 + \delta^{-1}\left(\nabla(\lim_{n\to\infty}\zeta_n) + \frac{1}{2}\left[\lim_{n\to\infty}\zeta_n, \lim_{n\to\infty}\zeta_n\right]\right)$$

$$= \zeta_0 + \delta^{-1}\left(\nabla\zeta + \frac{1}{2}[\zeta,\zeta]\right),$$

since the operators that respect the total degree commute with the limit. Next,

$$\delta^{-1}\zeta = \delta^{-1}\left(\zeta_0 + \delta^{-1}\left(\nabla\zeta + \frac{1}{2}[\zeta,\zeta]\right)\right) = \delta^{-1}\zeta_0 = 0,$$

since $(\delta^{-1})^2 = 0$. Finally, $\zeta_{00} = 0$ because the image of δ^{-1} consists of the elements of strictly positive bi-degrees.

It remains to check that taking $\Delta\Gamma = \omega(y) + \zeta$ where ζ solves the Problem (13.20) we shall get the necessary result. Put $A = \Omega_\zeta - (-\frac{\sqrt{-1}}{\hbar}\omega + \varphi)$, then we compute:

$$\begin{aligned}\delta^{-1}(A) &= \delta^{-1}\left(\Omega_0 - \delta\zeta + \nabla\zeta + \frac{1}{2}[\zeta,\zeta]\right) \\ &= \zeta_0 + \delta^{-1}\left(\nabla\zeta + \frac{1}{2}[\zeta,\zeta]\right) - \delta^{-1}\delta\zeta \\ &= \zeta - \delta^{-1}\delta\zeta = 0,\end{aligned}$$

since $\delta^{-1}\delta\zeta = \zeta$ under the assumptions we made. Similarly

$$D(A) = D\left(\Omega_\zeta - \left(-\frac{\sqrt{-1}}{\hbar}\omega + \varphi\right)\right) = D(\Omega_\zeta) - d\left(-\frac{\sqrt{-1}}{\hbar}\omega + \varphi\right) = 0,$$

where we used the Bianchi identity (see (12.3)) for Ω_ζ and the fact that D coincides with de Rham differential d on the center of Weyl algebra (since all the additional terms are given by commutators). But $D(A) = -\delta A + \nabla A + [\zeta, A]$, so $\delta A = \nabla A + [\zeta, A]$. Also observe that by definition $A_{00} = 0$ (in fact A is a 2-form with values in Weyl algebra), so

$$A = (\delta\delta^{-1} + \delta^{-1}\delta)A = \delta^{-1}\delta A = \delta^{-1}(\nabla A + [\zeta, A]). \qquad (13.21)$$

But the operator δ^{-1} increases the total filtration, see 13.3, and all the other operations involved in the right-hand side of this equality do not change it. Hence Equation (13.21) implies that $A = 0$, and hence $\Omega_\zeta = -\frac{\sqrt{-1}}{\hbar}\omega + \varphi$.

14 Fedosov quantisation and its properties

In this section, we will continue the study of Fedosov's construction of deformation quantisation, which we started in Section 13. In particular, we will use all notation and agreements, introduced in that section: in particular, (M, ω) will denote a symplectic manifold with symplectic form ω; $\Omega^*(M)$ is the de Rham complex of M and d the de Rham operator, etc.

Recall that in previous section we constructed the fibre bundle $\widehat{W}_\hbar^{\mathbb{C}}(M, \omega)$ (the bundle of completed complex Weyl algebras on (M, ω)) whose space of (smooth) sections $\widehat{W}_\hbar(M)$ is an associative (but noncommutative) algebra with respect to the pointwise product; we shall denote the product of two sections $a, b \in \widehat{W}_\hbar(M)$ either by $a \cdot b$ or simply by ab. We also denoted by $\widehat{W}_\hbar^*(M)$ the space of $\widehat{W}_\hbar^{\mathbb{C}}(M, \omega)$-valued differential forms on M, i.e. $\widehat{W}_\hbar^*(M)$ is the space of sections of the tensor product bundle $\Lambda^*(T^*M) \otimes \widehat{W}_\hbar^{\mathbb{C}}(M, \omega)$; as before $\widehat{W}_\hbar^*(M)$ is a graded algebra with respect to the point-wise product (we shall use the same notation for it).

After this we defined Abelian (or *Fedosov's*) connections on $\widehat{W}_\hbar(M, \omega)$ as connections D, whose curvature operator on $\widehat{W}_\hbar(M, \omega)$ is equal to the commutator with a central element in $\widehat{W}_\hbar^2(M)$ and hence vanishes identically. We proved that *for any $\varphi \in \Omega^2(M)[[\hbar]]$, $d\varphi = 0$, there exists an Abelian connection with curvature equal to the commutator with* $-\frac{\sqrt{-1}}{\hbar}\omega + \varphi$.

In this section we shall finalise the construction of deformation quantisation and study some of its properties.

14.1 Fedosov's construction: The isomorphism

First of all let us conclude the proof of Fedosov quantisation theorem by identifying the space of D-flat sections of $\widehat{W}_\hbar^{\mathbb{C}}(M, \omega)$

$$\widehat{W}_D(M) \subseteq \widehat{W}_\hbar(M) = \widehat{W}_\hbar^0(M), \ \widehat{W}_D(M) = \{a \in \widehat{W}_\hbar(M), Da = 0\}$$

(where D is a fixed Fedosov connection) with $C^\infty(M)[\hbar^{-1}, \hbar]]$. To this end, we consider two cochain complexes:

$$0 \to \widehat{W}_\hbar^0(M) \xrightarrow{D} \widehat{W}_\hbar^1(M) \xrightarrow{D} \widehat{W}_\hbar^2(M) \xrightarrow{D} \ldots$$
$$\ldots \xrightarrow{D} \widehat{W}_\hbar^{2n-1}(M) \xrightarrow{D} \widehat{W}_\hbar^{2n}(M) \to 0; \quad (14.1)$$

$$0 \to \widehat{W}_\hbar^0(M) \xrightarrow{\delta} \widehat{W}_\hbar^1(M) \xrightarrow{\delta} \widehat{W}_\hbar^2(M) \xrightarrow{\delta} \ldots$$
$$\ldots \xrightarrow{\delta} \widehat{W}_\hbar^{2n-1}(M) \xrightarrow{\delta} \widehat{W}_\hbar^{2n}(M) \to 0, \quad (14.1')$$

where $\delta a = [\omega(y), a]$ is the operator we studied in Sections 13.3.3, 13.3.4 (loosely speaking $\delta = dx^i \wedge \frac{\partial}{\partial y^i}$, see Equation (13.14)); we know that $\delta^2 = 0$. The cohomology of the complex (14.1) in degree 0 clearly coincides with $\widehat{W}_D(M)$; and due to the decomposition (13.19) we see that the cohomology of the complex (14.1') can be easily computed. Indeed, for every $a \in \widehat{W}_\hbar^p(M)$ with $p > 0$ we have $a_{00} = 0$, so $a = (\delta\delta^{-1} + \delta^{-1}\delta)a$. Hence if $\delta a = 0$, then $a = \delta(\delta^{-1}a)$; and if $a \in \widehat{W}_\hbar^0(M)$, $\delta a = 0$, then for the same reason $a - a_{00} = \delta(\delta^{-1}a)$. Summing up we get:

$$H^p(\widehat{W}_\hbar^*(M), \delta) = \begin{cases} \widehat{W}_\hbar^0(M)_{00}, & p = 0, \\ 0, & p > 0. \end{cases} \quad (14.2)$$

Here $\widehat{W}_\hbar^0(M)_{00}$ is the set of elements in Weyl algebra of bi-degree $(0,0)$; it's easy to see that $\widehat{W}_\hbar^0(M)_{00} = C^\infty(M)[\hbar^{-1}, \hbar]]$; clearly replacing δ by $-\delta$ does not change this result. We are now going to construct an isomorphism of complexes

$$Q : (\widehat{W}_\hbar^*(M), -\delta) \xrightarrow{\cong} (\widehat{W}_\hbar^*(M), D),$$

which will show that the cohomology of $(\widehat{W}_\hbar^*(M), D)$ is also concentrated in degree 0 and is equal to $C^\infty(M)[\hbar^{-1}, \hbar]]$; moreover, restricting Q to the cohomology we get a linear isomorphism

$$Q : C^\infty(M)[\hbar^{-1}, \hbar]] \to \widehat{W}_D(M)$$

and this completes the construction of Fedosov's $\star$-product, see Section 13.1.

So let $b \in \widehat{W}_\hbar^*(M)$; recall that for all $a \in \widehat{W}_\hbar^*(M)$ we have $Da = \nabla a - \delta a + [\zeta, a]$. For a given b consider the equation

$$a = b + \delta^{-1}(D + \delta)a. \tag{14.3}$$

Observe that the operator $D + \delta = \nabla + [\zeta, \cdot]$ does not change the total grading (see Definition 13.3) and δ^{-1} increases it by 1. Now we can use an iterative process, similar to the one employed in construction of ζ (see Section 13.3.5) to solve Equation (14.3). Namely:

$$a_0 = b,$$
$$a_1 = b + \delta^{-1}(D + \delta)a_0,$$
$$a_2 = b + \delta^{-1}(D + \delta)a_1,$$
$$\dots\dots\dots$$
$$a_n = b + \delta^{-1}(D + \delta)a_{n-1},$$
$$\dots\dots\dots$$

Then $a_n - a_{n-1} = \delta^{-1}(D+\delta)(a_{n-1} - a_{n-2})$, and by induction the total degree of the difference $a_n - a_{n-1}$ tends to infinity. Hence we conclude that there exists $a = \lim_{n\to\infty} a_n$, which solves Equation (14.3). This a is uniquely defined for any b, since otherwise the difference $a - a'$ of any two solutions of (14.3) would satisfy the equation

$$a - a' = \delta^{-1}(D + \delta)(a - a'),$$

and comparison of the total degrees on both sides shows that $a - a' = 0$. We now put $Q(b) = a$; here Q can be regarded as a set of $\hbar$-linear automorphisms $Q^p : \widehat{W}_\hbar^p(M) \to \widehat{W}_\hbar^p(M)$ for all $p \geq 0$ (just observe that Equation (14.3) is $\hbar$-linear).

Let us prove the equality $Q^{-1}D + \delta Q^{-1} = 0$ then $D = -Q\delta Q^{-1}$, so Q is the isomorphism of complexes (14.1) and (14.1'), which we sought; Equation (14.3) means $Q^{-1}a = a - \delta^{-1}(D + \delta)a$, so we compute

$$\begin{aligned}(Q^{-1}D + \delta Q^{-1})(a) &= Da - \delta^{-1}(D+\delta)Da + \delta a - \delta\delta^{-1}(D+\delta)a\\ &= (D+\delta)a - (\delta\delta^{-1} + \delta^{-1}\delta)Da - \delta\delta^{-1}\delta a\\ &= (D+\delta)a - (\delta\delta^{-1} + \delta^{-1}\delta)(D+\delta)a\\ &= (D+\delta)a - (D+\delta)a = 0,\end{aligned}$$

where we used the equalities $D^2 = \delta^2 = 0$ and the decomposition (13.19) applied to the element $(D+\delta)a$, which has trivial component of bidegree $(0,0)$.

Remark 14.1. A reader, familiar with the homological perturbation theory will easily recognise the main result of this theory, *homology perturbation lemma* in the construction of map Q. In effect, we used the decomposition $D = -\delta + D'$, where the map $D' = D + \delta$ is regarded as an infinitesimal perturbation of the differential $-\delta$. Then δ^{-1} is the cochain homotopy operator for δ, which satisfies the condition $(\delta^{-1})^2 = 0$ and Q^{-1} is obtained by perturbation of the identity map corresponding to D', as it is prescribed by the perturbation lemma; here the identity map is regarded as a self-isomorphism of the complex $(\widehat{W}_\hbar^*(M), -\delta)$. This approach matches with the one we used earlier in this section; in particular the following formula for Q can be obtained from either of them:

$$Q = (\mathbb{1} - \delta^{-1}D')^{-1} = \sum_{k=0}^{\infty} (\delta^{-1}D')^k. \qquad (14.4)$$

Similarly, we have $Q^{-1} = \mathbb{1} - \delta^{-1}D'$. Finally, observe that for any flat $a \in \widehat{W}_D(M)$, we have

$$Q^{-1}(a) = a - \delta^{-1}(D+\delta)(a) = a - \delta^{-1}\delta(a) = \delta\delta^{-1}a + a_{00} = a_{00}.$$

Thus in this case Q^{-1} is given just by projection to the bidegree $(0,0)$, or equivalently by setting all noncommutative variables y^i to 0.

Remark 14.2. Recall, that any Fedosov connection has the form

$$D(a) = \nabla a + [\varDelta\varGamma, a],$$

where $\varDelta\varGamma$ is a global 1-form with values in $\widehat{W}_\hbar^\mathbb{C}(M, \omega)$; we further represented $\varDelta\varGamma$ in the form

$$\varDelta\varGamma = \omega(y) + \zeta, \qquad (14.5)$$

where ζ is a Weyl-algebra valued 1-form, with the condition that $\zeta_{00} = 0$ and $\delta^{-1}\zeta = 0$. It turns out (see Exercise E.5.7) that the inverse is also true: if $D^2 = 0$, then one has the formula (14.5).

This observation allows us assume that the symplectic connection ∇ is the same in any two Fedosov connections: the difference of any two connections is given by a global tensorial 1-form, which can be merged with $\Delta\Gamma$, see formula (13.10).

14.2 Properties of Fedosov's deformation quantisation

Summing up the results from Sections 13.2, 13.3 and 14.1, we come to the following conclusion:

Proposition 14.3 (Fedosov's $\star$-product). *For every symplectic manifold M there exist Abelian (Fedosov's) connections on the completed Weyl bundle $\widehat{W}_\hbar^\mathbb{C}(M,\omega)$; for any such connection D, there exists an $\hbar$-linear isomorphism Q of the space $C^\infty(M)[\hbar^{-1},\hbar]]$ of formal Laurent series in $\hbar$ with coefficients in smooth $\mathbb{C}$-valued functions and the space $\widehat{W}_D(M)$ of D-flat sections of $\widehat{W}_\hbar^\mathbb{C}(M,\omega)$, so that the formula*

$$f \star g = Q^{-1}(Q(f) \cdot Q(g)), \quad f,g \in C^\infty(M)[\hbar^{-1},\hbar]] \qquad (14.6)$$

determines an associative $\hbar$-linear product in $C^\infty(M)[\hbar^{-1},\hbar]]$.

We shall call the product, determined by formula (14.6), *Fedosov product*. It is clear that the construction depends on the choice of Fedosov connection D. Let us describe the properties of this $\star$-product and also the classification of these products up to the equivalence relation. The reader should compare these results with the properties and classification of the $\star$-products, given by Lecomte and de Wilde's construction.

14.2.1 Fedosov product is a $\star$-product

First of all let us check, that this product is indeed a $\star$-product, i.e. that all the conditions of the deformation quantisation problem 1 hold (except maybe for the technical condition 4.).

(i) Formula (14.4) shows that the map Q is equal to the sum of iterations of the operator $\delta^{-1}D'$, where $D' = \nabla + [\zeta,\ ]$. Since locally

$$\nabla = \sum_{i=1}^{2n} dx^i \wedge \frac{\partial}{\partial x^i} + \frac{\sqrt{-1}}{\hbar}[\Gamma_{ijk}y^j y^k dx^i,\ ],\ \delta^{-1} = \frac{1}{p+q}\sum_{k=1}^{2n} y^k \circ \frac{\partial}{\partial dx^k},$$

where (p,q) is the bidegree of the argument, see Equations (13.10), (13.15) and (13.17), and since the product · in Weyl algebra is just the fibre-wise Moyal product we conclude that

$$Q(f) = f + \sum_{k=1}^{\infty} Q_k(f) \qquad (14.7)$$

where Q_k, $k = 1, 2, \ldots$ are some fixed by the choice of D Weyl-algebra valued differential operators on functions, and k denotes the total filtration of the image, see Definition 13.3. Similarly, since the Moyal product is formulated as a formal sum of bidifferential operators and from the description of Q^{-1} (see Remark 14.1) we conclude that the general formula for Fedosov $\star$-product (14.6) is expressed as a formal sum of bidifferential operators. Also it is evident that the "constant term" of the formula (14.6) is equal just to fg.

(ii) At degree 1 in $\hbar$ the $\star$-product (14.6) is obtained from $Q^{-1}((Q_0 + Q_1)(f) \cdot (Q_0 + Q_1)(g))$, where we use the symbol Q_k to denote the degree k part of Q, since the terms with higher filtration will end up with higher degrees in $\hbar$ (recall that the filtration is given by putting $\deg y^k = 1$, $\deg \hbar = 2$). But $Q_0(f) = f$ and

$$Q_1(f) = \delta^{-1} D'(f) = \delta^{-1}(\nabla + [\zeta,\ ])(f) = \delta^{-1} df = \sum_{k=1}^{2n} \frac{\partial f}{\partial x^k} y^k.$$

Hence the commutator of f and g with respect to the product (14.6) in degree 1 with respect to $\hbar$ is determined by the following equation (where we denote by ... the terms of degrees 2 and higher in $\hbar$ and use the fact that f, g commute with everything)

$$f \star g - g \star f = Q^{-1}\left(\left[f + \frac{\partial f}{\partial x^i} y^i, g + \frac{\partial g}{\partial x^j} y^j\right]\right) + \ldots$$

$$= Q^{-1}\left(\left[\frac{\partial f}{\partial x^i} y^i, \frac{\partial g}{\partial x^j} y^j\right]\right) + \ldots$$

$$= \frac{\partial f}{\partial x^i} \frac{\partial g}{\partial x^j} Q^{-1}([y^i, y^j]) + \cdots = \frac{\hbar}{\sqrt{-1}} \pi^{ij} \frac{\partial f}{\partial x^i} \frac{\partial g}{\partial x^j} + \ldots$$

$$= \frac{\hbar}{\sqrt{-1}} \{f, g\} + \ldots.$$

Here the square brackets stand for the commutator in Weyl algebra and we use the generating relation for Weyl algebra, see (13.1). A bit more accurate analysis, which we skip, shows that in effect the first term of the $\star$-product (14.6) is equal to $\frac{\hbar}{2\sqrt{-1}}\{f,g\}$, as prescribed by the condition 2 of Problem 1, see Exercise E.5.9.

(iii) Since $d1 = 0$ and since 1 is in the center of Weyl algebra, we have $Q(1) = 1$. Hence

$$f \star 1 = Q^{-1}(Q(f) \cdot Q(1)) = Q^{-1}(Q(f) \cdot 1) = Q^{-1}(Q(f)) = f,$$

and similarly $1 \star f = f$ for all $f \in C^\infty(M)[\hbar^{-1}, \hbar]]$.

14.2.2 Classification of Fedosov products

After the existence of $\star$-products has been demonstrated, it is natural to ask, how many different products of this sort there exist. Recall that two $\star$-products are equivalent if there exists a formal power series of linear differential operators $T = \mathbb{1} + \hbar T_1 + \hbar^2 T_2 + \ldots$, such that

$$T(f \star_1 g) = T(f) \star_2 T(g).$$

As we have already mentioned, the construction of Fedosov product depends on the choice of Abelian (Fedosov's) connection D. According to Theorem 13.4, the connection D is constructed when a closed form $\varphi \in \Omega^2(M)[[\hbar]]$ is fixed; following Fedosov's terminology we shall call the form φ *(Weyl's) curvature* of D. It is natural to expect that the equivalence class of Fedosov product depends on the cohomology class of φ. And this is the case, as shows the following theorem:

Theorem 14.4.

(i) *Two Fedosov products $\star_1$ and $\star_2$ constructed for the Abelian connections D_1 and D_2 are equivalent iff the connections D_1 and D_2 are gauge equivalent, i.e. if there exists an invertible section $c \in \widehat{W}_\hbar^0(M)$ of the form $c = 1 + O(\hbar)$, such that $D_1(\alpha^c) = D_2(\alpha)^c$ for any $\alpha \in \widehat{W}_\hbar^*(M)$.*[39]

---

[39] Here x^c denotes the conjugation by c, i.e. $x^c = c \cdot x \cdot c^{-1}$.

(ii) Fedosov connections D_1 and D_2 are equivalent if and only if the classes of their curvatures coincide, i.e. iff $[\varphi_1] = [\varphi_2] \in H^2(M)[[\hbar]]$.

The element $\frac{\sqrt{-1}}{\hbar}[\omega] + [\varphi] \in \frac{\sqrt{-1}}{\hbar}[\omega] \oplus H^2(M)[[\hbar]]$ is called *the characteristic class of Fedosov connection* or just *Fedosov characteristic class*. It follows from Theorem 14.4 that equivalence class of Fedosov product is uniquely determined by the characteristic class of the corresponding connection; in particular, equivalence classes of Fedosov's $\star$-products correspond bijectively to the elements of $H^2(M)[[\hbar]]$.

Let us sketch the *proof* of Theorem 14.4. An interested reader can find details in the original book by Fedosov, see [Fed96].

(i) It is clear, that conjugation by an invertible element sends a Fedosov connection to another Fedosov connection; in effect, it is clear that for any c and any Fedosov connection $D_1 = \nabla_1 + [\Delta\Gamma_1,\ ]$, the formula
$$D_2(x) = c^{-1} D_1(cxc^{-1})c,$$
gives again a connection of the form $D_2 = \nabla_2 + [\Delta\Gamma_2,\ ]$, for which $D_2^2 = 0$, hence a Fedosov connection, see Remark 14.2.

Recall that locally every symplectic connection can be written as $\nabla = d + [\Gamma(y),\ ]$ for some local Weyl-algebra valued differential 1-forms $\Gamma(y)$; so Fedosov connections D_1, D_2 locally are given by the formula
$$D_i = d - \delta + [\tilde{\zeta}_i,\ ], \, i = 1, 2$$
for some locally defined Weyl-algebra valued differential forms $\tilde{\zeta}_i$; then straightforward computations (c.f. Exercise E.5.10), show that if $D_2(x) = c^{-1} D_1(cxc^{-1})c$, (below we shall denote this relation $D_2 = D_1^c$) then
$$\tilde{\zeta}_2 = c^{-1}\tilde{\zeta}_1 c + c^{-1}dc - c^{-1}\delta c. \tag{14.8}$$

Clearly, conjugation by c^{-1} gives an isomorphism of the algebra $\widehat{W}_{D_1}(M)$ of D_1-flat sections with the algebra $\widehat{W}_{D_2}(M)$ of D_2-flat sections. Since the morphism Q of Section 14.1 is uniquely defined by the choice of Fedosov connection D, we conclude that
$$Q_2(f) = (Q_1(f))^{c^{-1}}, \text{ for all smooth functions } f \in C^\infty(M),$$

i.e. $c \cdot Q_2(f) \cdot c^{-1} = Q_1(f)$, for all f; on the other hand a simple unravelling of definitions shows that $c \cdot Q_2(f) \cdot c^{-1} = Q_2(T(f))$ for a formal sum of differential operators T beginning with identity $\mathbb{1}$, so

$$\begin{aligned}Q_2(T(f \star_1 g)) &= cQ_1(f \star_1 g)c^{-1} = c \cdot (Q_1(f) \cdot Q_1(g)) \cdot c^{-1} \\ &= (cQ_1(f)c^{-1})(cQ_1(g)c^{-1}) = Q_2(T(f)) \cdot Q_2(T(g)) \\ &= Q_2(T(f) \star_2 T(g)),\end{aligned}$$

hence T does indeed conjugate $\star_1$ with $\star_2$. The inverse statement (i.e. that equivalent products correspond to equivalent Fedosov connections) is based on the fact that locally any Fedosov product is isomorphic to the Moyal product, i.e. to Weyl algebra on an open neighbourhood of a point equipped with Darboux's coordinates, see below Remark 14.5. On the other hand restriction of the composition $Q_2TQ_1^{-1}$ to such an open neighbourhood $U \subseteq M$ gives an automorphism $\widehat{W}_{D_1}(U) \to \widehat{W}_{D_2}(U)$, which is equal to a conjugation by some element c_U; using a partition of unity we can restore the global element $c \in \widehat{W}_\hbar^0(M)$ such that the operator T is induced by conjugation with c. In particular, conjugation by c induces an isomorphism of the flat sections $\widehat{W}_{D_1}(U) \to \widehat{W}_{D_2}(U)$ and since locally every section of a vector bundle is uniquely represented as a linear combination of flat sections with functional coefficients, it means that conjugation by c intertwines the connections: $D_2 = D_1^c$.

(*ii*) Observe that since $c = 1 + O(\hbar)$, replacing $\hbar$ with $t\hbar$ we obtain a smooth family $c(t)$ of invertible elements, $c(0) = 1$, $c(1) = c$. We're going to prove a more general statement now: *for any smooth family $c(t)$, $t \in [0,1]$ of invertible elements the characteristic class of $D(t) = D^{c(t)}$ is constant: $\varphi(t) \equiv \varphi(0) \in H^2(M)[[\hbar]]$.*

To this end we let $\Omega_D(t)$ denote the curvature of the connection $D(t) = \nabla + [\Delta\Gamma(t), \ ]$; for our purposes it is enough to show that

$$\frac{d\Omega_D(t)}{dt} \in d\Omega^1(M)[[\hbar]] \text{ for all } t \in [0,1].$$

We consider the derivation $d(t) = \frac{d}{dt}D(t)$ at some moment $t \in [0,1]$. Since $D(t)$ was a differentiation of the algebra $\widehat{W}_\hbar^*(M)$, so is $d(t)$; since $d(t)$ acts trivially on $\Omega^*(M)[\hbar^{-1}, \hbar]] \subseteq \widehat{W}_\hbar^*(M)$ (in effect all

Fedosov connections act the same way on this space[40]), it is given by a commutation with an element $\gamma(t) = \frac{d\Delta\Gamma(t)}{dt} \in \widehat{W}_\hbar^1(M)$. Finally, since the map $d(t)$ preserves the space of flat sections, $\gamma(t)$ commutes with flat sections and so (again, we use the local isomorphism of Fedosov quantisation with Weyl algebra, see Remark 14.5) $\gamma(t)$ is equal to an element in $\Omega^1(M)[[\hbar]]$. Next, since (see Exercise E.5.11)

$$\frac{d\Omega_D(t)}{dt} = \nabla\gamma(t) + [\Delta\Gamma(t), \gamma(t)] = d\gamma(t),$$

the conclusion follows.

The opposite statement (that equal characteristic classes imply equivalent Fedosov connections) is proved in two steps. Let $\lambda \in \Omega^1(M)[[\hbar]]$ be a formal power series of 1-forms on M such that $\varphi_2 = \varphi_1 - d\lambda$. First, we construct a homotopy of Fedosov connections $D(t)$ on $\widehat{W}_\hbar^*(M)$, such that $D(0) = D_1$, and for every $t \in [0,1]$ the connection $D(t)$ is determined by the closed form $\varphi_t = \varphi_1 - td\lambda$. The construction is obtained by a modification of the original construction of the Abelian connections. Then $D(1) = D_2$, which is due to the fact that Fedosov's construction gives a unique Abelian connection for any closed form φ.

Thus we end up with a smooth homotopy $D(t)$ of Abelian connections between D_1 and D_2. The next step is to show that every smooth homotopy of Abelian connections is of the form $D(t) = D^{c(t)}$ for a 1-parameter family of invertible elements $c(t) \in \widehat{W}_\hbar(M)$, $t \in [0,1]$, $c(0) = 1$. So we fix such a homotopy $D(t)$, $t \in [0,1]$; let us denote by $\Delta\Gamma(t)$ the corresponding correction terms, so that $D(t)(a) = \nabla a + [\Delta\Gamma(t), a]$, see Remark 14.2 again.

Let $\Omega_D(t)$ denote the curvature form of $D(t)$; here it is just $\frac{\sqrt{-1}}{\hbar}\omega + td\lambda$. Let $\lambda(t) \in \Omega^1(M \times [0,1])[[\hbar]]$ be the smooth 1-parameter family of formal power series of 1-forms such that

$$\frac{d\Omega_D(t)}{dt} = -d\lambda(t).$$

In the example we consider above this equation holds trivially with $\lambda(t) \equiv \lambda$. We set the "quantum Hamiltonian function" $H(t) \in$

---
[40]One can also prove this by direct computation.

$\widehat{W}_\hbar(M \times [0,1])$, to be the solution of the equation

$$D(t)H(t) = \lambda(t) + \frac{d_\Delta \Gamma(t)}{dt}. \qquad (14.9)$$

The existence of $H(t)$ follows from the equality

$$D(t)\left(\lambda(t) + \frac{d_\Delta \Gamma(t)}{dt}\right) = -\frac{d\Omega_D(t)}{dt} + \frac{d\Omega_D(t)}{dt} = 0,$$

so that the iterative process can be used to find $H(t)$ (see Exercise E.5.12). Given $H(t)$ we can solve the (formal) Heisenberg equation:

$$\frac{da(t)}{dt} + \frac{\sqrt{-1}}{\hbar}[H(t), a(t)] = 0. \qquad (14.10)$$

Taking for $t = 0$ the initial condition $a(0)$ to be a $D(0)$-flat section, $a(0) \in \widehat{W}_{D(0)}$ we see that $a(t)$ is a $D(t)$-flat section. On the other hand, it turns out that $a(t)$ can be written in the form of a conjugation by a family $c(t)$ of invertible elements in $\widehat{W}_\hbar(M)$, which solves the problem, i.e. $D(t) = D^{c(t)}$; see Fedosov's book [Fed96] for details. □

Let us conclude this section by two important observations.

Remark 14.5. (*i*) The whole construction of isomorphisms of Fedosov's algebras can be extended to incorporate a nontrivial flow of diffeomorphisms of the base manifold, $f_t : M \to M$; if f_t are not symplectomorphisms, then the symplectic structure on M gives rise to a flow $(f_t^{-1})^*\omega = \omega(t)$ of symplectic structures on M. In this case one should consider a flow of symplectic connections $\nabla(t)$ with curvatures $\Omega_\nabla(t)$. Then, the following theorem is proved in Fedosov's book [Fed96]:

Theorem 14.6. *Let $f_t : M \to M$ be a diffeomorphism flow on M, for which $f_0 = \mathrm{id}$; let $\omega(t)$ be the corresponding family of symplectic forms and $\nabla(t)$ be a 1-parameter family of symplectic connections over $\omega(t)$. Suppose*

(*i*) $(f_t^{-1})^*(\Omega_\nabla(0)) = \Omega_\nabla(t)$;

(*ii*) $\frac{d\Omega_{D(t)}}{dt} = -d\lambda(t)$ *for a family of 1-forms $\lambda(t) \in \Omega^1(M)[\hbar^{-1}, \hbar]]$.*

Then there exists a 1-parameter family $c(t)$ of invertible elements in $\widehat{W}_\hbar(M)$ such, that the conjugation by $c(t)$ is an isomorphism between $\widehat{W}_{D(0)}(M)$ and $\widehat{W}_{D(t)}(M)$.

The proof is obtained by suitable modification of the reasonings we used in part (ii) of Theorem 14.4; in particular we should change the Heisenberg equation by its suitable modification, in which the vector field X_t, whose flow induces the diffeomorphism f_t plays an important role.

Applying this theorem to the case $M = U$, a small open neighbourhood in M, and f_t a flow of diffeomorphisms, induced by Darboux's coordinates change on U we can prove that locally Fedosov quantisation coincides with the Moyal one, and hence to the (completed) Weyl algebra, the fact that we used few times in this section.

(ii) Geometrically speaking, the construction, which we sketched in Theorem 14.4 can be described as follows: we have a flat bundle $(\widehat{W}_\hbar^{\mathbb{C}}(M,\omega), D)$ (here D is the flat connection) over a manifold M, which we extend to a family of flat bundles $(\widehat{W}_\hbar^{\mathbb{C}}(M,\omega(t)), D(t))$, $D(0) = D$ over $M \times [0,1]$.

We then extend the connections $D(t)$ to a flat connection over $M \times [0,1]$ by adding to $D(t)$ the term $\frac{d}{dt} \wedge dt + [H(t),\ ]$; Equation (14.9) is just the condition of flatness of this connection. Then the Heisenberg equation amounts to finding flat sections over the Cartesian product, which is just the parallel transport of the flat sections for $t = 0$ in t direction. Finally, just like in the usual Differential Geometry solution of the parallel transport equation can be obtained with the help of the time-ordered exponent of the connection, acting on the initial value of the section, which is precisely the conjugation we seek.

15 Properties, generalisations and applications of Fedosov quantisation (a survey)

Although Fedosov's construction of deformation quantisation that we have just discussed is quite interesting and deserves much attention on its own, its true value becomes clear from various and numerous applications that the ideas involved in this construction find in other branches of Mathematics. Giving account of all possible generalisations and results related with Fedosov quantisation would make these notes too long, so we shall only give few examples of our choice. Even in this case going deep into the details of proofs would be out of reach of our lecture course, so in all cases we will only sketch the main ideas related with the theorems and give a reference to the original works for an interested reader.

15.1 Trace and algebraic index theorem

The first application of Fedosov's construction was given by Fedosov himself. In effect, the original motivation of Fedosov lay within the scope of *Index theory*: he used his construction of deformation quantisation to derive an algebraic analog of the Atiyah–Singer index theorem, see [Fed91] for example. So in order to explain his results let us briefly recall the main ideas of the Index theory. Of course giving a comprehensive sketch of this old and renowned branch of modern Mathematics if at all feasible, is far beyond the modest purposes of our lecture notes. The reader interested in getting a more complete information on this subject should consult the book of Palais [Pa65] or any other textbook on this subject.[41]

15.1.1 Outline of the index theory

Let $E, F \to X$ be two (complex) vector bundles over a manifold; *differential operator of degree p* between these bundles, is any linear operator $D : \Gamma^\infty(E) \to \Gamma^\infty(F)$ on the space of smooth sections

---

[41] One can skip this section at the first reading; just pay attention to the formulas (15.2) and (15.3).

of E, which in local coordinates is expressible as the sum of partial derivatives of degrees less or equal to p with matrix-valued coefficient functions, i.e. in notation of Section 2.1.3 (here $I = (i_1, \ldots, i_n)$ is a multiindex):

$$D(s) = \sum_{|I| \leq p} A_I(x) \partial^I(s),$$

where $A_I(x)$ are matrix-valued functions in local coordinates $x^1, \ldots, x^n$. These operators can be extended to continuous operators $D : H^r(E) \to H^{r-p}(F)$, where $H^r(E)$, $H^r(F)$, $r \in \mathbb{R}$ are the *Soboloev spaces*, the completions of the spaces of smooth sections of E and F with respect to certain functional norms, called the Sobolev's norms. Thus the ideas and methods from Functional Analysis are introduced into this area. In particular, one can look for the Fredholm operators, i.e. the operators $S : V \to W$ between the Banach spaces, for which $\dim \ker S < \infty$, $\dim \operatorname{coker} S < \infty$, and try to compute their indices:

$$\operatorname{ind} S = \dim \ker S - \dim \operatorname{coker} S. \qquad (15.1)$$

Index of an operator is an important analytic invariant: it depends continuously on the operator and hence is invariant under homotopies and small perturbations. It is used in many situations; in particular it can be used to evaluate the number of solutions of partial differential equations.

It turns out that in the case of a differential operator $D : E \to F$ there exists a geometric characterisation of Fredholm operators: let

$$\sigma(D)(x, \xi) = \sum_{|I|=p} A_I(x) \xi^I$$

be the *principal symbol* of D; it is a polynomial in the direction of fibres function on T^*X with values in the sections of $Hom(E, F)$-bundle. One says that D is *elliptic*, if $\sigma(D)$ is invertible outside of the zero section of T^*X (i.e. if $(\xi_1, \ldots, \xi_n) \neq (0, \ldots, 0)$). Then *elliptic operators naturally extend to Fredholm operators between Sobolev spaces; their indices depend continuously only on their principal symbols.*

One theoretical way to calculate the index (15.1) of a Fredholm operator D on a Hilbert space is to find its quasi-inverse operator T, i.e. such an operator for which both compositions DT and TD are equal to 1 modulo compact[42] operators. In particular, both differences $1 - DT$ and $1 - TD$ are trace class (i.e. the notion of trace is well defined for them) and we have:

$$\operatorname{ind} D = tr(1 - TD) - tr(1 - DT). \tag{15.2}$$

In case of elliptic operators the operator T is called *parametrix* of D; on informal level it can be constructed as a differential operator, whose symbol is inverse to the principal symbol of D outside the zero section of T^*X. In order to give a rigorous meaning to this idea, we have to expand the domain of our theory so that it will include the calculus of *pseudo-differential operators*; a reader not familiar with the strict definitions might regard these operators as the generalisation of differential operators, obtained by a method, similar to the way we introduced the canonical quantisation on $\mathbb{R}^{2n}$ in Section 2.1.3, i.e. by taking composition of the Fourier transform, multiplication by a function and the inverse Fourier transform. However, here we need to do some work so as to extend the construction from Euclidean space to a general manifold and from Schwarz functions to the Sobolev space of sections of a vector bundle; in these notes we completely omit this theory.

It follows from the continuity property of the index that $\operatorname{ind} D$ is a homotopy invariant of the deformations of manifold and the fibre bundles (as long as these deformations preserve its ellipticity), thus we come up with the problem of finding a formula that would express this invariant in the terms of classical topological invariants of E, F and X. This problem was originally formulated in 1950s and was successfully resolved by Atiyah and Singer in the beginning of 1960s, see the paper [AS63] (see also the papers [AS68$_1$], [AS68$_2$], [AS71] where details of the proof and some generalisations of the result are given). The answer is given by the famous *Atiyah–Singer index formula*.

---

[42] A simple way to define compact operators is to say that they are the operators lying in the completion of the space of operators with finite-dimensional image.

Let us briefly explain the main ingredients of this formula. First of all, since the principal symbol $\sigma(D)$ is invertible outside the zero section of T^*X, it determines an isomorphism of vector bundles

$$\sigma(D)|_{S(T^*X)} : p^*(E) \to p^*(F),$$

where $p : S(T^*X) \to X$ is the spherical subbundle inside T^*X, i.e. the space of unit covectors in T^*X (unit with respect to a Riemannian metric) and p^*E, p^*F are the pull-backs of the bundles E and F onto the spherical bundle. One can now use this isomorphism to "sew together" the bundles p^*E and p^*F along the boundary $S(T^*X)$ of the disc bundle $D(T^*X)$ (i.e. the set of all covectors $v \in T^*X$, $|v| \le 1$). This construction gives a vector bundle over the space $D(T^*X)/S(T^*X)$, which is homeomorphic to the *Thom space* $Th(T^*X)$ of T^*X:

$$Th(T^*X) = T^*X/T_0^*X,$$

where T_0^*X is the set of all nonzero covectors.

This bundle is called *the Atiyah difference construction* or *Atiyah difference element*, denoted by $d(D; E, F)$. We then compute the Chern character[43] of $d(D; E, F)$, which gives a (nonhomogenous) class $ch(d(D; E, F))$ in the even de Rham cohomology (or even rational cohomology) of $Th(T^*X)$. Now, according to the Thom's theorem, there is a degree-shifting isomorphism of cohomology φ : $\tilde{H}^*(X) \cong H^{*+\dim X}(Th(T^*X))$ (here $\tilde{H}^*(X) = H^*(X, pt)$). We put

$$ch(D) = \varphi^{-1}(ch(d(D; E, F))).$$

It is a nonhomogeneous cohomology class, whose parity coincides with the parity of $\dim X$.

Another ingredient in the index formula is the so called Todd genus, or Todd class of the complexified tangent bundle $T_\mathbb{C}X$: to define it we consider the formal power series

$$Q(x) = \frac{x}{1-e^{-x}} = 1 + \frac{x}{2} + \sum_{n=1}^{\infty}(-1)^{n-1}\frac{B_{2n}}{(2n)!}x^{2n},$$

---

[43]Chern character is a special formal power series of the Chern classes of a complex vector bundle, which plays an important role in Toplogy, see for instance [MS74].

where B_k, $k \geq 1$ are the Bernoulli numbers. We further define for the variables $x_1, \ldots, x_N$, $N = \dim X$ the product $td(x_1, \ldots, x_N) = \prod_{i=1}^{N} Q(x_i)$. This is a symmetric in the variables x_i formal power series with rational coefficients and hence it can be written as a formal series in elementary symmetric polynomials of these variables, $\sigma_k(x_1, \ldots, x_N)$, $k = 1, \ldots, N$, i.e.

$$td(x_1, \ldots, x_N) = Td(\sigma_1, \ldots, \sigma_N).$$

We put

$$Td(X) = Td(T_{\mathbb{C}}X) = Td(c_1(T_{\mathbb{C}}X), \ldots, c_N(T_{\mathbb{C}}X)),$$

where $c_k(E)$ is the k-th rational or real Chern class[44] of a complex bundle E. By a slight abuse of notation we shall denote by the same symbols $ch(D)$, $Td(X)$ the (nonhomogenous) closed differential forms, representing these classes.

The final version of the index formula due to Atiyah and Singer can be now written as follows: *for any elliptic operator D on X its index is computed by the following integral:*

$$\mathrm{ind} D = \int_X ch(D) Td(X). \tag{15.3}$$

15.1.2 Quantum symbol calculus and the algebraic index

As we have mentioned above (see Remark 10.2), if symplectic manifold M is equal to the cotangent bundle T^*X of a smooth compact manifold X, then its quantisation can be identified with the algebra of (usual) differential operators on X.[45] Thus, the whole construction of Fedosov can be regarded as the generalised calculus of (pseudo) differential operators.

We can make this observation more specific. To this end we will first of all generalise Fedosov's construction to include the symbols of pseudo-differential operators between vector bundles;

---

[44]In effect, if X is orientable, only even Chern classes $c_{2k}(T_{\mathbb{C}}X)$ will be nontrivial; in this case everything is expressible in terms of Pontryagin classes of X.

[45]The analogy can be made more clear if we regard functions on T^*X as the symbols of operators on X equipped with a kind of Moyal product.

this amounts to considering (locally) matrix-valued functions on M and the corresponding (matrix-valued) Weyl algebras. Namely, if $E, F \to M$ are two vector bundles, we shall consider the tensor product $\widehat{W}_\hbar^\mathbb{C}(M,\omega) \otimes \mathrm{Hom}(E,F)$ or just $\widehat{W}_\hbar(M;E,F)$. Then the sections of this bundle make up a bimodule over $\widehat{W}_\hbar(M)$ (and even left/right module over the sections of $\widehat{W}_\hbar^\mathbb{C}(M,\omega) \otimes \mathrm{End}(E,E)$ and $\widehat{W}_\hbar^\mathbb{C}(M,\omega) \otimes \mathrm{End}(F,F)$, respectively).

Now for any connection ∇_{Hom} in the bundle $\mathrm{Hom}(E,F)$ and any symplectic connection ∇ in $\widehat{W}_\hbar^\mathbb{C}(M,\omega)$ their tensor product $\nabla \otimes 1 + 1 \otimes \nabla_{\mathrm{Hom}}$ induces a connection in $\widehat{W}_\hbar(M;E,F)$ which satisfies the Leibniz rule with respect to the $\widehat{W}_\hbar(M)$-bimodule structure. In the same manner as before, we can extend the connection $\nabla \otimes 1 + 1 \otimes \nabla_{\mathrm{Hom}}$ to a flat (Abelian) connection on $\widehat{W}_\hbar(M;E,F)$; then the space of flat sections in $\widehat{W}_\hbar(M;E,F)$, which we denote by $\widehat{W}_D(M;E,F)$ is a bimodule over the quantum algebra $\widehat{W}_D(M)$ (here and below we identify this algebra with $C^\infty(M)[[\hbar]]$ equipped with Fedosov's $\star$-product). We shall call $\widehat{W}_D(M;E,F)$ *quantum symbols algebra*.

One further defines *elliptic symbols* in $\widehat{W}_D(M;E,F)$. To this end for any $\hat{a} \in \widehat{W}_D(M;E,F)$, we consider its "classical part" $Q^{-1}(\hat{a}) = \hat{a}_{00}$; one can regard it as the section of $\mathrm{Hom}(E,F)[[\hbar]]$ given by setting all y^i equal to 0 in the local expression for $\hat{a}$: this operation is well-defined for flat sections. Let

$$\hat{a}_{00} = \hbar^{-N} a(x) + \ldots,$$

where dots denote the terms with higher degrees in $\hbar$. Then we shall say that $\hat{a}$ is elliptic, *if $a(x)$ is invertible outside of a compact set in M*. Equivalently, $\hat{a}$ is elliptic, if it admits "quantum parametrix", i.e. *if there exists such element $\hat{b} \in \widehat{W}_D(M;F,E)$ that both $1 - \hat{a}\hat{b} \in \widehat{W}_D(M;F,F)$ and $1 - \hat{b}\hat{a} \in \widehat{W}_D(M;E,E)$ have compact support*. We shall call this element $\hat{b}$ *quantum parametrix* of $\hat{a}$.

In order to bear further the analogy with the index theory, we will look at the formula (15.2): index of any operator can be obtained from traces of the operators $1 - TD$ and $1 - DT$. In quantum case this is equivalent to finding traces[46] on the subspace

---

[46]Recall that if A is an associative k-algebra, one says that a map $\nu : A \to V$, where V is a k-linear vector space, is a trace, iff $tr(ab) = tr(ba)$ for all $a, b \in A$; the space of traces is the dual space of the 0 degree Hochschild homology of A.

$\widehat{W}_D^{comp}(M)$ of flat sections with compact support on M: indeed, if this is done we can extend this trace to compactly supported sections in $\widehat{W}_D(M; E, E), \widehat{W}_D(M; F, F)$ by taking the sum of traces of diagonal entries in the matrices, and apply this construction to $1 - \hat{a}\hat{b}$, $1 - \hat{b}\hat{a}$ thus imitating the formula (15.2):

Definition 15.1. *Let $\hat{a}$ be an elliptic symbol in $\widehat{W}_D(M; E, F)$. We shall call the expression*

$$\mathrm{ind}_{alg}(\hat{a}) = tr(1 - \hat{b}\hat{a}) - tr(1 - \hat{a}\hat{b}) \tag{15.4}$$

where $\hat{b}$ is a parametrix of $\hat{a}$ and tr is a trace on $\widehat{W}_D^{comp}(M; E, E)$ or $\widehat{W}_D^{comp}(M; F, F)$ extended from $\widehat{W}_D^{comp}(M)$ as explained above, the **algebraic index** *of $\hat{a}$.*

15.1.3 The trace on $\widehat{W}_D^{comp}(M)$

In order to justify Definition 15.1, we need to describe the traces in the algebra $\widehat{W}_D^{comp}(M)$ (observe that this algebra is an ideal in the flat sections algebra $\widehat{W}_D(M)$). This was done by Fedosov in his original paper. First of all we shall give the definition of trace, used by Fedosov himself:

Definition 15.2.

(i) *We shall say that the functional $tr : \widehat{W}_D^{comp}(M) \to \mathbb{C}[\hbar^{-1}, \hbar]]$ (for all Abelian connections D) is **local**, if it is given by formula*

$$tr(\hat{a}) = \int_M G(\hat{a}, D). \tag{15.5}$$

Here $G(\hat{a}, D)$ is some density on M, depending on the coefficients of $\hat{a}$ and the connection D. In other words, the formula (15.5) determines a family of functionals on the family of algebras $\widehat{W}_D^{comp}(M)$, parametrised by Abelian connections.

(ii) *We shall say that the local functional tr is **invariant**, if for any automorphism A of the Weyl algebra bundle $\widehat{W}_\hbar^{\mathbb{C}}(M, \omega)$ we have:*

$$\int_M G(\hat{a}, D) = \int_M G(A(\hat{a}), D_A), \tag{15.6}$$

where D_A is the Abelian connection associated with D with the help of A.

Observe that since every infinitesimal automorphism of the Weyl algebra is given by commutator with a fixed element H (this follows from the description of Hochschild cohomology of Weyl algebra), one can show that any local invariant functional is a trace on $\widehat{W}_D^{comp}(M)$ with values in $\mathbb{C}[\hbar^{-1}, \hbar]]$ in the usual sense: $tr([\hat{a}, \hat{b}]) = 0$. Also observe that local functionals can be constructed from functionals on (Darboux) coordinate charts in M with the help of a partition of unity, and since

$$[x^i, \hat{a}] = \pi^{ij} \frac{\partial \hat{a}}{\partial x^j},$$

where $x^1, \ldots, x^{2n}$ are the local Darboux coordinates, we see that the trace vanishes on partial derivations of the symbol.

Starting with these observations, Fedosov proved the following

Proposition 15.3. *There exists a unique local invariant trace* $tr : \widehat{W}_D^{comp}(M) \to \mathbb{C}[\hbar^{-1}, \hbar]]$, *given by the formula*

$$tr(\hat{a}) = \int_M Q^{-1}(\hat{a}) \frac{\omega^n}{n!}. \tag{15.7}$$

We shall call this functional *canonical* or *Fedosov trace* on $\widehat{W}_D(M)$.

15.1.4 Fedosov's algebraic index formula

Now that we have at our disposal the canonical trace on Fedosov quantisation, we can eventually use it to define the algebraic index $\mathrm{ind}_{alg}(\hat{a})$ of an elliptic symbol $\hat{a} \in \widehat{W}_D(M; E, F)$, see Definition 15.1.

On the other hand, one can ask, if an analog of the Atiyah–Singer formula (15.3) is valid in this situation. Indeed, this happens to be the case: first of all since the symbol $\hat{a}$ is elliptic, the lowest degree (in $\hbar$) term $a(x)$ of $\hat{a}_{00}$ is invertible outside a compact subset in M. Similarly to the construction of Atiyah's difference element we can now "glue" E and F outside this set with the help of $a(x)$ to get an element

$d(\hat{a})$ in $K^{comp}(M)$, the *compactly supported K-theory of* M.[47] The Chern character can be applied to the elements of $K^{comp}(M)$ and takes values in the even de Rham cohomology with compact support on M; we shall abbreviate $ch(d(\hat{a}))$ to simply $ch(a)$.

We further introduce the A-genus of the tangent bundle of M: to this end we consider the formal power series:

$$A(x) = \frac{x/2}{\sinh(x/2)}; \text{ observe that } A(-x) = A(x).$$

Let further $A(x_1, \ldots, x_{2n}) = \prod_{i=1}^{2n} A(x_i)$ (here $2n$ is the dimension of the symplectic manifold M); this is an even symmetric formal power series and as such can be represented as a formal series in elementary symmetric polynomials in $x_1^2, \ldots, x_{2n}^2$:

$$A(x_1, \ldots, x_{2n}) = \mathscr{A}(\tilde{\sigma}_1, \ldots, \tilde{\sigma}_{2n}), \text{ where } \tilde{\sigma}_i = \sigma_i(x_1^2, \ldots, x_{2n}^2).$$

Finally, one puts

$$\mathscr{A}(M) = \mathscr{A}(p_1(TM), \ldots, p_{2n}(TM)),$$

where $p_i(TM) = c_{2i}(T_{\mathbb{C}}M)$ are the *Pontryagin classes* of M. It is an inhomogeneous even cohomology class on M; by a slight abuse of notation we shall denote de Rham form representing this class by the same symbol.

In his original papers Fedosov proved the following theorem, sometimes called *the algebraic index theorem*

Theorem 15.4 (Fedosov's algebraic index formula). *The algebraic index* $\text{ind}_{alg}\hat{a}$ *of an elliptic symbol* $\hat{a}$ *can be calculated by the following formula*

$$\text{ind}_{alg}\hat{a} = \int_M ch(a) e^{-\frac{1}{2\pi\sqrt{-1}}\omega} \mathscr{A}(M); \qquad (15.8)$$

here the right-hand side is well-defined since the form $ch(a)$ has compact support. If the manifold M is equal to the cotangent bundle of a

---

[47]Loosely speaking, classes of $K^{comp}(M)$ are generated by pairs of complex vector bundles over M, isomorphic outside of a compact set in M, and the homotopy class of the isomorphism is a part of the data; reader unfamiliar with this concept can look it up in the well-known Atiyah's book [At64], or Karoubi's book [Ka78].

smooth manifold X, then right-hand side of the formula (15.8) turns into that of the classical Atiyah–Singer formula (15.3).

Proof of this theorem is rather technical, and consists of unravelling the definitions, since both sides of the equation are in effect given by integrals over M.

Remark 15.5. Theorem 15.4 was one of the first instance of results pertaining to a large area of research, somewhat loosely called "the algebraic index theory". It evolved from the attempts to better understand the Atiyah–Singer formula and give a more direct proof thereof. The original reasoning due to Atiyah and Singer consists of meticulous check of the properties of expressions on both sides of the equality (15.3): since both these formulas satisfy the same set of axioms, one can deduce that they give the same result for any elliptic operator D from the fact that they agree in just a single well-chosen example, see [Pa65]; no solid explanation as to the origin of this construction was originally given. This reasoning although perfectly rigorous leaves behind lots of questions, concerned, first of all, with the way one may hit on a formula like (15.3). Thus there have been many attempts to find a method to derive this formula in a more direct way from some basic principles, see for instance the papers by Atiyah, Patody and Singer [APS75$_1$], [APS75$_1$], [APS76].

Algebraic index theorems can be regarded as an attempt to derive the analog of index theorem from purely algebraic premises; to this end one has to replace the differential operators (or pseudo-differential operators) with some objects of algebraic nature, like the elements of Fedosov algebra in the case of Theorem 15.4. Other generalisations included in particular the further loosening of the notion of differential operators, allowing in particular operators on sheaves rather than on vector bundles, working with general Poisson instead of symplectic structures, etc., see [BNT02$_1$] and [BNT02$_2$] for example.

15.2 Group action, quantisation and quantum momentum maps

As we remarked in the end of Section 12 (see Remark 12.8), the existence of symplectic invariant connections on a symplectic manifold with a group action, guarantees the existence of invariant $\star$-products on these manifolds. In the context of Fedosov theory we can obtain similar results.

15.2.1 Invariant connections and equivariant $\star$-products

Let a group G act on symplectic manifold M from the left by diffeomorphisms: $x \mapsto g \cdot x$. This action induces the action of G by isomorphisms on the algebra of smooth functions: for any $x \in M$, $f \in C^\infty(M)$, $g \in G$ we put

$$f^g(x) = f(g \cdot x),$$

(observe that this transposed action of G on $C^\infty(M)$ is from the right) and one can ask if this action survives under quantisation, i.e. if there exists such $\star$-product on $C^\infty(M)[[\hbar]]$ that

$$(f \star g)^h = f^h \star g^h, \qquad (15.9)$$

for all functions $f, g \in C^\infty(M)$ and all elements h of the group (we extend the action to $C^\infty(M)[[\hbar]]$ by $\hbar$-linearity). We shall call the $\star$-product, satisfying Equation (15.9), *equivariant* or *invariant*; if invariant $\star$-product exists, we will say that manifold admits *equivariant deformation quantisation*.

It is clear that the necessary condition imposed on the group action that admits equivariant $\star$-product is that *the action of G preserves the Poisson structure*, i.e. that

$$\{f, g\}^h = \{f^h, g^h\}, \ \forall f, g \in C^\infty(M), \ h \in G. \qquad (15.10)$$

Indeed, this is the degree 1 (in $\hbar$) part of Equation (15.9).

In many cases, condition (15.10) is sufficient. To explain this, consider the Fedosov quantisation construction again: it is clear that Fedosov $\star$-product will be invariant, as long the corresponding Abelian connection D commutes with the group action. On the other hand D depends in a substantial way on symplectic connection ∇ on M: if the connection ∇ is fixed the choice of correction terms $\Delta\Gamma = \omega(y) + \zeta$ is more or less automatic, as soon as the form φ is given: recall that ζ is obtained by the iterative process (13.20_n),

applied to $\omega(y)$, ∇ and φ. In particular, if φ is invariant with respect to the action of G (e.g. $\varphi = 0$) then all the terms in ζ will also be G-invariant as long as the symplectic connection ∇ satisfies the equality
$$\nabla^h = \nabla, \ \forall h \in G; \qquad (15.11)$$
in this case it is called *G-invariant symplectic connection*. Thus, the existence of a G-invariant symplectic connection ∇ is a key element for the equivariant quantisation. It is clear that if the group G is compact, then there always exist invariant symplectic connections on any symplectic manifold M with a G-action: it is sufficient to do the averaging over G trick on any symplectic connection on M. More generally, this is the case if the action of G is proper.[48] However, in many important cases this assumption is false; for instance it is usually so if the group is equal to $\mathbb{R}^p$ and the action is induced by Poisson-commuting elements in $C^\infty(M)$.

Observe that symplectic connections can be obtained by adding correction terms to usual symmetric connections, moreover these correction terms are obtained from symplectic form and the connection in question, see Proposition 11.6. Thus, if the symplectic form is G-invariant, it is enough to choose G-invariant symmetric connection on M. In order to tell, if there exist such connections, one can consider the following construction: let us assume that the group G is simply connected, and let $\mathfrak{g}$ be the Lie algebra of G; then the conditions (15.9) and (15.10) are equivalent to the equations

$$X(f \star g) = X(f) \star g + f \star X(g) \qquad (15.9')$$

and

$$X(\{f, g\}) = \{X(f), g\} + \{f, X(g)\} \qquad (15.10')$$

for all $f, g \in C^\infty(M)$ and all $X \in \mathfrak{g}$, respectively. Then in order to find the equivariant $\star$ product we will need a symmetric connection

---

[48]The action of a topological group G on topological space X is called proper if the map
$$H : G \times X \to X \times X, \ (g, x) \mapsto (g \cdot x, x)$$
is proper, i.e. if $H^{-1}(K)$ is compact for all compact $K \subseteq X \times X$.

on M that would satisfy the condition $\mathcal{L}_X(\nabla) = 0$ for all $X \in \mathfrak{g}$, where the Lie derivative of ∇ was defined earlier in Section 11.3.2, Equation (11.15). We will call the connections, satisfying this equality $\mathfrak{g}$-*invariant*.

In general, it is difficult to say, whether $\mathfrak{g}$-invariant connections on M exist or not; an abstract way to answer the question, whether a $\mathfrak{g}$-invariant connection exists is based on considering the map:

$$\vartheta : \mathfrak{g} \to \operatorname{Hom}(S^2(TM), TM), \ \vartheta(X)(A,B) = \mathcal{L}_X(\nabla)(A,B),$$

where ∇ is an arbitrary symmetric connection, see formula (11.15) and discussion after it. Then, since $\mathcal{L}_{[X,Y]} = \mathcal{L}_X \mathcal{L}_Y - \mathcal{L}_Y \mathcal{L}_X$, ϑ determines a cohomology class in the Chevalley-Eilenberg cohomology of $\mathfrak{g}$ with values in the sections of the bundle $\operatorname{Hom}(S^2(TM), TM)$. This class vanishes iff there exists $\mathfrak{g}$-invariant connections.[49] Unfortunately, this cohomology rarely vanishes and is pretty hard to compute in a general situation.

15.2.2 Quantum momentum maps

An important particular case of the group action on a symplectic or Poisson manifold is the case of *Hamiltonian action*:

Definition 15.6. *One says that the action of G is* **Hamiltonian** *if the corresponding Lie algebra action is by the Hamiltonian vector fields:*
$$U(f) = X_{J(U)}(f) = \{f, J(U)\}, \ \forall U \in \mathfrak{g},$$
where $J : \mathfrak{g} \to C^\infty(M)$ is a Lie algebra homomorphism with respect to the Poisson brackets in $C^\infty(M)$:

$$J([U,V]) = \{J(U), J(V)\}. \tag{15.12}$$

Dualising this construction, we can replace J by a continuous map μ from M to $\mathfrak{g}^*$:

$$\mu(x)(U) = J(U)(x), \ \forall x \in M, U \in \mathfrak{g}. \tag{15.13}$$

---
[49]Observe the relation of this class and the Vey class S_Γ^3 of the symplectic structure, which is also determined with the help of a symplectic connection.

In this context the map μ is called *momentum map* of the action, and the map J is restored from μ in an evident way. If we choose a basis $T_1, \ldots, T_N$ in $\mathfrak{g}$, then the map μ can be identified with the continuous map from M into the Euclidean space $\mathbb{R}^N$:

$$\mu(x) = (J(T_1)(x), \ldots, J(T_N)(x)).$$

By a slight abuse of notation, we shall refer to J as momentum map too.

Let us now assume that the action under consideration admits equivariant $\star$-product; a little more generally, let us assume that there is a representation of $\mathfrak{g}$ by derivations on the quantum algebra $\mathcal{A}(M) = (C^\infty(M)[\hbar^{-1}, \hbar]], \star)$. This means that there is a homomorphism of Lie algebras from $\mathfrak{g}$ to the Lie algebra of derivations of the quantum algebra, i.e. for every $U \in \mathfrak{g}$ there is a derivation

$$\widehat{U} : C^\infty(M)[\hbar^{-1}, \hbar]] \to C^\infty(M)[\hbar^{-1}, \hbar]],$$

so that $\widehat{U+V} = \widehat{U} + \widehat{V}$,

$$\widehat{[U,V]} = [\widehat{U}, \widehat{V}],$$

and Equation (15.9') holds for $\widehat{U}$:

$$\widehat{U}(f \star g) = \widehat{U}(f) \star g + f \star \widehat{U}(g).$$

It is also natural to assume that the map $\widehat{U}$ is given by the action of $U \in \mathfrak{g}$ up to terms of degrees 1 and higher in $\hbar$:

$$\widehat{U}(f) = U(f) + \sum_{i=1}^{\infty} \hbar^i U_i(f).$$

The purely equivariant case here corresponds to the situation when $U_i = 0$, $i > 1$. The following definition now is a "quantum analog" of Definition 15.2.2:

Definition 15.7. *One says that the action of $\mathfrak{g}$ on $\mathcal{A}(M)$ is **quantum Hamiltonian** if all maps $\widehat{U}$ are equal to inner derivations:*

$$\widehat{U}(f) = \frac{1}{\hbar}[f, \widehat{J}(U)], \forall U \in \mathfrak{g}, \text{ where } [a, b] = a \star b - b \star a,$$

and where $\hat{J} : \mathfrak{g} \to \mathcal{A}(M)$ is a Lie algebra homomorphism with respect to the commutator brackets in $\mathcal{A}(M)$:

$$\hat{J}([U,V]) = \frac{1}{\hbar}[\hat{J}(U), \hat{J}(V)]. \tag{15.14}$$

In this case the map $\hat{J}$ is called **quantum momentum map**; one has to use the factor $\frac{1}{\hbar}$ because of the formula

$$f \star g - g \star f = \hbar\{f, g\} + o(\hbar).$$

Summing up, suppose that an equivariant $\star$-product on a symplectic manifold M is given; this means that we have a linear representation $\hat{\rho} : \mathfrak{g} \to Der(\mathcal{A}(M))$, $U \mapsto \widehat{U}$. Then finding a quantum momentum map for it consists of two independent problems:

(i) Find a linear map $\hat{J} : \mathfrak{g} \to \mathcal{A}(M)$ such that $\widehat{U}(f) = \frac{1}{\hbar}[f, \hat{J}(U)]$;

(ii) Check Equation (15.14) for the map $\hat{J}$; modify this map, if necessary.

These questions can be resolved separately by the deformation theory methods, independently of the way the $\star$-product is constructed. However, we shall only concentrate on the case when the product in question is given by Fedosov's construction. Our purpose is to describe the obstructions that allow the construction of quantum momentum maps for a given equivariant $\star$-product and to classify the equivariant Fedosov's $\star$-products and quantum momentum maps, associated with them (provided there exist invariant symplectic connections). There exist different approaches that lead to this end; in our lectures we shall roughly follow the exposition of the paper [RW16].

15.2.3 Classification of quantum momentum maps and equivariant cohomology

Let us fix a Hamiltonian action of the Lie algebra $\mathfrak{g}$ on the symplectic manifold (M, ω). We assume that this action allows invariant connections, so fixing a $\mathfrak{g}$-invariant closed 2-form $\varphi \in \Omega^2(M)[[\hbar]]$ we have an action of the Lie algebra $\mathfrak{g}$ on the quantised algebra $\mathcal{A}_\varphi(M)$

by derivations (here we make explicit the dependence of Fedosov product on the choice of φ). Let

$$\hat{J} : \mathfrak{g} \to \mathcal{A}_\varphi(M)$$

be a quantum momentum map, corresponding to this action (clearly, every such quantum momentum map extends some classical momentum map J of the action). As a linear map $\hat{J}$ is a map to $C^\infty(M)[[\hbar]]$. It turns out that in terms of ω and φ the condition that the given linear map $\hat{J}$ is a quantum momentum map can be written in the form of the following two equations:[50]

$$d(\hat{J}(U)) = i_{\widehat{U}} \left(\frac{1}{\hbar} \omega + \varphi \right), \qquad (15.15)$$

$$\hat{J}([U,V]) = \left(\frac{1}{\hbar} \omega + \varphi \right)(\widehat{U},\widehat{V}). \qquad (15.16)$$

Here and below we denote by $\widehat{U}, \widehat{V}$ the vector fields on M, corresponding to $U, V \in \mathfrak{g}$ under the representation of $\mathfrak{g}$. The first Equation (15.15), means that the commutator with $\hat{J}(U)$ coincides with $\widehat{U}$, and the second one (15.16) is responsible for the condition (15.14).

These two equations can be written down compactly with the help of equivariant cohomology. Recall that for a Lie algebra $\mathfrak{g}$ acting on a manifold M one defines *Cartan equivariant complex* as the graded linear space

$$C_\mathfrak{g}^*(M) = (\Omega^*(M) \otimes S(\mathfrak{g}^*[2]))^G,$$

where V^G denotes the G-invariant part of a G-module V (G is the 1-connected Lie group, corresponding to $\mathfrak{g}$) and $S(\mathfrak{g}^*[2])$ is the graded symmetric algebra generated by the vector space $\mathfrak{g}[2]$, i.e. by the dual space of $\mathfrak{g}$ shifted into dimension 2. One can regard $C_\mathfrak{g}^*(M)$ as the (graded) space of G-invariant functions on $\mathfrak{g}$ with values in $\Omega^*(M)$. Forgetting about the invariance condition, we allow $\mathfrak{g}^*$ act on such functions by multiplication when we identify $\mathfrak{g}^*$ with the space of linear functions on $\mathfrak{g}$. The differential $d_\mathfrak{g}$ of this complex is given

---

[50] Here and below in this section we shall omit the $\sqrt{-1}$ factor in front of ω in the formulas.

then by the formula:

$$d_{\mathfrak{g}}(\alpha) = d(\alpha) + \sum_{k=1}^{N} i_{\widehat{X}_k}(\alpha) \cdot X^k, \qquad (15.17)$$

where d is de Rham differential, $N = \dim \mathfrak{g}$ and $X_1, \ldots, X_N \in \mathfrak{g}$, $X^1, \ldots, X^N \in \mathfrak{g}^*$ are some dual basis. The formula (15.17) in effect does not depend on the choice of basis, and the equation $d_{\mathfrak{g}}^2 = 0$ follows from the Cartan formula $\mathcal{L}_\xi = i_\xi d + d i_\xi$.

Now $\hat{J}$ being a map $\mathfrak{g} \to C^\infty(M)[[\hbar]]$ we can reinterpret it as a (linear) function on $\mathfrak{g}$ with values in $C^\infty(M)[\hbar^{-1}, \hbar]]$; similarly we regard $\frac{1}{\hbar}\omega + \varphi$ as an element in $C_{\mathfrak{g}}^2(M)[\hbar^{-1}, \hbar]]$ (recall that both ω and φ are $\mathfrak{g}$-invariant). If we now consider the difference $\theta_{\mathfrak{g}} = \frac{1}{\hbar}\omega + \varphi - \hat{J}$, then the condition (15.15) is equivalent to the equation $d_{\mathfrak{g}}\theta_{\mathfrak{g}} = 0$ and the invariance of $\theta_{\mathfrak{g}}$ then follows from (15.16): indeed the G-invariance of a function $\alpha \in C(\mathfrak{g}, \Omega^*(M))$ is equivalent to its $\mathfrak{g}$-invariance and this is given by

$$\alpha(ad_U(V)) = -\mathcal{L}_{\widehat{U}}\alpha(V),$$

where $ad_U(V) = [U, V]$ in $\mathfrak{g}$; so $\theta_{\mathfrak{g}}(ad_U(V)) = \theta_{\mathfrak{g}}([U, V]) = -\hat{J}([U, V])$ and on the other hand, since both ω and φ are $\mathfrak{g}$-invariant

$$\mathcal{L}_{\widehat{U}}\theta_{\mathfrak{g}}(V) = \mathcal{L}_{\widehat{U}}\hat{J}(V) = i_{\widehat{U}}d\hat{J}(V) = i_{\widehat{U}}i_{\widehat{V}}\left(\frac{1}{\hbar}\omega + \varphi\right) = \left(\frac{1}{\hbar}\omega + \varphi\right)(\widehat{U}, \widehat{V}),$$

where we used Cartan's formula again and Equation (15.15). Thus we can reformulate the conditions (15.15), (15.16) as the following statement:

Proposition 15.8. *Let a Lie algebra $\mathfrak{g}$ act on a symplectic manifold (M, ω) in a Hamiltonian way; assume that this action admits an invariant symmetric connection. Then there exists a quantum momentum map $\hat{J}$ for an equivariant Fedosov $\star$-product on M, determined for an invariant closed 2-form $\varphi \in \Omega^2(M)^G$ iff the sum $\frac{1}{\hbar}\omega + \varphi$ can be extended to a closed element $\theta_{\mathfrak{g}} \in C_{\mathfrak{g}}^2(M)[\hbar^{-1}, \hbar]]$.*

The class of $\theta_{\mathfrak{g}} = \theta_{\mathfrak{g}}(\varphi, \hat{J})$ in equivariant cohomology should be regarded as the analog of the class $[\frac{\omega}{\hbar} + \varphi] \in \frac{[\omega]}{\hbar} + H^2(M)$ of Fedosov quantisation; its lowest term can be written separately, just as we did it in the usual (non-equivariant) case, so the class $[\theta_{\mathfrak{g}}(\varphi,\hat{J})]$ belongs to the affine subspace $[\frac{\omega+J}{\hbar}]+H_{\mathfrak{g}}^2(M)[[\hbar]]$ in equivariant cohomology (here J is the classical momentum map, associated with the action). In effect a more detailed analysis of the construction shows that the following statement, analogous to Theorem 14.4 holds:

Proposition 15.9.
(i) *Let $\hat{J}$, $\hat{J}'$ be two quantum momentum maps defined for the same $\mathfrak{g}$-equivariant quantum algebra $\mathcal{A}_\varphi(M)$; then these maps are equivalent (i.e. are mapped to each other by some self-equivalence of $\mathcal{A}_\varphi(M)$) iff the difference $\hat{J}-\hat{J}'$ determines a trivial class in $H_{\mathfrak{g}}^2(M)[\hbar^{-1},\hbar]]$.*

(ii) *More generally, if $(\mathcal{A}_\varphi(M),\hat{J})$, $(\mathcal{A}_{\varphi'}(M),\hat{J}')$ are two equivariant Fedosov deformations, determined by invariant closed 2-forms φ, φ' respectively, equipped with quantum momentum maps $\hat{J}$, $\hat{J}'$, then there exists an equivalence of $\mathcal{A}_\varphi(M)$ and $\mathcal{A}_{\varphi'}(M)$ that intertwines $\hat{J}$ and $\hat{J}'$ iff the class of the difference*

$$\theta_{\mathfrak{g}}(\varphi,\hat{J})-\theta_{\mathfrak{g}}(\varphi',\hat{J}')=(\varphi-\varphi')-(\hat{J}-\hat{J}')$$

is trivial in $H_{\mathfrak{g}}^2(M)$.

Proofs of Propositions 15.8 and 15.9 are obtained by suitable modifications of the iterative process used in the proof of Fedosov Theorems 13.4, 14.4 and others. We omit it here.

Remark 15.10. The equivariant complex $C_{\mathfrak{g}}^*(M)$ is a special choice of complex that is used to calculate *equivariant cohomology*. This cohomology plays central role in many modern physical theories, where the group action is considered. It can be defined in purely topological terms (and with integer coefficients, or with coefficients in any other commutative ring), as the cohomology of the *homotopy quotient space*, see for instance [Tu20] for details.

15.3 Deformation of holomorphic symplectic structures

In the previous part of this lecture we described the way one can extend the action of a Lie group or a Lie algebra from symplectic manifold to its quantised function algebra. This can be regarded as an instance of "quantisation with additional structures", i.e. a search for the quantisation procedure that would somehow preserve an additional properties that exist on original manifold. In the case we considered there this property was the group action. Another important instance of such additional structures is the complex structure on manifolds, i.e. the choice of complex coordinate systems on M with holomorphic change of coordinates. In this case, one can introduce the notion of holomorphic functions, holomorphic differential forms (in particular, holomorphic symplectic form), holomorphic vector fields, holomorphic differential operators etc. on M. If now M is endowed with a holomorphic symplectic structure, the natural question that one can ask is: "Is it possible to find such deformation series that the product of any two holomorphic functions is again holomorphic?" This is the question that we are going to briefly discuss in this section. In our exposition, we closely follow the paper of Nest and Tsygan [NT01].

15.3.1 Few words about complex geometry: Dolbeault cohomology, Atiyah class

Before we can even formulate the results, we need briefly introduce the notions from complex geometry that will be used here. A reader unfamiliar with the basic results of this field can look it up in any of the numerous textbooks on Complex Differential Geometry, for instance the first two chapters of the famous Griffiths and Harris course [GrHa78] are a perfect introduction into the subject.

So recall that an even-dimensional manifold M is said to be given a *complex structure* (speaking simply, it is called *complex manifold*) if one can choose an atlas of local complex coordinate charts $(z^1, \ldots, z^n)$ on it with holomorphic coordinate changes from one chart to another; here and below holomorphicity of a function of many complex variables can be regarded as the condition that all partial Cauchy–Riemann operators $\bar{\partial}_i = \frac{\partial}{\partial \bar{z}^i}$ on it vanish; if $z^i = x^i + \sqrt{-1} y^i$ is the i-th complex coordinate on M (regarded as the complex combination of real coordinates (x^i, y^i)), then we can

write
$$\frac{\partial}{\partial z^i} = \frac{\partial}{\partial x^i} - \sqrt{-1}\frac{\partial}{\partial y^i}, \quad \frac{\partial}{\partial \bar{z}^i} = \frac{\partial}{\partial x^i} + \sqrt{-1}\frac{\partial}{\partial y^i}.$$

Similarly, a complex vector bundle $E \to M$, where M bears a complex structure, is called *holomorphic vector bundle* if one can choose bundle coordinate charts on E with holomorphic gluing cocycle between them; complexified tangent and cotangent bundles of M are examples of such holomorphic bundles.

If M is a complex manifold, one can define (locally) holomorphic functions on it as such complex-valued functions $f \in C^\infty(M)$ for which $\bar{\partial}_i f = 0$ for all $i = 1, \ldots, n$; this notion is well-defined (does not depend on the local coordinates from the atlas). However, due to the maximal module principle any globally holomorphic function on a compact complex manifold should be constant. In this case, one is compelled to consider *sheaf of (local) holomorphic functions* $\mathscr{O}_M$ rather than the algebra of global holomorphic functions; this sheaf is often called *the structure sheaf of M*. In simple terms $\mathscr{O}_M$ is the functor that associates with every open set $U \subseteq M$ the ring $\mathscr{O}_M(U)$ of holomorphic functions on U; then global holomorphic functions on M are just the elements of $\mathscr{O}_M(M)$, the space of global sections of $\mathscr{O}_M$.

On the whole, the language of sheaf theory is more suitable for many purposes in the theory of complex holomorphic manifolds. For instance, one can define holomorphic vector fields, differential forms and more generally holomorphic sections of any holomorphic bundle $E \to M$. As before, globally holomorphic objects of this sort are rare so it is more reasonable to consider *sheaves of local holomorphic sections* $\mathscr{E}_M$ of such bundles; these are the functors that send open sets $U \subseteq M$ to the corresponding spaces of local sections $\mathscr{E}_M(U)$. We will distinguish this sheaf from the usual sheaves of smooth sections $\mathscr{C}^\infty(M, E)$. Generally speaking *sheaf on a manifold M* is a contravariant functor from the category of open sets on M (with arrows given by inclusions) to the category of abelian groups, rings, vector spaces etc., satisfying certain additional properties; we are not going to describe these properties here, it suffices to know that in the case of the sheaves of holomorphic sections they hold automatically.

For every sheaf $\mathscr{A}$ on a topological space one can define its *sheaf cohomology*. There are many different ways to introduce it, the most appropriate one being the derived functor of the functor of global sections of $\mathscr{E}$, see [Wei94], [Go58] or [Wa71]. A more constructive way to define it is with the help of Čech *complex*: if $M = \bigcup_\alpha U_\alpha$ is an open cover of M, we define

$$\check{C}^p(\{U_\alpha\}, \mathscr{A}) = \bigoplus_{(\alpha_0,\ldots,\alpha_p)} \mathscr{A}(U_{\alpha_0} \cap \cdots \cap U_{\alpha_p})$$

with differential given by the alternate sum of the restrictions from p-fold intersections of sets U_α to $p+1$-fold intersections. It turns out that when the cover $\{U_\alpha\}$ is sufficiently refined, the cohomology of this complex does not depend on it. It is these stable (i.e. independent on the choice of sufficiently fine open cover) groups that are called the cohomology of M with coefficients in $\mathscr{A}$, denoted $H^*(M, \mathscr{A})$.

In complex case de Rham differential d on the space of (complex-valued) differential forms on M is naturally represented as a sum of two differentials: if

$$dz^i = dx^i + \sqrt{-1}dy^i, \ d\bar{z}^i = dx^i - \sqrt{-1}dy^i,$$

then any complex-valued form φ on M can be written as

$$\varphi = \sum_{I,J} \varphi_{I,J} dz^I \wedge d\bar{z}^J,$$

where $I = (i_1, \ldots, i_p)$, $J = (j_1, \ldots, j_q)$ and $dz^I = dz^{i_1} \wedge \cdots \wedge dz^{i_p}$, $d\bar{z}^J = d\bar{z}^{j_1} \wedge \cdots \wedge d\bar{z}^{j_q}$. In this way, we obtain a double grading on differential forms, so we write $\varphi \in \Omega^{p,q}(M)$ and the de Rham differential d is equal to the sum $d = \partial + \bar{\partial}$,

$$\partial : \Omega^{p,q}(M) \to \Omega^{p+1,q}(M), \ \bar{\partial} : \Omega^{p,q}(M) \to \Omega^{p,q+1}(M);$$

in local coordinates:

$$\partial \varphi = \sum_{i,I,J} \partial_i(\varphi_{I,J}) dz^i \wedge dz^I \wedge d\bar{z}^J, \ \bar{\partial}\varphi = \sum_{i,I,J} \bar{\partial}_i(\varphi_{I,J}) d\bar{z}^i \wedge dz^I \wedge d\bar{z}^J.$$

Then

$$\partial^2 = 0, \ \bar{\partial}^2 = 0, \ \partial\bar{\partial} + \bar{\partial}\partial = 0.$$

Clearly $\varphi \in \Omega^{p,0}(M)$ is holomorphic iff $\bar{\partial}\varphi = 0$. The complex

$$0 \to \Omega^{p,0}(M) \xrightarrow{\bar{\partial}} \Omega^{p,1}(M) \xrightarrow{\bar{\partial}} \ldots \xrightarrow{\bar{\partial}} \Omega^{p,n-1}(M) \xrightarrow{\bar{\partial}} \Omega^{p,n}(M) \to 0$$

is called the q-th *Dolbeault complex of M* and denoted

$$\mathcal{A}^p_{\bar{\partial}}(M, \mathbb{C}) = \bigoplus_{q \geq 0} \mathcal{A}^{p,q}_{\bar{\partial}}(M, \mathbb{C}).$$

Its cohomology is denoted by $H^{p,q}_{\bar{\partial}}(M, \mathbb{C})$. One can show that this cohomology is equal to the *sheaf cohomology* of $\Omega^p(M)$, the sheaf of degree p holomorphic forms on M (it is the sheaf of sections of the holomorphic bundle of p-forms on M). Since $(\Omega^*(M), d)$ can be identified with bicomplex $(\bigoplus_{p+q=*} \Omega^{p,q}(M), \partial, \bar{\partial})$, there is a bicomplex spectral sequence

$$E^{pq}_r(M) \Rightarrow H^k_{dR}(M, \mathbb{C}) \qquad (15.18)$$

that connects the de Rham cohomology $H^k_{dR}(M, \mathbb{C})$ of M with the Dolbeault cohomology $H^{p,q}_{\bar{\partial}}(M, \mathbb{C})$ for $p + q = k$ (in many cases this sequence converges to $H^*_{dR}(M, \mathbb{C})$). In particular the "boundary map" of this sequence is a map $H^k_{dR}(M, \mathbb{C}) \to H^k(M, \mathcal{O}_M) = H^{0,k}_{\bar{\partial}}(M, \mathbb{C})$. Also observe that in many important cases, e.g. if M is a Kähler manifold,[51] there is an isomorphism

$$H^k_{dR}(M, \mathbb{C}) = \bigoplus_{p+q=k} H^{p,q}_{\bar{\partial}}(M, \mathbb{C}).$$

More generally, for a holomorphic vector bundle $E \to M$ one defines an operator

$$\bar{\partial}_E : \mathscr{C}^\infty(M, E) \to \Omega^{0,1}(M, E)$$

where $\Omega^{0,1}(M, E)$ is the (sheaf of) differential $(0,1)$-forms on M with values in E. One first defines $\bar{\partial}_E$ in terms of local coordinates; then one can show that it is well-defined (does not depend on the choice of coordinates) because the bundle is holomorphic; moreover $\bar{\partial}_E^2 = 0$.

---

[51] *Kähler manifolds* is an important class of complex manifolds, which satisfy certain analytic condition on the complex structure. They have many nice additional properties.

So iterating $\bar{\partial}_E$, we obtain a complex $\mathcal{A}_{\bar{\partial}}(M, E)$, whose cohomology is equal to the sheaf cohomology of $\mathscr{E}(M)$.

Further for a holomorphic bundle E any connection on it is a map

$$\nabla : \mathscr{C}^\infty(M, E) \to \Omega^1(M, E) = \Omega^{1,0}(M, E) \oplus \Omega^{0,1}(M, E)$$

and it is natural to demand that the $(0,1)$ part of ∇ should coincide with $\bar{\partial}_E$. Then every connection is completely determined by its $(1,0)$ connection form $A_\nabla \in \Omega^{1,0}(M, \text{End}(E))$. Connections of this sort are called *complex linear connections*. One says that *the connection ∇ is holomorphic* iff one can choose A_∇ to be holomorphic. In this case, the curvature of ∇ is a $(2,0)$ form with coefficients in $\text{End}(E)$, which has many topological and geometrical consequences. However just like any other global holomorphic object, holomorphic connections are rather rare: in fact holomorphic bundles seldom admit holomorphic connections. In fact there exists a homological obstruction for the existence of holomorphic connections. One way to define it is to consider any complex connection ∇ on E; then it is easy to see that the $(1,1)$-part of its curvature is closed with respect to the operator $\bar{\partial}$. One can show that the class of this $(1,1)$-part of the curvature in Dolbeault cohomology of the sheaf $\text{End}(E)$ vanishes if and only if there exist holomorphic connections on E. This class is called *the Atiyah class of E* because it has been first considered by Michael Atiyah, see [At57].

15.3.2 Nest and Tsygan's construction

We will say that a complex manifold bears *complex holomorphic symplectic structure* if there is a $(2,0)$ form ω on M, which is nondegenerate (i.e. its matrix is invertible complex matrix at every point of M), holomorphic (i.e. $\bar{\partial}\omega = 0$) and holomorphically closed (i.e. $\partial\omega = 0$). Equivalently, $d\omega = 0$.

We are to find *holomorphic $\star$-product* on M, i.e. a product that sends (formal power series with coefficients in) holomorphic functions on M again to series with holomorphic functions as coefficients. However as we explained earlier, global holomorphic objects (in particular functions) are rare, so instead we shall speak about sheaves. Thus one has to find an associative product on the sheaf $\mathscr{O}_M[[\hbar]]$, (the sheaf of formal power series with coefficients in the structure sheaf $\mathscr{O}_M$). In this case instead of quantised algebra of functions $\mathcal{A}_\varphi(M)$ we will obtain a sheaf of associative algebras on M.

Of course, if the holomorphic tangent bundle $T^{1,0}(M)$, i.e. the subbundle of the complexified bundle $T_\mathbb{C} M$, spanned by the local fields $\frac{\partial}{\partial z^i}$, admits symmetric holomorphic connection (and hence symplectic holomorphic connection, since the latter can be obtained from the former by adding a correction term equal to a combination of the coefficients of symplectic form), there is no problem to construct the product, since Fedosov's construction can be extended to complex case in a word-for-word manner, yielding a family of $\star$-products, classified by the elements of the Dolbeault cohomology $H^{2,0}_{\bar{\partial}}(M,\mathbb{C})$. However, Nest and Tsygan show that similar constructions can exist under a far less restrictive assumptions.

Namely, let us suppose that *the boundary maps*

$$H^i_{dR}(M,\mathbb{C}) \to H^{0,i}_{\bar{\partial}}(M,\mathbb{C}) = H^i(M, \mathscr{O}_M) \qquad (15.19)$$

of the spectral sequence $E^{pq}_r(M)$ (see (15.18)) are epimorphic for $i = 1, 2$. This condition holds in a wide range of examples, where Atiyah class of the holomorphic tangent bundle is nontrivial, and hence no holomorphic connections exist, for example for all Kähler manifolds. In this case, if we take $i = 2$, it follows from the spectral sequence arguments that one can choose a splitting $H^2(M, \mathscr{O}(M)) \to H^2_{dR}(M,\mathbb{C})$, which leads to an isomorphism

$$H^2_{dR}(M,\mathbb{C}) = H^2(F^1\Omega^{*,*}(M), d) \oplus H^2(M, \mathscr{O}_M). \qquad (15.20)$$

Here $F^k\Omega^{*,*}(M)$ denotes the k-th term of filtration in bicomplex $(\Omega^{p,q}(M), \partial, \bar{\partial})$ that determines the spectral sequence $E^{pq}_r(M)$, so

$$F^k\Omega^{*,*}(M) = \bigoplus_{p \geq k, q} \Omega^{p,q}(M).$$

Then Nest and Tsygan proved the following theorem:

Theorem 15.11.

(i) *If the boundary map (15.19) is epimorphic for $i = 1$, then there exist holomorphic $\star$-products for all holomorphic symplectic forms on M.*

(ii) If in addition the boundary map (15.19) is epimorphic for $i = 2$, the set of equivalence classes of holomorphic $\star$-products associated with a given holomorphic symplectic form is in a bijective correspondence with the set $H^2(F^1\Omega^{*,*}(M), d)[[\hbar]]$.

The idea behind this result is to use a slightly more general approach, than that of Fedosov. Let us briefly outline it here. Recall, that a vector bundle $E \to M$ is called *Lie algebroid*, if it is equipped with two additional structures:

(i) A skew-symmetric bilinear bracket $[,]$ on the space of sections $\Gamma(E)$, which satisfies the Jacobi identity.

(ii) A bundle morphism $\rho : E \to TM$, called *the anchor map*, such that the following analog of the Leibniz rule holds:

$$[\xi, f\eta] = f[\xi, \eta] + \rho(\xi)(f)\eta.$$

In this case one can define E-differential forms on M as polylinear skew-symmetric functions of sections of E and use Cartan's formula to define the differential; the corresponding cohomology are denoted by $H_E^*(M, \mathbb{C})$.

Further, one can define E-symplectic form on M as a E-differential 2-form ω_E on M, nondegenerate and closed (with respect to E). It turns out that all the machinery of Weyl algebra bundles and Fedosov connections can be used in this case just like in the usual Fedosov's construction yielding a $\star$-product on M with coefficients in E-differential operators, i.e. such $\star$-product that all the bidifferential operators B_k in the formal series for this product are equal to compositions of elements in the image of the anchor map ρ. Moreover, the equivalence classes of such products are in a bijective correspondence with the elements in $H_E^2(M, \mathbb{C})[[\hbar]]$. In holomorphic case, this construction can be easily "sheafified", i.e. pulled from the space of global sections of E, functions on M, differential forms and bidifferential operators on M to the case where one deals with sheaves of all these objects; in this case we shall have to consider sheaf cohomology version of the space $H_E^2(M, \mathbb{C})[[\hbar]]$ for classification.

In the situation we discuss here, one should take E to be equal to $T^{1,0}M$; holomorphic symplectic form then is reinterpreted as the E-symplectic structure. Then the main difficulty it is to identify the classifying space of holomorphic $\star$-products, the sheaf cohomology as above, with $H^2(F^1\Omega^{*,*}(M), d)[[\hbar]]$, which Nest and Tsygan do by careful inspection of definitions.

One more question, addressed by Nest and Tsygan is the following: "Every holomorphic $\star$-product is *par excellence* a usual $\star$-product on smooth functions (not necessarily holomorphic). Thus there is a unique cohomology class in $H^2_{dR}(M, \mathbb{C})[[\hbar]]$ corresponding to it. On the other hand, any $\star$-product on M will also correspond to an element in this set. Then how does one identify the classes inside this set, that come from holomorphic $\star$-products from all others?"

It turns out, that holomorphic quantisations correspond to a certain affine subvariety inside the space $H^2_{dR}(M, \mathbb{C})[[\hbar]]$. More accurately, here's a theorem, proved by Nest and Tsygan:

Theorem 15.12. *Suppose that the assumptions of the part (ii) of Theorem 15.11 hold. Fix a splitting* (15.20), *then there exists a sequence of (polynomial) maps*

$$\tau_n : H^2(F^1\Omega^{*,*}(M), d)^{\times n} \to H^2(M, \mathcal{O}(M)), \tag{15.21}$$

such that for any sequence $\{\alpha_k\}_{k \geq 0}$ *of elements* $\alpha_k \in H^2(F^1\Omega^{*,*}(M), d)$, *the class (in* $H^2_{dR}(M, \mathbb{C})[\hbar^{-1}, \hbar]]$*) of Fedosov quantisation associated with* $\alpha_0 + \hbar\alpha_1 + \ldots$ *by part (ii) of Theorem 15.11 is given by*

$$\varphi = \frac{1}{\hbar}\omega + \alpha_0 + \sum_{k \geq 1} \hbar^k(\alpha_k + \tau_k(\alpha_0, \ldots, \alpha_{k-1})).$$

The classes $\tau_k(\alpha_0, \ldots, \alpha_{k-1})$ are sometimes called the *Nest-Tsygan classes*.

Observe that if the holomorphic tangent bundle on M admits holomorphic connections, then the class of $\star$-product is determined by a class of closed holomorphic form $\varphi \in \Omega^{2,0}(M)[[\hbar]]$ and there's no need to add elements from $H^2(M, \mathcal{O}(M))$. Thus the maps τ_n should vanish if the Atiyah class of M is trivial. However, in the original paper of Nest and Tsygan the existence of maps τ_n was proved by induction and no explicit formulas were given so the geometric nature of τ_n and their exact relation to Atiyah class remain mysterious till today.[52]

---

[52] Up to the knowledge of the author, January 2022.

Exercises 5: Fedosov quantisation

E.5.1 Prove that the center of Weyl algebra $\widehat{W}_\hbar(\mathbb{R}^{2n}, \omega)$ consists only of the constants $\mathbb{R}[[\hbar]]$.

E.5.2 Prove that the action of the group $Sp_{2n}(\mathbb{R})$ on $\mathbb{R}^{2n}$ induces automorphisms on the Weyl algebra $W_\hbar(\mathbb{R}^{2n}, \omega_0)$.

E.5.3 Prove that $F_{ij,kl} = F_{ij;lk}$, if $F_{ij;lk} = F^m_{ij;k}\omega_{ml}$, where we denote the coefficients of the curvature matrix of a symplectic connection ∇ on M by the symbols $F^m_{ij;k}dx^i \wedge dx^j$.

E.5.4 Let $F^m_{ij;k}dx^i \wedge dx^j$ be the coefficients of the curvature matrix of a symplectic connection ∇ on M; let $F_{ij;lk} = F^m_{ij;k}\omega_{ml}$. Prove that the formal sum

$$\Omega(y) = \Omega_{ij}(y)dx^i \wedge dx^j = F_{ij;lk}y^l y^k dx^i \wedge dx^j$$

(here we omit the constant factor $\frac{\sqrt{-1}}{\hbar}$) does not depend on the choice of coordinates and hence determines a global section in $\widehat{W}^2_\hbar(M)$.

E.5.5 Prove that $\nabla(\alpha) = d\alpha + [\Gamma(y), \alpha]$ for all $\alpha \in \widehat{W}^p_\hbar(M)$, where $\Gamma(y) = \frac{\sqrt{-1}}{\hbar}\Gamma^l_{ij}\omega_{lk}y^j y^k dx^i$ (here Γ^k_{ij} are the Christoffel symbols of a symplectic connection ∇ and the brackets $[,]$ denote the graded commutator).

E.5.6 Put $\omega(y) = \frac{\sqrt{-1}}{\hbar}\omega_{ij}y^j dx^i$. Prove that this expression does not depend on the choice of local coordinates, so $\omega(y) \in \widehat{W}^1_\hbar(M)$ is a global differential 1-form with values in Weyl algebras. Check by direct computation that the square of the graded commutator with $\omega(y)$ is up to a constant factor equal to the commutator with symplectic form ω:

$$[\omega(y), [\omega(y), \alpha]] = -\frac{\sqrt{-1}}{2\hbar}[\omega, \alpha],$$

and hence it vanishes identically.

E.5.7 Let $D' = \nabla + [\varDelta\Gamma,]$ be an Abelian connection, (here ∇ is a symplectic connection and $\varDelta\Gamma \in \widehat{W}_\hbar^1(M)$ is a global 1-form with values in $\widehat{W}_\hbar^{\mathbb{C}}(M,\omega)$), in particular $D'^2(a) = 0$ for any $a \in \widehat{W}_\hbar(M)$. Prove that the correction term $\varDelta\Gamma$ can be represented as $\varDelta\Gamma = \omega(y) + \zeta$, where

$$\omega(y) = \frac{\sqrt{-1}}{\hbar}\omega_{ij}y^i dx^j, \ \zeta \in \widehat{W}_\hbar^1(M), \ \zeta_{00} = 0$$

and $\delta^{-1}\zeta = 0$.

E.5.8 Choose a local trivialisation of the Weyl bundle $\widehat{W}_\hbar^{\mathbb{C}}(M,\omega)$ on an open set $U \subseteq M$ and use it to extend the de Rham operator d to $\widehat{W}_\hbar^*(U)$. Prove that the operator δ anti-commutes with this extension.

E.5.9 Prove, that the formal power series that determines the product (14.6) begins with

$$f \star g = fg + \frac{\hbar}{2\sqrt{-1}}\{f,g\} + \ldots.$$

Hint: try unravelling the definition of the symmetrised product $a \circ b$.

E.5.10 Let D_1, D_2 be two Fedosov connections, given by

$$D_i = \nabla_i - \delta + [\zeta_i, \ ], \ i = 1, 2$$

for differential 1-forms ζ_i with values in Weyl algebras (see Section 13.3.3). Suppose, that there exists an invertible section c of $\widehat{W}_\hbar^{\mathbb{C}}(M,\omega)$ such, that

$$D_2(x) = c^{-1}D_1(cxc^{-1})c.$$

Prove the formulas (compare with (14.8)):

$$\nabla_2 = \nabla_1^c, \ \zeta_2 = c^{-1}\zeta_1 c - c^{-1}\delta c,$$

where $\nabla^c(f) = c^{-1}\nabla(cxc^{-1})c$.

Exercises 5: Fedosov quantisation   237

E.5.11 Recall (see Remark 14.2) that any Fedosov connection has the form $D = \nabla + [\Delta\Gamma, \ ]$ for some fixed symplectic connection ∇ and a Weyl-algebra valued 1-form $\Delta\Gamma$. Prove that for any $\alpha \in \widehat{W}_\hbar^*(M)$ we have:

$$D^2(\alpha) = \left[\Omega(y) + \nabla\Delta\Gamma + \frac{1}{2}[\Delta\Gamma, \Delta\Gamma], \alpha\right],$$

where $\Omega(y)$ is the curvature of ∇ (considered as a section in $\widehat{W}_\hbar^2(M)$, see formula (13.9)). We shall call the sum $\Omega(y) + \nabla\Delta\Gamma + \Delta\Gamma^2$ the curvature Ω_D of D; by definition it is an element in the center of $\widehat{W}_\hbar^2(M)$. Find the way the curvature of D changes under the conjugation $D \mapsto D^c$, where c is an invertible element.

E.5.12 Let $D(t) = \nabla + [\Delta\Gamma(t), \ ]$ be a family of Fedosov connections and $\lambda(t) \in \Omega^1(M)[[\hbar]]$ be a family of 1-forms. Prove that if $D(t)(\lambda(t) + \Delta\dot{\Gamma}(t)) = 0$ (where $\dot{f}$ denotes the differentiation with respect to the parameter t), then there exists $H(t)$, determined by the equation

$$H(t) = \delta^{-1}(\lambda + \Delta\dot{\Gamma}(t) - (\delta + D(t))H(t)),$$

and that this $H(t)$ satisfies Equation (14.9). **Hint:** use the iterative process, analogous to the one used in Sections 13.3.5 and 14.1.

16 Higher homotopy algebras: Topological background and definitions

In the last few sections of this course we are going to talk about the solution of the general deformation quantisation problem for Poisson structures on manifolds. This solution was first given by Maxim Kontsevich in September 1997: this is the date when the paper appeared in electronic archive (https://arxiv.org/pdf/q-alg/9709040.pdf). An updated version of this text later appeared as the IHES preprint (https://www.ihes.fr/~maxim/TEXTS/DefQuant_final.pdf), and the official publication date in a refereed journal is in 2003, see [Kon03].

Unlike previously known solutions, even including Fedosov quantisation, the main idea behind Kontsevich's theorem was to consider the deformation series as a single object in certain homological context rather than as a collection of separate bidifferential operators, satisfying some homological equations as in traditional deformation theory, see Section 9. In other words, we should find a homological interpretation of the associativity condition for the deformation series $B = \sum_{k \geq 1} \hbar^k B_k$ and then prove the existence of such objects: in this way we will prove that *all equations on B_k hold simultaneously*.

It turned out that a proper way to interpret this equation is within the theory of higher homotopy algebras, so our first purpose is to give a brief introduction into this theory. We begin with a short excursion into Topology, where the ideas of strong homotopy algebraic structures originated. In this section we assume that the reader is familiar with the basic notions of homotopy theory, such as the notion of homotopy, homotopy equivalence etc., see [Sp66] for example.

16.1 Topological preliminaries: Loop spaces and loop products

If one has to put it roughly, Topology is the study of various properties of geometric objects (*topological spaces*), which survive under more or less wide range of continuous deformations of the spaces.

In this study topologists often apply various functorial geometric constructions to spaces and compare properties of the new object, obtained by these constructions; these objects can be of more or less any reasonable nature, such as groups, algebras, topological spaces or elements in some sets. One of popular constructions of this kind is the *loop space* construction. So we let X be a connected CW-complex[53] and let $x_0 \in X$ be a point (0-dimensional cell) in it. The pair (X, x_0) is often called *pointed space*, and the point x_0 is called *the base point* of this space; any map $(X, x_0) \to (Y, y_0)$ between pointed spaces should send the base point x_0 into the base point y_0.

Definition 16.1. *The loop space of X at x_0 is the set*

$$\Omega_{x_0} X = \{\gamma : [0,1] \to X \mid \gamma(0) = \gamma(1) = x_0\},$$

*endowed with the **compact-open topology**, i.e. where the base of topology consists of subsets*

$$\mathscr{U}(K, V) = \{\gamma \in \Omega_{x_0} X \mid \gamma(K) \subseteq V\},$$

where $K \subseteq [0,1]$ ranges through all compact subsets in the segment $[0,1]$, and V goes through all open subsets in X.

Putting it short, one says that $\Omega_{x_0} X$ is a particular example of *mapping space*: for any two topological spaces C and X one puts

$$C^X = \{f : X \to C\},$$

which is also usually endowed with the compact-open topology. Below we shall pay very little attention to the point set topology of X and $\Omega_{x_0} X$; just observe that if X is a metric space, then the compact-open topology of the loop space coincides with the pointwise convergence topology induced from the metric. Also below we will use the

---

[53] *Cell complexes* or *CW complexes* is a wide class of topological spaces studied by Algebraic Topology. Loosely speaking they are topological spaces obtained by "glueing" together Euclidean discs (called *cells*) of various dimensions in such a way that the boundary sphere of any such cell is inside the union of cells of lower dimensions. The purpose of using this class of spaces is that many results in this case can be obtained by induction on dimension of the cells. The reader nonfamiliar with basic properties of CW-complexes can think of triangulated manifolds or simplicial complexes instead.

"exponential property" of the mapping spaces: under mild assumptions on the topology of X, Y and C there is a homeomorphism of topological spaces
$$(C^X)^Y = C^{X \times Y}.$$

The actual identification is given by associating with a map $F : X \times Y \to C$ the map $f : Y \to C^X$, given by
$$(f(y))(x) = F(x, y).$$

Using this property one can, for example, identify the double loop space $\Omega_{\omega_0}(\Omega_{x_0} X)$ (here ω_0 is the constant path at the point x_0) with the space of maps
$$\Omega^2_{x_0} X = \left\{ \alpha : I^2 \to X \mid \alpha\left(\{0,1\} \times I \bigcup I \times \{0,1\}\right) = x_0 \right\};$$

here I denotes the unit segment: $I = [0,1]$. Before we proceed, observe that in many cases, when the marked point is fixed it is dropped from notation, so we shall often write ΩX, $\Omega^2 X$, $\Omega^3 X$ etc. instead of $\Omega_{x_0} X$, $\Omega^2_{x_0} X$, $\Omega^3_{x_0} X$ etc. The spaces $\Omega^k X$ are called *iterated loop spaces*.

One important additional structure that comes naturally with the loop space is the *composition* or *concatenation of loops*: for any two loops $\alpha, \beta : [0,1] \to X$ we define $\alpha * \beta \in \Omega_{x_0} X$ by the formula

$$\alpha * \beta(t) = \begin{cases} \alpha(2t), & 0 \le t \le \frac{1}{2}, \\ \beta(2t-1), & \frac{1}{2} < t \le 1. \end{cases} \qquad (16.1)$$

This construction can be regarded as a sort of product on ΩX (from now on we shall drop the base point from our notation):

$$m_0 : \Omega X \times \Omega X \to \Omega X, \quad m_0(\alpha, \beta) = \alpha * \beta. \qquad (16.2)$$

The "product" m_1 is neither commutative, nor even associative: clearly

$$m_0(m_0(\alpha, \beta), \gamma) = (\alpha * \beta) * \gamma \neq \alpha * (\beta * \gamma) = m_0(\alpha, m_0(\beta, \gamma)),$$

since on the right the loop α takes the first half of the segment, and on the left it spans only from $t = 0$ till $t = \frac{1}{4}$. Similarly, there is neither the unit element, nor the inverse element in ΩX. However, if we consider all the operations *up to homotopies*, then all these properties are restored: the unit is represented by the map

$$\omega_0 : [0,1] \to X, \; \omega_0(t) = x_0,$$

and the inverse element is given by $\bar{\alpha}(t) = \alpha(1-t)$. In effect, the reader familiar with basic notions of Topology will immediately identify the set of linear components of ΩX with the fundamental group $\pi_1(X, x_0)$ of the pointed space (X, x_0).

Observe, that the homotopy that restores the associativity of the product m_0 on ΩX can be given by a "simultaneous reparametrisation" of the segment $[0,1]$, i.e. there exists a map h from the unit square $I^2 \to [0,3]$ such that the composition of h with

$$\alpha \cdot \beta \cdot \gamma = \begin{cases} \alpha(t), & 0 \leq t < 1, \\ \beta(t-1), & 1 \leq t < 2, \\ \gamma(t-2), & 2 \leq t \leq 3 \end{cases}$$

gives a homotopy from $(\alpha * \beta) * \gamma$ to $\alpha * (\beta * \gamma)$. Fixing h we can regard this homotopy as the map:

$$m_1 : \Omega X \times \Omega X \times \Omega X \times I \to \Omega X,$$
$$m_1(\alpha, \beta, \gamma, 0) = (\alpha * \beta) * \gamma, \qquad (16.3)$$
$$m_1(\alpha, \beta, \gamma, 1) = \alpha * (\beta * \gamma)$$

for all $\alpha, \beta, \gamma \in \Omega X$. Figure 16.1 gives a graphic description of such a map: every quadrilateral into which the square is decomposed at this figure is mapped piecewise linearly into unit square. One can regard the map m_1 as a map that extends into the interior part of the segment I the maps $(\alpha * \beta) * \gamma$ and $\alpha * (\beta * \gamma)$ on its boundary; in this case the product m_0 is called *homotopy associative*. Similarly, there are maps

$$L_e, R_e : \Omega X \times I \to \Omega X,$$

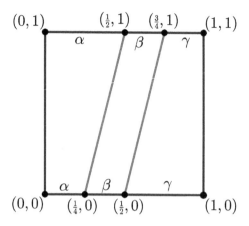

Figure 16.1: The map m_1.

such that
$$L_e(\alpha, 0) = \alpha * \omega_0, \quad L_e(\alpha, 1) = \alpha,$$
$$R_e(\alpha, 0) = \omega_0 * \alpha, \quad R_e(\alpha, 1) = \alpha$$
for any $\alpha \in \Omega X$. In other words, the maps
$$l_e, r_e : \Omega X \to \Omega X, \quad l_e(\alpha) = \alpha * \omega_0, \quad r_e(\alpha) = \omega_0 * \alpha$$
are *homotopic to identity*.

16.2 H-spaces and Stasheff's theorem

One of the main problems in Topology is to distinguish various topological spaces up to some equivalence relation (e.g. up to a homotopy equivalence or homeomorphism). This usually is achieved by comparing various algebraic and geometric structures (invariant under the equivalence transformations) associated to these spaces. Products analogous to m_0 are an example of such structures. Namely, one gives the following definition:

Definition 16.2. *A cell complex[54] Y is called **homotopy associative H-space** if there is a homotopy associative product map*

---
[54]For the sake of simplicity we restrict our attention to CW complexes, so all conditions on structure maps of an H-space can be relieved to what we write here; this is more convenient rather than using the most generic definition.

$\mu_0 : Y \times Y \to Y$, and an element $e \in Y$ such that both maps $\lambda_e, \rho_e : Y \to Y$, $\lambda_e(y) = \mu(y, e)$, $\rho_e(y) = \mu(e, y)$ are homotopic to identity.

The first condition (i.e. the associativity of m up to a homotopy) here means that there exists a map μ_1, similar to m_1 above:

$$\mu_1 : Y \times Y \times Y \times I \to Y,$$
$$\mu_1(a, b, c, 0) = \mu(\mu(a, b), c), \qquad (16.3')$$
$$\mu_1(a, b, c, 1) = \mu(a, \mu(b, c)).$$

Clearly, for any X, its loop space ΩX is a homotopy-associative H-space. The opposite is not clear and a long-standing problem in Topology was to *find a criterion for a homotopy-associative H-space Y to be homotopy equivalent to the loop space ΩX of some X*. Here we assume that the maps that induce this homotopy equivalence should respect the product "up to a homotopy", i.e. that the composition

$$m_0 \circ (f \times f) : Y \times Y \to \Omega X \times \Omega X \to \Omega X$$

is should be homotopic to

$$f \circ \mu : Y \times Y \to Y \to \Omega X,$$

where $f : Y \to \Omega X$ is the homotopy equivalence map and similarly for the homotopy inverse map $g : \Omega X \to Y$. Similar conditions should hold for the units $e \in Y$, $\omega_0 \in \Omega X$ and the homotopies L_e, R_e. The first consistent answer to this question was given in 1963 by Jim Stasheff [Sta63₁], [Sta63₂] (see also the book [Sta70]). Now we are going to explain it briefly, since it is closely related with the algebraic structures that we will use later.

Consider the pentagon P_2 at Fig. 16.2: its vertices are spanned by various expressions that involve the product (composition) $*$ of four elements (loops) $\alpha, \beta, \gamma, \delta$ in ΩX or in any other H-space Y. Since this product is not associative, the way we put parentheses in these expressions matters, and it is easy to see that there are exactly 5 different formulas of this sort (see Fig. 16.2). On the other hand, since the product is homotopy associative, one can connect the vertices in this diagram by segments, and extend the map from

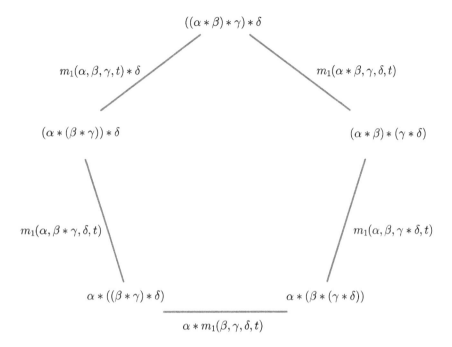

Figure 16.2: The map m_2.

$(\Omega X)^{\times 4} \times P_2^{(0)}$ (or $Y^{\times 4} \times P_2^{(0)}$) into ΩX (or into Y), where $P_2^{(0)}$ denotes the vertices of P_2, to the sides of this pentagon, i.e. to a map $(\Omega X)^{\times 4} \times \partial P_2 \to \Omega X$ (resp. $Y^{\times 4} \times \partial P_2 \to Y$). To this end, we will use the maps that "connect" those different expressions; these maps are determined by the homotopy m_1 (or μ_1) and are given by explicit formulas (Fig. 16.2).

Now we have a continuous mapping $(\Omega X)^{\times 4} \times \partial P_2 \to \Omega X$ (and similarly for any homotopy associative H-space Y). It turns out that in case of the loop space this map can be extended from the boundary ∂P_2 of the pentagon (called *Stasheff's pentagon* sometimes) to its interior. Speaking a little loosely, this is done by putting the map

$\tilde{m}_0^4 : (\Omega X)^{\times 4} \to \Omega X$,

$\tilde{m}_0^4(\alpha_0, \alpha_1, \alpha_2, \alpha_3)(t) = \alpha_i(4t - i)$, if $i \leq 4t \leq i+1$, $i = 0, 1, 2, 3$

in the centre of P_2 and connecting it with the boundary by piecewise-linear reparametrisations of the segment.

Reasoning by induction Stasheff further defines for all natural $n \geq 1$ the n-dimensional Euclidean polytopes P_n, whose vertices are spanned by all possible ways to form the (nonassociative) product of $n+2$ loops $\alpha_0, \ldots, \alpha_{n+1}$ in ΩX; the boundary of P_n can be decomposed into the union of polytopes P_k, $k < n$ and their Cartesian products; in particular $P_1 = I$ and P_2 is the pentagon and we can put $P_0 = *$ (the one-point set). Assuming by induction that there exist maps

$$m_k : (\Omega X)^{\times k+2} \times P_k \to \Omega X, \ k \geq 1, \tag{16.4}$$

for $k \leq n$ one can define a map $(\Omega X)^{\times n+2} \times \partial P_n \to \Omega X$. Then Stasheff observes that this map can now be extended into the interior of P_n by the same trick as before: we put a map $\tilde{m}_0^{n+2} : (\Omega X)^{\times n+2} \to \Omega X$ into the center of P_n and connect it with the boundary by piecewise-linear reparametrisations.

Thus we obtain a sequence of maps m_k (see (16.4)) for all $k \geq 1$; if now an H-space Y is homotopy equivalent to ΩX, then these maps can be modified to give maps $\mu_k : Y^{\times k+2} \times P_k \to Y$ (we can use the maps that define homotopy equivalence to do this) and we see that *the existence of μ_k is a necessary condition for Y to be homotopy equivalent to a loop space*. The big achievement of Stasheff is that he proved that this structure is *not only necessary but also sufficient condition for the homotopy equivalence of H-spaces*. Summing up, we obtain the following theorem:

Theorem 16.3 (Stasheff). *There exists a series of Euclidean polytopes P_n, $n = 0, 1, 2, \ldots$, called **Stasheff's polytopes** or **associahedra**, such that*

(i) $\dim P_n = n$; *in low dimensions $P_0 = *$, $P_1 = I$ and P_2 is pentagon;*

(ii) *The set of vertices $P_n^{(0)}$ of P_n is in bijective correspondence with the set of all possible non-associative products of $n+2$ elements;*

(iii) *Boundary ∂P_n of P_n consists of Cartesian products of various pairs of P_k, $k < n$;*

(iv) For every n and for every cell complex X there is a map
$$m_n : (\Omega X)^{\times n+2} \times P_n \to \Omega X,$$
which coincides with the maps induced from P_k, $k < n$ on the boundary; in low dimensions these maps are given by the concatenation of loops m_0 for $n = 0$ and the associativity homotopy m_1 for $n = 1$;

(v) An H-space Y is homotopy equivalent to the loop space ΩX of a cell space X iff there exists a series of maps
$$\mu_n : Y^{\times n+2} \times P_n \to Y,$$
satisfying the same conditions as m_n above.

This theorem justifies the following important definition:

Definition 16.4. *A cell space Y is called A_∞-space if it can be equipped with a series of maps*
$$\mu_n : Y^{\times n+2} \times P_n \to Y$$
(here P_n, $n \geq 0$ are Stasheff's polytopes) which match on the boundary ∂P_n with the maps induced from μ_k, $k < n$.

Clearly, it follows from the part (v) of Theorem 16.3 that all A_∞-spaces are homotopy equivalent to loop spaces of cell complexes. One can define morphisms of A_∞-spaces as maps which somehow preserve the structure maps, homotopies of such maps and homotopy equivalences of A_∞-spaces, so that the homotopy category of such spaces appears. We are not going to explain it here, but let us briefly describe one of the main properties of this category:

Proposition 16.5. *If a cell-complex A is homotopy equivalent to an A_∞-space B, then A is an A_∞-space itself and A and B are homotopy equivalent as A_∞-spaces.*

The idea behind this statement is that every homotopy equivalence of cell spaces can be given by a deformation retraction $i : A \to B$, $r : B \to A$ such that $ri = \mathrm{id}_A$ and $ir : B \to B$ is homotopic to id_B via a homotopy h relative to A. Then taking compositions of i, r and h we obtain the structure maps; one can regard this construction as a topological analog of the homology perturbation lemma.

16.3 From topology to algebra: Strong homotopy algebras

We have just seen that keeping trace of all the homotopies that ensure associativity of the product of loops proved to be quite useful in Topology. If we now want to pass from Topology to Algebra, we can use one of homology constructions: in effect, if we replace any cell space X by its chain complex Λ_X, we can use various maps to and from tensor products of complexes (Eilenberg-Sielber and Alexander-Whitney morphisms etc.) and functoriality of morphisms to obtain maps

$$\Lambda_Y^{\otimes n+2} \otimes \Lambda_{P_n} \to \Lambda_Y, \ n \geq 0$$

for any A_∞-space Y. Now P_n being a polytope, its cell chain complex is acyclic and consists of just one top-dimensional cell and the boundary complex. Thus we can omit this unique cell from the notation and get maps from tensor products $\Lambda_Y^{\otimes n+2}$ to Λ_Y that change the degrees of the elements by $-n$.

One can work out all details in this case, but we shall skip this now since we only need Topology as an inspiration. So we are going to pass straight to the algebraic definition. But first let us recall the basic notation and terminology from homological algebra: a $\mathbb{Z}$-*graded* (or just *graded*) vector space[55] V is a collection of vector spaces $V = \{V_n\}_{n \in \mathbb{Z}}$ often thought of as the direct sum $\bigoplus_{n=0}^\infty V_n$; in this case elements $v \in V_n \subseteq V$ are called *homogenous of degree n*. Also we will usually assume that $V_n = 0$, $n < 0$ (or at least for sufficiently large negative values of n): this ensures that one can define tensor product of two graded spaces as the following graded space:

$$(V \otimes W)_n = \bigoplus_{p+q=n} V_p \otimes W_q,$$

otherwise this direct sum would contain infinitely many summands (this can sometimes be useful, especially if one considers direct product of graded components instead of the direct sum). A map $f : V \to W$ between graded spaces is called *homogenous of degree k* for some $k \in \mathbb{Z}$ if it is a map between the direct sums and

---

[55] In all examples that we consider in these lecture notes our vector spaces will be defined over characteristic zero fields, even only over $\mathbb{R}$ or $\mathbb{C}$.

$f(V_n) \subseteq W_{n+k}$ for all $n \in \mathbb{Z}$. Also recall that *suspension* or *shift* of a graded space V by $k \in \mathbb{Z}$ is the graded space $V[k]$ given by

$$V[k]_n = V_{n+k}.$$

In particular $V_n = V[1]_{n-1} = V[-1]_{n+1}$. All algebraic structures on V (if any) change their signs and are otherwise altered when the shift is used, in accordance with the *Koszul signs convention* or *Koszul signs rule* (see the discussion following Definition 16.6), below we will see this on several occasions.

Before we go on, let us mention that there exists a large number of books in which the basic notions of the strong homotopy algebraic structures are explained, see for instance [Sm01] or [MSS02]. These books are also basic references for the Appendix A.

16.3.1 A_∞-algebras

Let us begin with the notion, that is a straightforward generalisation from the study of loop spaces

Definition 16.6 (A_∞-algebras). *A graded vector space V is said to bear an A_∞-algebra or strong homotopy associative algebra structure if it is equipped with a sequence of maps $m_n : V^{\otimes n} \to V$, $n \geq 1$ of degree $2 - n$, satisfying the following system of equations:*[56]

$$\sum_{p+q=n} \sum_{k=1}^{p} (-1)^{|a_1|+\cdots+|a_{k-1}|-k+1} m_p(a_1, \ldots$$

$$\ldots, a_{k-1}, m_q(a_k, \ldots, a_{k+q-1}), a_{k+q}, \ldots, a_n) = 0. \quad (16.5)$$

Here $a_1, \ldots, a_n$ are arbitrary homogenous elements in V of degrees $|a_1|, \ldots, |a_n|$ and we use comma instead of the tensor product sign to make the formulas look more compact.

The sign $(-1)^{|a_1|+\cdots+|a_{k-1}|-k+1}$ that appears in formula (16.5) is typical for homological algebra, it comes from the rule that whenever

---

[56]Here we regard A_∞-algebras cohomologically rather than homologically; in the homological case the degrees of maps m_n would be $n - 2$ and there can be changes in some signs.

we move an element of degree a across an element of degree b, we should multiply our term by $(-1)^{ab}$. However, to use it here we should consider the maps m_n as maps of degree 1, rather than as maps of degree $2-n$ and change the degrees of a_i to $|a_i|-1$: this comes from the fact that the definition of A_∞-algebras involves the $V[1]$ rather than V, c.f. Definitions 16.8 and 16.9. Thus here we move an element of degree 1 across the elements $a_1, \ldots, a_{k-1}$ of degrees $|a_1|-1, \ldots, |a_{k-1}|-1$ hence the sign. This rule of signs is usually referred to as the *Koszul rule* or *Koszul sign convention*, and it is ubiquitous in homological algebra, although sometimes to see it one needs to shift the gradings first. In future we shall often abbreviate the signs obtained by this rule to simply $(-1)^\varepsilon$ and similar expressions.

Let us consider the first few equations from the system (16.5): taking $n=1$ we see that m_1 is a map $m_1 : V \to V$ of degree $+1$, for which
$$m_1^2(a) = 0 \tag{16.5_1}$$
for all $a \in V$. Thus *V is equipped with the cochain differential*. More accurately, since we assume the shift $V[1]$ instead of V, we need to change the sign of the differential, so we put
$$\partial a = (-1)^{|a|} m_1(a).$$

Moving forward we obtain the map $m_2 : V \otimes V \to V$ of degree 0, for which we have the following equation: for all homogenous elements $a, b \in V$
$$m_1(m_2(a,b)) + m_2(m_1(a), b) + (-1)^{|a|-1} m_2(a, m_1(b)) = 0. \tag{16.5_2}$$

Thus m_2 is a graded product in V, such that the differential m_1 satisfies graded Leibniz rule with respect to m_2. More precisely, since we work with shifted space $V[1]$ rather than with V, we need to change the signs (and the order of terms) again: we put
$$a \cdot b = (-1)^{|a|} m_2(b, a)$$
then Equation (16.5$_2$) takes form
$$(-1)^{|b|} \partial(b \cdot a) = b \cdot \partial a + (-1)^{|b|} \partial b \cdot a,$$

which is precisely the graded Leibniz rule for ∂ and $\cdot$. However, this product needs not be either commutative or associative. The associativity question is somehow resolved at the next step: take $n = 3$ and consider the next equation in the system (16.5); it involves the map m_3 of degree -1, and looks as follows:

$$m_3(m_1(a), b, c) + (-1)^{|a|-1} m_3(a, m_1(b), c) + (-1)^{|a|+|b|} m_3(a, b, m_1(c))$$
$$+ m_1(m_3(a,b,c)) + m_2(m_2(a,b), c) + (-1)^{|a|-1} m_2(a, m_2(b,c)) = 0. \quad (16.5_3)$$

Working out the signs and other conventions we see that m_3 *gives a cochain homotopy between the products* $a \cdot (b \cdot c)$ *and* $(a \cdot b) \cdot c$. Moving forward, one can say that all the remaining maps m_n, $n \geq 4$ are *higher homotopies*, which connect various combinations of m_k, $k < n$. In particular, if $V = A$ is a differential graded algebra, i.e. if the product $\cdot$ is associative "on the nose", then putting $m_3 = m_4 = \cdots = m_n = \cdots = 0$, we get an example of A_∞-algebra. In other words, *every differential graded algebra is A_∞-algebra*. The opposite is not true: counterexamples can be obtained for instance, from topological considerations above.

Similarly to the definition of A_∞-algebras one can give definition of A_∞-*morphisms* between them: as above it is a series of maps $f_k : V^{\otimes k} \to V$, $\deg f_k = 1 - k$, satisfying certain equations (these equations involve the maps m_n for all n). However, writing down all these equations is not very enlightening and we will not do it; instead we are going to take a more "global" point of view on this subject, see the next section, Remark 17.3.

16.3.2 Cofree coalgebras cogenerated by a graded space

The notion of A_∞-algebras is closely related with the study of loop spaces. It also appears in many other situations in Topology. However, in our pursuit of deformation quantisation we shall need another higher homotopy structure, called L_∞-*algebras*. To make the definition easier, let us recall few constructions from algebra before we define this new structure.

Let V be a graded vector space; let ΛV be the graded exterior algebra of V

$$\Lambda V = \bigoplus_{n \geq 0} \Lambda^n V$$

where

$$\Lambda^n V = V^{\otimes n}/\langle x \otimes y = -(-1)^{|x||y|} y \otimes x \rangle$$

is the n-th graded exterior power of V. We shall denote by $x \wedge y$ the image of $x \otimes y \in V \otimes V$ under the natural projection to $\Lambda^2 V$. Then a generic element of ΛV can be written as a linear combination of elementary exterior products:

$$x = \sum_\alpha v_1^\alpha \wedge \cdots \wedge v_n^\alpha.$$

Observe that since tensor product of graded spaces is graded, and since the relations that define the exterior powers are homogenous, ΛV is a graded vector space. However, unless $V_k = 0$, $k \leq 0$ graded components of ΛV can be infinite-dimensional even if V is finite-dimensional itself and we might need the additional grading that comes from the tensor powers to work with this space.

Also recall that a *coalgebra* over $\mathbb{C}$ is a complex vector space C equipped with *coproduct* $\Delta_C : C \to C \otimes C$ and *counit* $\eta_C : \mathbb{C} \to C$, which satisfy the axioms dual to the associativity and unit axioms for algebras (one obtains them by switching the directions of all the arrows in the corresponding diagrams). A coalgebra is called *cocommutative* if the map that swaps the legs of tensor product does not change Δ_C; of course this swap should satisfy the Koszul rule of signs. If we let the swap change signs in the way opposite to the Koszul rule, we shall get the graded cocommutativity condition.

The following theorem is a classical result from the theory of coalgebras (see, for example [Sw69]):

Proposition 16.7. *The graded exterior algebra ΛV of V bears the structure of cofree graded cocommutative coalgebra, generated by V, when we equip it with the coproduct*

$$\Delta(v_1 \wedge \cdots \wedge v_n)$$
$$= \sum_{p+q=n} \sum_{\sigma \in Sh_{p,q}} (-1)^\varepsilon (v_{\sigma(1)} \wedge \cdots \wedge v_{\sigma(p)}) \otimes (v_{\sigma(p+1)} \wedge \cdots \wedge v_{\sigma(n)})$$
$$+ 1 \otimes (v_1 \wedge \cdots \wedge v_n) + (v_1 \wedge \cdots \wedge v_n) \otimes 1$$

and counit, given by the inclusion of $\mathbb{C}$ into ΛV as $\Lambda^0 V$. Here the summation is over all (p,q)-shuffles $Sh_{p,q}$ and the sign $(-1)^\varepsilon$ is given by the Koszul rule.

Of all the ideas that appear in this proposition the most exotic probably is the seemingly straightforward concept of *cofree coalgebra*. To clarify it, recall that the *free algebra* $A(V)$ generated by a vector space V is characterised by the following condition: *the space V is canonically included into $A(V)$ and for every linear map $f: V \to A$ where A is an algebra, there exists a unique homomorphism $\varphi(f): A(V) \to A$, which makes the following diagram commute*:

$$V \xrightarrow{\subseteq} A(V)$$
$$f \searrow \quad \downarrow \varphi(f)$$
$$A$$

Now to define *cofree coalgebra* $C(V)$, cogenerated by V we must invert all the arrows in this definition: thus $C(V)$ is a coalgebra, equipped with a canonical projection Π onto V such that *for every coalgebra C and every linear map $g: C \to V$ there exists a unique homomorphism of coalgebras*[57] $\phi(g): C \to C(V)$ *such that the following diagram commutes*:

(16.6)

If we add the condition that $C(V)$ is cocommutative and correspondingly add the clause that the diagram (16.6) exists for all *cocommutative* coalgebras C, then we will get the definition of *cofree cocommutative coalgebra, cogenerated by V*.

Seemingly simple, the concept of cofreeness can be a source of confusion for an unprepared reader; in particular it comes somewhat

---

[57]Recall that a linear map $a: C \to K$ of coalgebras is *homomorphism of coalgebras* if (i) $(a \otimes a) \circ \Delta_C = \Delta_K \circ a$; (ii) $\eta_C = \eta_K \circ a$.

unexpected that coalgebra homomorphisms in the cofree case are defined by maps *into* the space of generators, and not *from* it. We recommend that the reader should spend some time thinking about it and solving problems from the Exercise section, in particular Exercises E.6.1 (try writing down the coalgebra homomorphism in terms of its "Taylor coefficients") and E.6.3.

16.3.3 L_∞-algebras

Now one can take advantage of the constructions we introduced in Section 16.3.2. Let C be a coalgebra; recall that a map $D : C \to C$ is called *coderivation* if

$$\Delta_C \circ D = (D \otimes \mathbb{1} + \mathbb{1} \otimes D) \circ \Delta_C.$$

Recall also that cofree coalgebra cogenerated by a vector space V coincides as linear space with the tensor algebra TV. Then the following definition is equivalent to the definition of A_∞-algebras 16.6 (see Exercises E.6.3 and E.6.4):

Definition 16.8 (Second definition of A_∞-algebras). *One says that a graded vector space V is equipped with the structure of A_∞-algebra if the (graded) cofree coalgebra $TV[1]$, generated by the shifted space $V[1]$, is endowed with a coderivation $M : TV[1] \to TV[1]$ of degree 1, such that $M^2 = 0$.*

This definition is convenient since it reduces all the infinite series of maps m_n and relations on them to just one "global" map M with "Taylor coefficients" m_n satisfying a single equation that looks like the definition of cochain differential. We leave it to the reader to check that Equations (16.5) (in particular (16.5$_1$)–(16.5$_3$)) follow from the equality $M^2 = 0$. Now the definition of L_∞-algebras is obtained just by replacing general cofree coalgebras by graded cocommutative ones:

Definition 16.9 (L_∞-algebras). *One says that a graded vector space V is equipped with the structure of L_∞-**algebra** or **strong homotopy Lie algebra** if the cofree graded cocommutative coalgebra*

$\Lambda V[1]$, (co)generated by the shifted space $V[1]$ is endowed with a coderivation $L : \Lambda V[1] \to \Lambda V[1]$ of degree 1, such that $L^2 = 0$.[58]

As before (see Exercise E.6.3) we can replace L by its "Taylor coefficients" $\ell_n : \tilde{\Lambda}^n V \to V$, where

$$\tilde{\Lambda}^n V = \Lambda^n V[1] = V^{\otimes n}/\langle a \otimes b = -(-1)^{(|a|-1)(|b|-1)} b \otimes a \rangle.$$

The degree of map ℓ_n is $2 - n$ and one can rewrite the equation $L^2 = 0$ in terms of the maps ℓ_n. We leave this exercise to the reader, see Exercise E.6.5; let us only write down the first three equations from this series:

$$\ell_1(\ell_1(a)) = 0, \qquad (16.7_1)$$
$$\ell_1(\ell_2(a,b)) + \ell_2(\ell_1(a),b) \pm \ell_2(a,\ell_1(b)) = 0, \qquad (16.7_2)$$

and finally

$$\ell_1(\ell_3(a,b,c)) + \ell_3(\ell_1(a),b,c) \pm \ell_3(a,\ell_1(b),c) \pm \ell_2(a,b,\ell_1(c))$$
$$= \pm\ell_2(\ell_2(a,b),c) \pm \ell_2(\ell_2(b,c),a) \pm \ell_2(\ell_2(c,a),b). \qquad (16.7_3)$$

Here a, b, c are arbitrary homogenous elements in V and $\pm$ replaces the generic signs that appear from Koszul sign conventions. The first of these equalities shows that ℓ_1 is a cochain differential on V; similarly Equation (16.7$_2$) means that ℓ_2 defines a bilinear graded skew-symmetric bracket $V \wedge V \to V$ of degree 0. Finally, Equation (16.7$_3$) says that ℓ_3 is a cochain homotopy, that makes the bracket, induced by ℓ_2 on the cohomology of V (i.e. the cohomology of V with respect to ℓ_1) satisfy the Jacobi identity. In particular, *every differential graded Lie algebra is an L_∞ algebra for which $\ell_n = 0, n \geq 3$*. Some examples of L_∞-algebras which are not Lie algebras are given in the exercise section.

---

[58] In Physics literature one often uses still another definition of L_∞-algebras, based on the notions of formal geometry; in these terms *L_∞-algebra is a formal affine space equipped with a cohomological vector field Q of degree 1, such that $Q^2 = 0$*. This approach has the advantage of allowing nonaffine formal manifolds with similar structures on them; such objects are called simply *Q-manifolds*. We are not going to use this approach here; an interested reader can find it in the literature, for instance in [MSS02].

17 Maurer–Cartan equations and Kontsevich's theorem

We are now going to take advantage of the structures introduced in the previous section; namely we shall show that the solution of the deformation quantisation problem follows from an answer to some abstract question, connected with L_∞-algebras. Material of this section is somewhat scattered through the literature and is treated by different authors under different names, see the books [Sm01] and [MSS02], mentioned earlier. The results that we discuss here appear under different disguises in most modern texts on the deformation theory, including the paper of Kontsevich [Kon03] or part I of the book [CKTB06].

17.1 Maurer–Cartan equations in differential Lie algebras

We begin with reformulating the deformation quantisation question in terms of the differential Lie algebras. To this end we recall that the Hochschild complex of an algebra A can be equipped with an important algebraic structure, the *Gerstenhaber bracket*:

$$[,]: C^p(A,A) \otimes C^q(A,A) \to C^{p+q-1}(A,A),$$

fort all $p, q \geq 0$. This bracket is graded skew-symmetric, satisfies a variant of graded Jacobi identity and graded Leibniz rule with respect to the Hochschild differential (see Section 8.3.2 for the definitions and proofs). It is not difficult to show (see Section 8.3.2 again), that signs and gradings in all formulas are such that the shifted space $C^\cdot(A,A)[1]$ becomes a differential graded Lie algebra in the usual sense (i.e. the Lie bracket of the elements of degrees p and q is of degree $p+q$ etc.).

Also recall that a $\star$-product on a Poisson manifold M with Poisson bracket $\{,\}$ is given by a formal series

$$f \star g = fg + \sum_{n \geq 1} \hbar^n B_n(f,g), \text{ where } B_1(f,g) = \frac{1}{2}\{f,g\},$$

where all B_n are bidifferential operators on M; the associativity of this product is equivalent to the series of Equations (9.6):

$$\delta B_n = \frac{1}{2} \sum_{p+q=n} [B_p, B_q],$$

where we consider the bidifferential operators as Hochschild 2-chains.

Let us put $B = \sum_{n\geq 1} \hbar^n B_n$, so that $B \in C^1_{diff}(A, A)[1][[\hbar]]$, where $A = C^\infty(M)$ and $C^\cdot_{diff}(A, A)$ denotes the complex of differentiable Hochschild cochains (see Definition 8.1). The complex of differentiable cochains is closed in $C^\cdot(A, A)$ with respect to the Gerstenhaber bracket; we shall denote this cochain complex by $\mathscr{D}^\cdot_{poly}(M)$ and call it (with a slight abuse of terminology) *the algebra of polydifferential operators*. Clearly shifted complex of differentiable cochains $\mathscr{D}^\cdot_{poly}(M)[1]$ is a DG Lie subalgebra in the DG Lie algebra $C^\cdot(A, A)[1]$.

Now the associativity relations for $\star$-product can be written as a single equation in the differential graded algebra $\mathscr{D}^\cdot_{poly}(M)[1][[\hbar]]$

$$\delta B + \frac{1}{2}[B, B] = 0, \qquad (17.1)$$

here the sign is changed due to the shift.

Equation (17.1) is called *Maurer–Cartan equation*; more accurately:

Definition 17.1. *Let $\mathfrak{g}^\cdot$ be a differential graded Lie algebra with differential d and Lie bracket $[,]$; then an element $\varpi \in \mathfrak{g}^1$ is said to satisfy* **the Maurer–Cartan equation** *if*

$$d\varpi + \frac{1}{2}[\varpi, \varpi] = 0.$$

In this case ϖ is called the **Maurer–Cartan element of the algebra** $\mathfrak{g}^\cdot$. *The set of all Maurer–Cartan elements in $\mathfrak{g}^\cdot$ is denoted* $\mathfrak{MC}(\mathfrak{g}^\cdot)$.

This type of equations seems to be ubiquitous in modern Mathematics: they appear under different disguises in the study of Lie groups and Lie algebras (as the usual Maurer–Cartan equation), in

Differential Geometry (as the flat connection condition), in Algebraic Topology (as twisting cochains and their generalisations), in Integrable systems (as various compatibility conditions) and in many other situations.

Summing up we say that *solutions of the deformation quantisation problem on M are given by the Maurer–Cartan elements in the differential graded Lie algebra $\mathscr{D}_{poly}^{\cdot}(M)[1][[\hbar]]$ (i.e. the shifted algebra of formal power series of polydifferential operators on M), which in degree 1 in variable $\hbar$ coincide with the Poisson bracket on M.* Below we shall extend this observation to the equivalence relations of $\star$-products.

17.2 L_∞-morphisms and Maurer–Cartan elements

We are now about to explain the relation of Maurer–Cartan elements and L_∞-algebra theory. One can say that this relation is the key fact, that motivates Kontsevich's theorem, the crucial mathematical idea that relates this theorem to the deformation theory. To this end we first introduce the notion of morphisms between L_∞-algebras and then apply these morphisms to Maurer–Cartan equations.

17.2.1 L_∞-morphisms: definitions

We begin with the following:

Definition 17.2 (L_∞-morphisms). *Let V, V' be two L_∞-algebras; let $L = \{\ell_n\}$, $L' = \{\ell'_n\}$ be the corresponding L_∞-structures (i.e. L, L' are degree 1 square-zero coderivations in cofree cocommutative coalgebras, cogenerated by the shifts of V and V' respectively, and ℓ_n, ℓ'_n are their "Taylor coefficients"). Then L_∞-(homo)morphism $F : V \Rightarrow V'$ is a homomorphism of coalgebras*[59]

$$F : \Lambda V[1] \to \Lambda V'[1],$$

such that
$$F \circ L = L' \circ F. \tag{17.2}$$

---
[59] Just like in the case of L_∞-algebras, L_∞-morphisms can be regarded as a particular case of the morphisms between formal Q-manifolds, that preserves the formal vector fields Q, see footnote 58 at page 254.

Remark 17.3. Similarly if A, A' are A_∞-algebras, with the corresponding A_∞-structure morphisms $\{m_n\}$ and $\{m'_n\}$, then A_∞-morphism $F : A \Rightarrow A'$ is a homomorphism of cofree coalgebras cogenerated by $A[1]$, $A'[1]$ commuting with the codifferentials. We are not going to use this notion here, but the reader is encouraged to look into the properties of A_∞-morphism, as they are perfectly similar, but somewhat easier to understand than those of L_∞-maps.

As before, any homomorphism F into the cofree cocommutative coalgebra $\Lambda V'[1]$ is uniquely determined by its "Taylor coefficients" $f_n : \Lambda^n V \to V'$. Here f_n are maps of degrees $1 - n$ and the condition (17.2) generates a series of equations satisfied by $f_1, f_2, \ldots$. The first of this equations is just

$$\ell'_1(f_1(a)) = (-1)^{|a|} f_1(\ell_1(a)), \qquad (17.2_1)$$

where the degree of f_1 is 0; i.e. it means that (after the shift) f_1 is a cochain map between V and V', equipped with the differentials induced by ℓ_1, ℓ'_1. Next, the map f_2 has degree -1 and satisfies the following equation

$$\ell'_1(f_2(a,b)) - f_2(\ell_1(a), b) - (-1)^\epsilon f_2(a, \ell_1(b))$$
$$= \pm \left(\ell'_2(f_1(a), f_1(b)) + (-1)^\eta f_1(\ell_2(a,b)) \right), \qquad (17.2_2)$$

where we use shorthand notation for the signs. If the signs are unravelled we will see that *f_2 is a cochain homotopy, which ensures that f_1 induces a homomorphism of Lie algebras in the cohomology of V and V' with respect to the differentials ℓ_1 and ℓ'_1 respectively*. All the other equations involving f_k, $k \geq 3$ can (and should) be regarded as higher homotopies between various maps, induced by f_k, ℓ_i and ℓ'_j; we leave it to the reader to describe the structure of these equations (see Exercise E.6.8).

For future references let us observe that in particular, every homomorphism of differential graded Lie algebras $\varphi : \mathfrak{g}^{\cdot} \to \mathfrak{h}^{\cdot}$ can be regarded as an L_∞-morphism by setting $f_1 = \varphi$, $f_k = 0$, $k \geq 1$ (here we regard DG Lie algebras as L_∞-algebras with trivial higher homotopies). The opposite, however, is not true: in many cases one needs to extend the set of morphisms between DG Lie algebras to account for the lack of identities in certain places. So it is worth describing

the set of L_∞-morphisms between Lie algebras. To this end we recall that if $\varphi : V \to V'$ is a homogenous map of degree p between two graded vector spaces and d, d' are degree $+1$ differentials on V, V' (so that $d^2 = d'^2 = 0$), then the *differential of φ* is given by the formula
$$d(\varphi) = d' \circ \varphi - (-1)^p f \circ d, \tag{17.3}$$
see Exercise E.6.7. Then *a collection of maps $f_n : \Lambda^n \mathfrak{g} \to \mathfrak{h}$ of degrees $1 - n$ defines an L_∞-morphism $F : \mathfrak{g} \Rightarrow \mathfrak{h}$ iff the following equations hold* for all n and all $g_1, \ldots, g_n \in \mathfrak{g}$

$$d(f_n)(g_1, \ldots, g_n)$$
$$= \sum_{1 \leq i < j \leq n} (-1)^{\epsilon_1} f_{n-1}([g_i, g_j], g_1, \ldots, \widehat{g_i}, \ldots, \widehat{g_j}, \ldots, g_n)$$
$$+ \frac{1}{2} \sum_{p+q=n} \sum_{\sigma \in Sh_{p,q}} (-1)^{\epsilon_2} [f_p(g_{\sigma(1)}, \ldots, g_{\sigma(p)}), f_q(g_{\sigma(p+1)}, \ldots, g_{\sigma(n)})]. \tag{17.4}$$

Here we as usual abbreviate the signs given by Koszul rule applied to a permutation of graded elements to just $(-1)^{\epsilon_i}$, $i = 1, 2$ and the hat $\widehat{}$ above a symbol denotes the omission of this symbol in the formula.

17.2.2 Properties of L_∞- and A_∞-morphisms

Before we proceed with the relation of L_∞-morphisms and Maurer–Cartan elements, let us briefly describe the main properties of L_∞-morphisms. These properties are mostly identical with the properties of A_∞-homomorphisms, and since in the A_∞ case formulas are a little easier to grasp, we shall speak rather about the latter than about the former. A reader interested in detailed proofs and discussion of properties of such maps is referred to the ample bibliography that exists on this subject, e.g. see [MSS02] and references therein (see also the "Further reading" section of the Introduction).

First of all, we remark that one can compose L_∞ and A_∞ morphisms, since the composition of two coalgebra homomorphisms is again a coalgebra homomorphism. However, the formula for such composition is rather complicated: for instance in A_∞ case we have the following formula:

$$(F \circ G)_n(a_1, \ldots, a_n)$$
$$= \sum_{k=1}^{n} \sum_{p_1+\cdots+p_k=n} (-1)^\epsilon f_k(g_{p_1}(a_1, \ldots, a_{p_1}), \ldots, g_{p_k}(a_{n-p_k+1}, \ldots, a_n))$$

for the "Taylor coefficients" of the composition $F \circ G$ of two A_∞-morphisms in terms of the "Taylor coefficients" of components. One can go further and define homotopy of A_∞- and L_∞-morphisms: to this end we introduce the A_∞-algebra $\Omega^\cdot(I)$ of differential forms on the unit segment (and similarly in L_∞ case, which we omit) and define a homotopy of two A_∞-morphisms $F, G : V \Rightarrow V'$ as the map $H : V \Rightarrow V' \otimes \Omega^\cdot(I)$, which coincides with F and G on the natural projections $\varepsilon_0, \varepsilon_1 : V' \otimes \Omega^\cdot(I) \to V'$ at the ends of the segment, i.e. such that

$$F = \varepsilon_0 \circ H, \; G = \varepsilon_1 \circ H.$$

These definitions allow one to speak about the homotopy equivalence of A_∞- and L_∞-algebras; in particular, one can ask, if an A_∞- or L_∞-map has homotopy inverse. It turns out that there is a simple criterion for an A_∞- or L_∞-map to have a homotopy inverse:

Proposition 17.4. *Let $F = \{f_n\} : V \Rightarrow V'$ be an A_∞-morphism between two A_∞-algebras (or L_∞-morphism between homotopy Lie algebras), and f_n its "Taylor coefficients". Then F has homotopy inverse iff the map $f_1 : V \to V'$ induces an isomorphism on the level of homology (i.e. $f_1^* : H^*(V, m_1) \to H^*(V', m_1')$ is an isomorphism).*

We are not going to give a complete proof of this fact here, let us sketch it in the important (for us) particular case:

Proposition 17.4′. *Let $f : \mathfrak{g}^\cdot \to \mathfrak{h}^\cdot$ be a homomorphism of differential graded Lie algebras (and so an L_∞-morphism of L_∞ algebras) over a characteristic zero field, which induces an isomorphism in cohomology. Then there exists a homotopy inverse L_∞-map $G : \mathfrak{h}^\cdot \Rightarrow \mathfrak{g}^\cdot$.*

Homomorphisms of differential Lie algebras, that induce isomorphisms in cohomology are called **quasi-isomorphisms**.

The *proof* of this fact is based on the following easy statement from Linear Algebra: *for every cochain map $f : C^{\cdot} \to D^{\cdot}$ of cochain complexes over a characteristic zero field that induces an isomorphism in cohomology, there exists a homotopy inverse cochain map $g : D^{\cdot} \to C^{\cdot}$* (see Exercise E.6.9). Also we remark that if the cochain complexes V, V' are over a characteristic zero field, a cocycle $\varphi \in \text{Hom}(V, V')$ induces a trivial element in the homology group $H^*(\text{Hom}(V, V'), d)$ (see Exercise E.6.7) iff the corresponding map $\varphi_* : H^*(V) \to H^*(V')$ vanishes: this follows from the identification $\text{Hom}(V, V') = V^* \otimes V'$ and the Künneth formula.[60]

Using these observations we can construct the homotopy inverse L_∞-morphism $G = \{g_n\}$ as follows. First of all, we will choose the homotopy inverse map g_1 of f on the level of cochains; for the sake of simplicity we will assume that $g_1 \circ f = \text{id}_{\mathfrak{g}^{\cdot}}$ (in general case this condition can be satisfied by passing to subcomplexes) and $f \circ g_1 = \text{id}_{\mathfrak{h}^{\cdot}} + d \circ h_1 + h_1 \circ d$ for some cochain homotopy h_1. The map g_1 will play the role of the first stage of the homotopy inverse L_∞-map G. In order to get the higher Taylor coefficients of G we consider the difference

$$\varphi(a, b) = g_1([a, b]) - [g_1(a), g_1(b)], \tag{17.5}$$

for any $a, b \in \mathfrak{h}^{\cdot}$; we can regard φ as a degree 0 element in the complex $\text{Hom}(\tilde{\Lambda}^2 \mathfrak{h}^{\cdot}, \mathfrak{g}^{\cdot})$. Then $d\varphi(a, b) = 0$ since every map in the formula (17.5) commutes with the differentials. On the other hand, if ξ, η are cohomology classes in $H^*(\mathfrak{h}^{\cdot})$, we have $\xi = f^*(\bar{x}), \eta = f^*(\bar{y})$ for some $\bar{x}, \bar{y} \in H^*(\mathfrak{g}^{\cdot})$ (here g_1^*, f^* are the homomorphisms in cohomology, induced by g_1 and f), then

$$\varphi_*(\xi, \eta) = g_1^*([f^*(\bar{x}), f^*(\bar{y})]) - [g_1^*(f^*(\bar{x})), g_1^*(f^*(\bar{y}))]$$
$$= g_1^*(f^*[\bar{x}, \bar{y}])) - [\bar{x}, \bar{y}] = 0,$$

since $g_1^* = (f_1^*)^{-1}$ and f is a homomorphism of Lie algebras. Hence the class of φ in $H^0(\text{Hom}(\tilde{\Lambda}^2 \mathfrak{h}^{\cdot}, \mathfrak{g}^{\cdot}))$ vanishes, so there exists $g_2 \in \text{Hom}^{-1}(\tilde{\Lambda}^2 \mathfrak{h}^{\cdot}, \mathfrak{g}^{\cdot})$ for which $dg_2 = \varphi$, so that $dg_2(a, b) = g_1([a, b]) -$

---

[60] This reasoning works well in finite-dimensional case; in general situation one needs use suitable modifications of the argument; e.g. by replacing the complexes by their finite-dimensional subcomplexes.

$[g_1(a), g_1(b)]$ for all $a, b \in \mathfrak{h}^{\cdot}$. Moreover for any $x, y \in \mathfrak{g}^{\cdot}$ we have the same identity
$$\varphi(f(x), f(y)) = 0,$$
and it follows that we can choose g_2 so that it vanishes on the image of $\tilde{\Lambda}^2 f : \tilde{\Lambda}^2 \mathfrak{g}^{\cdot} \to \tilde{\Lambda}^2 \mathfrak{h}^{\cdot}$.

Similarly if we consider the composition $f \circ g_2 : \tilde{\Lambda}^2 \mathfrak{h}^{\cdot} \to \mathfrak{h}^{\cdot}$ then again this map is closed and vanishes on cohomology, hence it is homotopic to 0 via a suitable homotopy h_2 (expressible in terms of h_1 and other structure maps).

At the second stage reasoning similarly we consider the map

$$\begin{aligned}\psi(a, b, c) = &\; g_2([a, b], c) + (-1)^{\epsilon_1} g_2([b, c], a) \\ &+ (-1)^{\epsilon_2} g_2([c, a], b) - ([g_2(a, b), g_1(c)] \\ &+ (-1)^{\epsilon_1} [g_2(b, c), g_1(a)] + (-1)^{\epsilon_2} [g_2(c, a), g_1(b)]).\end{aligned} \quad (17.6)$$

Then by a straightforward computation we see that $d\psi = 0$; and since g_2 vanishes on the image of $\tilde{\Lambda}^2 f$, we conclude that $\psi_* = 0$ (since the image of f coincides with the whole exterior square in cohomology) and we can find g_3 such that $dg_3 = \psi$ and g_3 vanishes on the image of $\tilde{\Lambda}^3 f$. Also we can find a chain homotopy $h_3 : \tilde{\Lambda}^3 \mathfrak{h}^{\cdot} \to \mathfrak{h}^{\cdot}$ of degree -3 between $f \circ g_3$ and 0.

Going on by induction, we obtain the maps $g_n : \tilde{\Lambda}^n \mathfrak{h}^{\cdot} \to \mathfrak{g}^{\cdot}$ for all n such that $\{g_n\}$ determines an L_∞-map $G : \mathfrak{h}^{\cdot} \Rightarrow \mathfrak{g}^{\cdot}$. Moreover g_n vanishes on $\tilde{\Lambda}^n f$ so we have

$$(G \circ f)_n = \begin{cases} \mathrm{id}_{\mathfrak{g}^{\cdot}}, & n = 1, \\ 0, & n \geq 2. \end{cases}$$

Hence $G \circ f = \mathrm{id}_{\mathfrak{g}^{\cdot}}$. At the same time we construct the maps $h_1, h_2, h_3, \ldots$ that determine homotopies[61]

$$f \circ g_n \sim_{h_n} \begin{cases} \mathrm{id}_{\mathfrak{h}^{\cdot}}, & n = 1, \\ 0, & n \geq 2. \end{cases}$$

Hence $f \circ G \sim \mathrm{id}_{\mathfrak{h}^{\cdot}}$ (here $\sim$ denotes the homotopy). □

---

[61]More accurately this series of maps can be used to determine the L_∞-homotopy $\mathfrak{h}^{\cdot} \Rightarrow \mathfrak{h}^{\cdot} \otimes \Omega^{\cdot}(I)$, as we defined it above; we omit the details here.

17.2.3 L_∞-morphisms and Maurer–Cartan elements

The importance of L_∞-morphisms for the deformation theory is explained by their property to move the Maurer–Cartan elements from the differential graded Lie algebra in their domain to the algebra in their range. Namely the following proposition holds:

Proposition 17.5. *Let $F = \{f_n\} : \mathfrak{g}^{\cdot} \Rightarrow \mathfrak{h}^{\cdot}$ be an L_∞-morphism between two differential graded Lie algebras, and $\varpi \in \mathfrak{MC}(\mathfrak{g}^{\cdot})$; then the element*[62]

$$F(\varpi) = \sum_{k=1}^{\infty} \frac{1}{k!} f_k(\underbrace{\varpi, \ldots, \varpi}_{k \text{ times}}) \in \mathfrak{h}^{\cdot} \qquad (17.7)$$

satisfies the Maurer–Cartan equation, see Definition 17.1.

Proof. This can be proved by a direct computation, see below. Observe, that there are no expressions $(-1)^\epsilon$ in our formulas, because in effect since ϖ has degree 1 in $\mathfrak{g}^{\cdot}$ the signs are precisely the ones that appear there.

We know that $\varpi \in \mathfrak{MC}(\mathfrak{g}^{\cdot})$, so

$$d\varpi + \frac{1}{2}[\varpi, \varpi] = 0.$$

Also F is an L_∞-morphism, so Equation (17.4) holds; thus we have

$$dF(\varpi) = \sum_{k=1}^{\infty} \frac{1}{k!} df_k(\varpi, \ldots, \varpi) = \sum_{k=1}^{\infty} \frac{1}{k!} k f_k(d\varpi, \varpi, \ldots, \varpi)$$

$$+ \sum_{k=2}^{\infty} \frac{1}{k!} \left(\binom{k}{2} f_{k-1}([\varpi, \varpi], \varpi, \ldots, \varpi) \right.$$

$$\left. - \frac{1}{2} \sum_{p+q=k} \binom{k}{p} [f_p(\varpi, \ldots, \varpi), f_q(\varpi, \ldots, \varpi)] \right).$$

---

[62] Here and below we will always mutely assume that the infinite sums in our formulas algebraically converge; e.g. this happens if $f_k = 0$ for sufficiently large k, or if $f_k(\pi, \ldots, \pi)$ belongs to I^k for some ideal I, so we can consider I-adic topology.

Here the first line is obtained by putting the differential d under the map f_k: all the terms of $d(\varpi \wedge \cdots \wedge \varpi)$ in $\tilde{\Lambda}\mathfrak{g}$ are equal to $d\varpi \wedge \varpi \wedge \cdots \wedge \varpi$ due to the sign agreements. Also the second line is just the right-hand side of the formula (17.4), where we observed that all the terms in the wedge product are the same, and hence we should obtain binomial coefficients from combinatorical considerations. Now we have:

$$k\frac{1}{k!} = \frac{1}{(k-1)!}, \quad \binom{k}{2}\frac{1}{k!} = \frac{1}{2}\frac{1}{(k-2)!}, \quad \binom{k}{p}\frac{1}{k!} = \frac{1}{p!}\frac{1}{q!}.$$

On the other hand f_k are poly-linear so changing the summation variable we get

$$\sum_{k=1}^{\infty} \frac{1}{(k-1)!} f_k(d\varpi, \varpi, \ldots, \varpi) + \frac{1}{2}\sum_{k=2}^{\infty} \frac{1}{(k-2)!} f_{k-1}([\varpi, \varpi], \varpi, \ldots, \varpi)$$

$$= \sum_{p=1}^{\infty} \frac{1}{(p-1)!} f_p(d\varpi + \frac{1}{2}[\varpi, \varpi], \varpi, \ldots, \varpi) = 0$$

and we obtain by changing the summation order:

$$dF(\varpi) = -\frac{1}{2}\sum_{k=2}^{\infty}\sum_{p+q=k} \frac{1}{p!}\frac{1}{q!}[f_p(\varpi, \ldots, \varpi), f_q(\varpi, \ldots, \varpi)]$$

$$= -\frac{1}{2}\left[\sum_{p=1}^{\infty}\frac{1}{p!}f_p(\varpi, \ldots, \varpi), \sum_{q=1}^{\infty}\frac{1}{q!}f_q(\varpi, \ldots, \varpi)\right]$$

$$= -\frac{1}{2}[F(\varpi), F(\varpi)].$$

□

Remark 17.6. This proposition is a particular case of a more general one, see Exercise E.6.11.

17.2.4 Equivalence of Maurer–Cartan elements

Before we proceed with Kontsevich's theorem, let us briefly discuss the equivalence relation on the set $\mathfrak{MC}(\mathfrak{g}\dot{})$: let $\varpi \in \mathfrak{g}^1$ be a Maurer–Cartan element; for arbitrary $\epsilon \in \mathfrak{g}^0$ we consider the analytic 1-parameter family of elements $\varpi_\epsilon(t)$, $t \in \mathbb{R}$, which is a solution of the following formal Cauchy problem:

$$\frac{d\varpi_\epsilon(t)}{dt} = -d\epsilon - [\varpi_\epsilon(t), \epsilon], \quad \varpi_\epsilon(0) = \varpi. \tag{17.8}$$

Observe that for any ϵ and any initial value ϖ we can solve this equation in a unique way. On the other hand, differentiating the Maurer–Cartan equation with respect to t and substituting (17.8) into the formula we get

$$\begin{aligned}\frac{d}{dt}\left(d\varpi_\epsilon + \frac{1}{2}[\varpi_\epsilon, \varpi_\epsilon]\right) &= d\frac{d\varpi_\epsilon}{dt} + \left[\frac{d\varpi_\epsilon}{dt}, \varpi_\epsilon\right] \\ &= d(-d\epsilon - [\varpi_\epsilon, \epsilon]) - [d\epsilon + [\varpi_\epsilon, \epsilon], \varpi_\epsilon] \\ &= -[d\varpi_\epsilon, \epsilon] + [\varpi_\epsilon, d\epsilon] - [d\epsilon, \varpi_\epsilon] - [\varpi_\epsilon, [\varpi_\epsilon, \epsilon]] \\ &= -\left[d\varpi_\epsilon + \frac{1}{2}[\varpi_\epsilon, \varpi_\epsilon], \epsilon\right].\end{aligned}$$

Now since $\varpi = \varpi_\epsilon(0)$ satisfies the Maurer–Cartan equation it follows by induction that

$$\frac{d^n}{dt^n}\bigg|_{t=0}\left(d\varpi_\epsilon + \frac{1}{2}[\varpi_\epsilon, \varpi_\epsilon]\right) = 0, \quad n \geq 0.$$

Hence, $\varpi_\epsilon(t)$ being analytic we conclude that $\varpi_\epsilon(t) \in \mathfrak{MC}(\mathfrak{g}\dot{})$ for all $t \in \mathbb{R}$. Now the following proposition is proved by a direct computation, similar to Proposition 17.5:

Proposition 17.7. *Choose any $\varpi \in \mathfrak{MC}(\mathfrak{g}\dot{})$ and $\epsilon \in \mathfrak{g}^0$, and let $F = \{f_n\} : \mathfrak{g}\dot{} \Rightarrow \mathfrak{h}\dot{}$ be an L_∞-morphism; put*

$$F(\epsilon) = \sum_{k \geq 0} \frac{1}{k!} f_{k+1}(\epsilon, \underbrace{\varpi, \ldots, \varpi}_{k \text{ times}}). \tag{17.9}$$

Then

$$F(\varpi_\epsilon) = F(\varpi)_{F(\epsilon)}.$$

We can now give the following definition

Definition 17.8. (i) *Let $\varphi, \psi \in \mathfrak{MC}(\mathfrak{g}\dot{})$. We shall say that φ and ψ are elementary equivalent, if there exists $\epsilon \in \mathfrak{g}^0$ such that $\varphi_\epsilon(0) = \varphi$, $\varphi_\epsilon(1) = \psi$; we will denote this as $\varphi \sim_\epsilon \psi$.*

(ii) Let $\alpha, \beta \in \mathfrak{MC}(\mathfrak{g}^\cdot)$ be two Maurer–Cartan elements; then we shall say that α **and** β **are equivalent** (denoted as $\alpha \sim \beta$) if there exist e finite sequence $\epsilon_0, \epsilon_1, \ldots, \epsilon_n \in \mathfrak{g}^0$ and $\psi_1, \ldots, \psi_n \in \mathfrak{MC}(\mathfrak{g})$ such, that

$$\alpha \sim_{\epsilon_0} \psi_1 \sim_{\epsilon_1} \cdots \sim_{\epsilon_{n-1}} \psi_n \sim_{\epsilon_n} \beta.$$

Clearly $\sim$ is an equivalence relation. We will denote by $\widehat{\mathfrak{MC}}(\mathfrak{g}^\cdot)$ the set of equivalence classes of the Maurer–Cartan elements in $\mathfrak{g}$. It follows from Proposition 17.7 that *every L_∞-map $F : \mathfrak{g}^\cdot \Rightarrow \mathfrak{h}^\cdot$ induces a map $\widehat{F} : \widehat{\mathfrak{MC}}(\mathfrak{g}^\cdot) \to \widehat{\mathfrak{MC}}(\mathfrak{h}^\cdot)$*. Moreover let $H : \mathfrak{g}^\cdot \Rightarrow \mathfrak{h}^\cdot \otimes \Omega^\cdot(I)$ (here $I = [0,1]$) be a homotopy of L_∞-maps, then it induces a map $\mathfrak{MC}(\mathfrak{g}^\cdot) \to \mathfrak{MC}(\mathfrak{h}^\cdot \otimes \Omega^\cdot(I))$; evaluating at the points of the segment gives a homotopy

$$H : \mathfrak{MC}(\mathfrak{g}^\cdot) \times [0,1] \to \mathfrak{MC}(\mathfrak{h}^\cdot).$$

Now with a little bit of work, which we omit (compare part (ii) of Theorem 14.4, especially Equations (14.9), (14.10) and part (ii) of Remark 14.5) one shows that the this homotopy can be realised as an elementary equivalence $\sim_\epsilon$ for some $\epsilon \in \mathfrak{g}^0$, which means that if the L_∞-morphisms $F, G : \mathfrak{g}^\cdot \Rightarrow \mathfrak{h}^\cdot$ are homotopic, then for any $\varpi \in \mathfrak{MC}(\mathfrak{g}^\cdot)$ the elements $F(\varpi)$ and $G(\varpi)$ are equivalent. Summing up we obtain the following result:

Theorem 17.9. *If the differential graded Lie algebras $\mathfrak{g}^\cdot$ and $\mathfrak{h}^\cdot$ are homotopy equivalent as L_∞-algebras and F is the corresponding L_∞-morphism, then $\widehat{F} : \widehat{\mathfrak{MC}}(\mathfrak{g}^\cdot) \to \widehat{\mathfrak{MC}}(\mathfrak{h}^\cdot)$ is a bijection.*

17.3 Formality of differential Lie algebras and Kontsevich's theorem

In order to eventually formulate Kontsevich's formality theorem we need to give one more definition:

Definition 17.10. *Let $\mathfrak{g}^\cdot, \mathfrak{h}^\cdot$ be two differential graded Lie algebras (DGL algebras). We will say, that $\mathfrak{g}^\cdot$ **is homotopy equivalent to** $\mathfrak{h}^\cdot$ **as Lie algebras** if there exist differential Lie algebras $\mathfrak{x}_0, \ldots, \mathfrak{x}_n$ and $\mathfrak{y}_1, \ldots, \mathfrak{y}_n$ which fit into the diagram*

where all the arrows represent quasi-isomorphisms of differential Lie algebras, see Proposition 17.4'. In particular, if $\mathfrak{h}^{\cdot}$ is equal to the cohomology of $\mathfrak{g}^{\cdot}$ with the bracket induced from $\mathfrak{g}^{\cdot}$, then one says that $\mathfrak{g}^{\cdot}$ is **formal**.

It is clear, that homotopy equivalence of Lie algebras is an equivalence relation. In fact, it follows directly from this definition and from Propositions 17.4', 17.5 and 17.7 that the next proposition holds

Proposition 17.11.

(i) *Two differential graded Lie algebras over a characteristic zero field are homotopy equivalent iff there exists an L_∞-morphism $F : \mathfrak{g}^{\cdot} \Rightarrow \mathfrak{h}^{\cdot}$, which induces an isomorphism on their cohomology (L_∞-quasi-isomorphism).*

(ii) *If $\mathfrak{g}^{\cdot}$ is homotopy equivalent to $\mathfrak{h}^{\cdot}$ and F is the corresponding L_∞-quasi-isomorphism, then $\widehat{F} : \widehat{\mathfrak{MC}}(\mathfrak{g}^{\cdot}) \to \widehat{\mathfrak{MC}}(\mathfrak{h}^{\cdot})$ is a bijection.*

We are going to apply this result to the deformation quantisation problem: we saw in the beginning of this section that $\star$-products are governed by the Maurer–Cartan elements in the differential Lie algebra of polydifferential operators $\mathfrak{g}^{\cdot} = \mathscr{D}_{poly}^{\cdot}(M)[1][[\hbar]]$. Moreover, one can show that two $\star$-products are equivalent (see the formulation of the deformation Problem 1, at page 50 and especially formulas (4.10) and (4.11)) iff the corresponding Maurer–Cartan elements are equivalent, see Exercise E.6.13.

On the other hand, we know that the cohomology of the Lie algebra $\mathscr{D}^{\cdot}(M)$ is equal to the differential graded Lie algebra of polyvector fields equipped with zero differential and the Schouten bracket (see the Hochschild–Kostant–Rosenberg Theorems 8.4, 8.9). Let us denote this algebra by $\mathscr{T}_{poly}^{\cdot}(M)[1][[\hbar]]$. Since the differential in this Lie algebra is trivial, it is easy to see that Maurer–Cartan elements in this algebra are given by the formal power series

$$\Pi = \pi_0 + \hbar\pi_1 + \hbar^2\pi_2 + \ldots, \qquad (17.10)$$

where π_k, $k \geq 0$ are bivectors, such that $[\Pi, \Pi] = 0$; one often refers to these elements as to *formal Poisson structures*. As we know (see Theorem 8.4) Hochschild–Kostant–Rosenberg map $\chi : \mathscr{T}_{poly}^{\cdot}(M) \to \mathscr{D}_{poly}^{\cdot}(M)$ induces an isomorphism of cohomology; in effect $\mathscr{T}_{poly}^{\cdot}(M)$ is just the cohomology of $\mathscr{D}_{poly}^{\cdot}(M)$. However, it is not a homomorphism of Lie algebras, so Proposition 17.11 is not applicable yet. Kontsevich's theorem provides a solution for this problem, at least at the local scale:

Theorem 17.12 (Kontsevich's formality theorem, '97). *The differential graded Lie algebra of polydifferential operators on Euclidean space $\mathbb{R}^d$ is formal. Moreover, one can choose the L_∞-quasi-isomorphism*

$$\mathcal{U} = \{\mathcal{U}_n\} : \mathscr{T}_{poly}^{\cdot}(\mathbb{R}^d)[1] \Rightarrow \mathscr{D}_{poly}^{\cdot}(\mathbb{R}^d)[1]$$

so that the first "Taylor coefficient" $\mathcal{U}_1$ of this morphism will coincide with the Hochschild–Kostant–Rosenberg map χ.

Now linearising this map over $\hbar$ we extend $\mathcal{U}$ to an L_∞-quasi-isomorphism

$$\mathcal{U} = \{\mathcal{U}_n\} : \mathscr{T}_{poly}^{\cdot}(\mathbb{R}^d)[1][[\hbar]] \Rightarrow \mathscr{D}_{poly}^{\cdot}(\mathbb{R}^d)[1][[\hbar]].$$

Observe that if the zeroth term π_0 in the formal Poisson structure Π (see formula (17.10)) vanishes, i.e. if $\Pi = \hbar\pi$ for a Poisson bivector π on $\mathbb{R}^d$ then the convergence condition for the formulas (17.7), (17.9) holds automatically: since the degree of $\hbar$ grows indefinitely, both series are convergent in $\hbar$-adic topology. Thus we come up with the following result:

Proposition 17.13 (Kontsevich quantisation of $\mathbb{R}^d$). *The formula $B = \mathcal{U}(\Pi)$ where*

$$\mathcal{U}(\Pi) = \sum_{n \geq 1} \frac{1}{n!} \mathcal{U}_n(\underbrace{\Pi, \ldots, \Pi}_{n \text{ times}}),$$

induces a bijection between the set of equivalence classes of formal Poisson structures $\Pi = \hbar\pi_1 + \hbar^2\pi_2 + \ldots$ on $\mathbb{R}^d$ and the set of equivalence classes of associative $\star$-products on $C^\infty(\mathbb{R}^d)[[\hbar]]$

$$f \star g = fg + \sum_{n \geq 1} \hbar^n B_n(f, g).$$

In particular, if taking $\Pi = \frac{\hbar}{2}\pi$ for a Poisson bivector π we see that the coefficients of the $\star$-product that solves the deformation quantisation problem for $(\mathbb{R}^d, \pi)$ are given by the formulas

$$B_n = \frac{1}{2^n n!} \mathcal{U}_n(\underbrace{\pi, \ldots, \pi}_{n \text{ times}}). \tag{17.11}$$

This approach is often referred to as *Kontsevich quantisation* or more specifically, *Kontsevich quantisation of $\mathbb{R}^d$*, especially, when the formality morphism $\mathcal{U}$ is given by the construction, that we will describe in the following section, see the formula (18.5). The $\star$-product obtained in this way (see formula 17.11) is called *Kontsevich $\star$-product*.

Remark, that in spite of the fact that topologically the space $\mathbb{R}^d$ is quite simple, this result is totally nontrivial: as one knows from the Weinstein's theorem (Theorem 4.11), Poisson structure on a manifold including $\mathbb{R}^d$ can have different ranks in different points and quite often one can not choose a regular coordinate system for this structure, similar to Darboux coordinates, in a generic point of $\mathbb{R}^d$. So we cannot apply a standard deformation procedure (e.g. Moyal formula) point-wise, like we did in Fedosov quantisation. Instead, we need to regard the deformation series on $\mathbb{R}^d$ as a whole. Further, in order to obtain the solution of the deformation quantisation problem on an arbitrary manifold, one can use Kontsevich's result on coordinate charts and then globalise this construction, glueing them together. This approach is similar to Fedosov's ideas; we shall talk about this procedure below.

18 Kontsevich's construction

In this section we will explain the main ideas of Kontsevich's original proof of Theorem 17.12: namely we shall construct in an explicit way the L_∞-map $\mathcal{U} : \mathcal{T}_{poly}^{\cdot}(\mathbb{R}^d)[1] \Rightarrow \mathcal{D}_{poly}^{\cdot}(\mathbb{R}^d)[1]$, which will coincide with the Hochschild–Kostant–Rosenberg map in degree 1. One should remember that this map is not unique: there exist infinitely many different L_∞-quasi-isomorphisms, extending the Hochschild–Kostant–Rosenberg map χ.

Kontsevich's construction is in a certain sense very intuitive and natural: if one thinks about all possible ways to compose polyvector fields into a polydifferential operator, one immediately comes up with the idea that any such composition can be uniquely encoded by an oriented graph. Roughly speaking, vertices of this graph will symbolise the polyvectors that are involved, and their edges will denote the substitution order (below we will elaborate this idea in detail, see Section 18.3). If $\mathcal{U}_\Gamma(P_1, \ldots, P_n)$ (where P_k are the polyvector fields and Γ is the graph) is the resulting operator, then it is natural to assume that the map $\mathcal{U}_n$ is given by a sum (see formula (18.5)):

$$\mathcal{U}_n(P_1, \ldots, P_n) = \sum_\Gamma W_\Gamma \mathcal{U}_\Gamma(P_1, \ldots, P_n),$$

where the summation is done over all suitable graphs and W_Γ are certain weights. Thus, the main problem here is to find the weights W_Γ which will turn the map $\mathcal{U} = \{\mathcal{U}_n\}$ into an L_∞-morphism with $\mathcal{U}_1 = \chi$. This has been done by Kontsevich by taking W_Γ equal to the integrals of certain differential forms over some configuration spaces. In spite of the deceiving simplicity of this construction, one should not overestimate the explicitness of Kontsevich's formula: the number of terms in it grows rapidly and computing each term is a big problem itself.

Exposition of Kontsevich's ideas naturally falls into two parts: in this section we describe the construction of the map $\mathcal{U} = \{\mathcal{U}_n\}$, after this we will sketch the proof of the fact that this map is indeed an L_∞-quasi-isomorphism, that begins with $\mathcal{U}_1 = \chi$ (see Section 18.4). We will give few details of the proof. However writing down a rigorous reasoning here would involve too much computations and thus render our text barely readable, so we prefer to sacrifice the rigorousness of our exposition for the sake of its intuitive clearness. Interested reader can refer to the original paper of Kontsevich https://arxiv.org/pdf/q-alg/9709040.pdf (see [Kon03], for the published version) or to one of the many expository texts about this subject (see Chapter I of [CKTB06] for example). On the other hand in order to fill this gap we will give the proof of Kontsevich's formality theorem due to Tamarkin (see Appendix B): this latter proof is a pure existence theorem, it does not give any explicit construction of an L_∞-morphism. As such it is based on a different circle of ideas, which might be helpful in many situations, and therefore we think it is worth explaining.

The rest of this section is dedicated to the exposition of results of Kontsevich's paper https://arxiv.org/pdf/q-alg/9709040.pdf; in particular we stick to the notation used there.

18.1 Configuration spaces and their compactifications

In order to get the weights W_Γ in the formula for Kontsevich's L_∞-quasi-isomorphism, we will need to integrate certain differential forms over configuration spaces; both objects will be encoded by appropriate graphs. The standard way to define the configuration spaces is to take certain quotients of open sets in Euclidean spaces, which means that these spaces are originally noncompact, and hence the integration over them is quite problematic. In order to circumvent this problem Kontsevich considers suitable compactifications of configuration spaces, a version of the Fulton–MacPherson compactifications [FM94], which we are about to describe.

18.1.1 Definitions

Let $\mathcal{H}$ denote the upper half-plane

$$\mathcal{H} = \{(x,y) \in \mathbb{R}^2 \mid y > 0\}$$

and

$$\bar{\mathcal{H}} = \{(x,y) \in \mathbb{R}^2 \mid y \geq 0\} = \mathcal{H} \bigcup \mathbb{R}^1$$

is its closure in $\mathbb{R}^2$. Let n, m be nonnegative integers with $2n + m \geq 2$ (equivalently $n \geq 1$ or $m \geq 2$). We are going to consider the configuration spaces $Conf_{n,m}$ of points in $\bar{\mathcal{H}}$, namely let

$$Conf_{n,m} = \{(p_1, \ldots, p_n; q_1, \ldots, q_m) \in \mathcal{H}^{\times n} \times (\mathbb{R}^1)^{\times m} \mid p_i \neq p_j, q_s \neq q_t\}.$$

Let us also denote by $Conf_n$ the configuration space of $n > 1$ points in $\mathbb{C}$:

$$Conf_n = \{(p_1, \ldots, p_n) \in \mathbb{C}^{\times n} \mid p_i \neq p_j\}.$$

Clearly $Conf_{n,m}$ and $Conf_n$ are open noncompact manifolds of dimension $2n + m$ and $2n$ respectively. Kontsevich's formula involves integration over such spaces so our primary purpose now is to describe their suitable compactification so that we will not have to worry about the convergence.

To this end we first factor out the evident action of the homotheties on $Conf_{n,m}$: let $G^{(1)}$ be the group of orientation-preserving affine transformations of $\bar{\mathcal{H}}$:

$$G^{(1)} = \{z \mapsto az + b \mid a, b \in \mathbb{R}, \, a > 0\}.$$

Clearly, $G^{(1)}$ sends $\bar{\mathcal{H}}$ to itself and acts freely on $Conf_{n,m}$ since $2n + m \geq 2$, so $C_{n,m} = Conf_{n,m}/G^{(1)}$ is a smooth noncompact manifold of dimension $2n + m - 2$. In case of $Conf_n$ we have a bit more freedom: we consider the group

$$G^{(2)} = \{z \mapsto az + b \mid a \in \mathbb{R}, \, b \in \mathbb{C}, \, a > 0\}.$$

Again the condition $n > 1$ ensures the fact that the action is free, so $C_n = Conf_n/G^{(2)}$ is a smooth noncompact manifold of dimension $2n - 3$. We will denote by the square brackets $[P]$ the image of $P \in Conf_{n,m}$ or $P \in Conf_n$ under this factorisation.

There exist many different methods to compactify a manifold; the one that we are going to use here is based on the following idea: closure of any subset in a compact space is compact. Thus fixing an embedding of $C_{n,m}$ and C_n into such space we obtain natural compactifications. In our case the role of compact space will be played by the torus $T^N = (\mathbb{R}/2\pi\mathbb{Z})^N$.

So let z be a complex number, and let $Arg\, z \in [0, 2\pi)$ denote the angle in its trigonometric representation: if $z = |z|e^{i\varphi}$, $\varphi \in \mathbb{R}$

mod 2π, then $Arg\, z = \varphi$ mod 2π. We put $N = 2n(n-1) + nm$; in this case the map
$$[(p_1,\ldots,p_n; q_1,\ldots,q_m)] \mapsto$$
$$\prod_{i,j=1}^{n}(Arg(p_i - p_j), Arg(p_i - \bar{p}_j)) \times \prod_{i=1}^{n}\prod_{j=1}^{m}(Arg(p_i - q_j)) \in \prod_{k=1}^{N}(\mathbb{R}/2\pi\mathbb{Z})$$

determines an embedding of $C_{n,m}$ into T^N. Indeed, if $m > 1$, then we can use the group action of $G^{(1)}$ to fix the positions of q_1 and q_2 on $\mathbb{R}^1$ (say, $q_1 = 0$, $q_2 = 1$ or the opposite); then positions of $p_1,\ldots,p_n$ will be uniquely determined by the angles $Arg(p_i - q_1)$, $Arg(p_i - q_2)$, and this allows one to restore the positions of remaining points q_j. If $m = 1$, we use the group to fix the position of q_1 (say, $q_1 = 0$) and the length of $p_1 - q_1$. Then p_1 is uniquely determined by $Arg(p_1 - q_1)$ and the remaining points are also fixed. We let the reader check the details and variants of this reasoning; similarly, one can embed C_n into the torus $T^{n(n-1)}$ by means of the angles $Arg(p_i - p_j)$.

Now we can put $\bar{C}_{n,m}$, $\bar{C}_n$ to be the compactifications of $C_{n,m}$, C_n inside the tori. The constructed spaces have many important combinatoric and geometric properties; first of all, they bear a natural structure of *manifolds with corners*, i.e. manifolds whose boundary have natural stratification, similar to the stratification of the coordinate octants in $\mathbb{R}^n$ by various coordinate hyperplanes; below we will describe various strata in this boundary. Observe that the permutation groups $S_n \times S_m$ act naturally on the configuration spaces $Conf_{n,m}$ (by reordering the points p_i and q_j), and this action extends to $C_{n,m}$ (same is true for C_n and the group S_n). We shall denote by $C_{A,B}$ (respectively C_A) the configuration space of points, indexed by the sets A, B (or just A) and by $\bar{C}_{A,B}$, $\bar{C}_A$ their compactifications. Here A, B are some subsets of $p_1,\ldots,p_n$, $q_1,\ldots,q_m$: the structure of $C_{A,B}$ depends only on the cardinality of A and B. Clearly, forgetting part of the data gives maps $C_{A,B} \to C_{A',B'}$, $C_A \to C_{A'}$ for any subsets $A' \subseteq A$, $B' \subseteq B$.

18.1.2 Examples

Let us describe the spaces $\bar{C}_{n,m}$, $\bar{C}_n$ for small n, m (satisfying the condition $2n + m \geq 2$ or $n > 1$):

$\bar{C}_{1,0}, \bar{C}_{0,2}$ Using the group $G^{(1)}$ we can fix the positions of the point p_1, for example put $p_1 = \sqrt{-1}$, hence $C_{1,0} = \{*\}$ is a one-point set, so the same is true for $\bar{C}_{1,0}$. Similarly, the positions of the points q_1, q_2 can be fixed uniquely as $\{0, 1\}$, where 0 corresponds to the left of these points, and 1 to the right one. Hence the set $C_{0,2}$ consists of two elements: when we fix the order of q_1, q_2, we fall into one of these elements. So $\bar{C}_{0,2} = \{*, *\}$ is a two-point set, whose components are enumerated by the orderings of q_1, q_2.

$\bar{C}_2$ Using the group $G^{(2)}$ we can fix the position of one of the points and the distance between them, e.g. put $p_1 = 2\sqrt{-1}$ and $|p_1 - p_2| = 1$, while the direction of the vector $p_1 - p_2$ is chosen freely. This shows that $C_2 = S^1$ is already compact and hence $\bar{C}_2 = S^1$.

$\bar{C}_{1,1}$ Again using the group $G^{(1)}$ we fix the position of q_1 (put $q_1 = 0$) and the length of $p_1 - q_1$ (put $|p_1 - q_1| = 1$). Hence the space $C_{1,1}$ is naturally identified with *open* upper half-circle $|z| = 1$, $\Im z > 0$. Its compactification hence is naturally identified with the *closed* upper half-circle with endpoints given by the two strata of $\bar{C}_{0,2}$ see Fig. 18.1: here $C_{0,2}^+$, $C_{0,2}^-$ denote the components of the two-point set $C_{0,2}$, corresponding to the direct $q_1 < q_2$ and inverse $q_1 > q_2$ order of the points q_i on the straight line.

$\bar{C}_{2,0}$ This example is more complicated than the previous ones: unlike there, we cannot use the group $G^{(1)}$ action to fix the position of two points to a reasonable extent. In effect, $\dim C_{2,0} = 2$; with the help of $G^{(1)}$ action we can move p_1 into the point $\sqrt{-1} \in \mathcal{H}$, thus $C_{0,2} \cong \mathcal{H} \setminus \{\sqrt{-1}\}$. To describe the boundary of the corresponding compactification $\bar{C}_{2,0}$ one should use accurate considerations of the possible ways to reach "infinity" inside $C_{2,0}$. It turns out, there are three different strategies for this (modulo the action of $G^{(1)}$:

(i) The distance from the points p_1, p_2 shrinks indefinitely in comparison with the distance to either of them from the boundary line of $\bar{\mathcal{H}}$. Since the only invariant of the $G^{(1)}$ action in this situation is the direction of $p_1 - p_2$, we can identify the corresponding stratum with $\bar{C}_2$; this gives us the internal circle at Fig. 18.2.

Kontsevich's construction 275

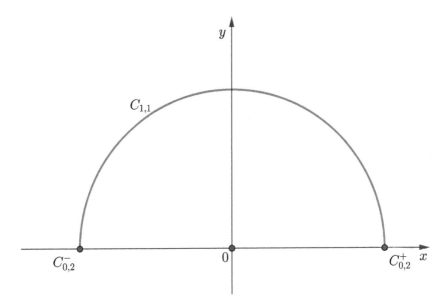

Figure 18.1: The space $\bar{C}_{1,1}$: $\bar{C}_{1,1} = C_{1,1} \bigcup C_{0,2}$.

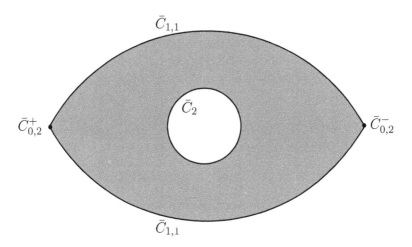

Figure 18.2: The space $\bar{C}_{2,0}$, "The Eye".

(ii) The distance from one of the points p_1, p_2 to the boundary $\partial \bar{\mathcal{H}}$ shrinks in comparison with the distance from the other point; passing to the limit we obtain a point in $\bar{C}_{1,1}$: there are two copies of this space in the boundary of $\bar{C}_{2,0}$ (see Fig. 18.2), depending on which of the points strikes the real line.

(iii) Both points go towards the real line simultaneously; this corresponds to the "corners" of Fig. 18.2, which represent the two components of the space $\bar{C}_{0,2}$.

The resulting shape was dabbed "the Eye" by Kontsevich, due to the striking similarity with Sauron's form from the "Lord of the Rings" movie, see Fig. 18.2.

18.1.3 The stratification of $\partial \bar{C}_{n,m}$, $\partial \bar{C}_n$, "magnifying glass"

Let us now briefly describe the structure of the boundary components of $\bar{C}_{n,m}$ or $\bar{C}_n$. We refer the interested reader to the original paper by Kontsevich for details.

First of all, we shall use the group action of $G^{(1)}$ and $G^{(2)}$ to put every configuration of points in a standard position, so that we don't need to distinguish the configurations that differ only by the group action. After this the size of the configuration will be fixed and we can obtain the boundary components of the compactification by sending certain points to the boundary $\mathbb{R}^1$ of $\bar{\mathcal{H}}$ or collapsing them into one point.

Boundary of $\partial \bar{C}_n$ We begin with the case C_n; here we shall say that a configuration is in standard position if

(i) The diameter of the convex hull of the points is equal to 1;

(ii) The centre of the circumcircle (minimal circle that contains the polygon) of this convex hull is at the point $\sqrt{-1}$.

This prescription fixes the size and position of the convex hull; clearly, using the transformations from $G^{(2)}$ one can put every configuration in a unique standard position of this sort. Now the additional points in compactification of C_n appear from letting some of the points in configuration collapse into one (while maintaining the standard position of the whole configuration).

More accurately let us denote by A the set of points in configuration (forgetting the order). Let $S \subset A$ be the subset of collapsing points. Considering S separately under "microscope" (using "magnifying glass" in Kontsevich's terminology, see figure) we see that the collapsing points contribute to the compactification of C_A a set, homeomorphic to $\bar{C}_S$. At the same time, varying the position of the point s, to which the points in S converge, as well as the positions of remaining points, we get the set $C_{A \setminus S \sqcup \{s\}}$. Summing up we see, that the boundary of $\bar{C}_A$ is made up of the sets

$$\partial_S \bar{C}_A \cong C_S \times C_{A \setminus S \sqcup \{s\}}$$

for various subsets $S \subseteq A$. Observe that the dimensions of these spaces are

$$(2|S| - 3) + (2(|A| - |S| + 1) - 3) = 2|A| - 4 = \dim C_A - 1,$$

thus we obtain a description of the codimension 1 strata in the boundary of $\bar{C}_A$.

Boundary of $\partial \bar{C}_{n,m}$ Here we say, that a configuration $C = (A, B) \in C_{n,m}$ of points in $\bar{\mathcal{H}}$ is in standard position, if

(i) The projection of the convex hull C of the points $p_1, \ldots, p_n$ onto the boundary line $\mathbb{R}^1$ of $\mathcal{H}$ is either a one-point set $\{0\}$, or an interval centred at 0;

(ii) The maximum of diameter of C and the distance from it to the line $\mathbb{R}^1$ is equal to 1.

After the standard position of the set has been fixed, we see that the boundary components of $\bar{C}_{n,m}$ appear in a way similar to the previous one. In fact, there are two principal possibilities, that give the strata of codimension 1 in the boundary of $\bar{C}_{n,m}$:

1. There is a subset $S \subseteq A$ of points in the upper half-plane $\mathcal{H}$, that collapse into one point that is still in the upper half-plane; in this case we get the component

$$\partial_S \bar{C}_{A,B} \cong C_S \times \bar{C}_{A \setminus S \sqcup \{s\}, B}, \qquad (18.1)$$

see Fig. 18.3.

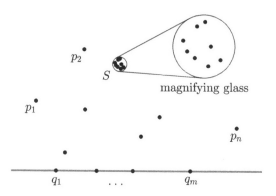

Figure 18.3: Magnifying glass: the set S collapses to a point in $\mathcal{H}$.

2. There are two subsets $S \subseteq A$, $S' \subseteq B$, which collapse together into a point s' that automatically lies on the line $\mathbb{R}^1$; this gives the stratum

$$\partial_{S,S'} \bar{C}_{A,B} \cong C_{S,S'} \times C_{A\setminus S, B\setminus S' \sqcup \{s'\}}. \tag{18.2}$$

For instance, as we saw above, the codimension 1 boundary components of $\bar{C}_{2,0}$ consist of three parts: if we take $S = \{p_1, p_2\}$, then $\partial_S \bar{C}_{2,0} = C_2 \times C_{1,0}$ which gives the internal circle $C_2 = S^1$, as $C_{1,0}$ is a one-point set; if we take $S = \{p_1\}$, $S' = \emptyset$, then $\partial_{S,S'} \bar{C}_{2,0} = C_{1,0} \times C_{1,1} = C_{1,1}$, which is the upper open arc of the boundary, as $C_{1,0}$ is one-point set again.

Of course, one can iterate these constructions and modify them, for instance by taking more than one collapsing subset: one easily sees that this amounts to descending deeper into the stratification of the boundary; for instance in $\bar{C}_{2,0}$ case in this way we get into the "eye's corners". In general, Kontsevich encodes the structure of the boundary set of $\bar{C}_{n,m}$ by the "tree of nested collapsing subsets". This in fact shows that $\bar{C}_{n,m}$ has the natural structure of *manifold with corners*. Our constructions allow us now define the differential forms on this space; moreover one can use Stokes formula in this case: integral of the form $d\omega$ against the space $\bar{C}_{n,m}$ is equal to the integral of ω against $\partial \bar{C}_{n,m}$.

18.2 Admissible graphs and integrals

In this section we will describe the set of all graphs that we will use to produce the polydifferential operators out of the polyvector fields. In addition to encoding the composition of polyvectors, these graphs will encode the differential forms that we will integrate over the spaces $\bar{C}_{n,m}$, $\bar{C}_n$ to obtain the weights in our formula; we will define these weights in Section 18.2.2.

18.2.1 The graphs $G_{n,m}$

Let n, m be two nonnegative integers for which $2n + m - 2 \geq 0$. Consider the set $V_{n,m} = \{1, \ldots, n\} \sqcup \{\bar{1}, \ldots, \bar{m}\}$ (V for short); we say that the elements $1, 2, \ldots, n$ are *the vertices of the first type*, and $\bar{1}, \bar{2}, \ldots, \bar{m}$ are *the vertices of the second type*.

Definition 18.1. *We shall say that an oriented graph[63] Γ on V (i.e. for which $V_\Gamma = V$ is the set of vertices) is* **admissible**, *if the following conditions hold:*

- *Source $s(e)$ of every edge $e \in E_\Gamma$ of Γ is in a vertex of the first type (target $t(e)$ of e can be of any type);*

- *There are no loops (i.e. edges $e \in E_\Gamma$ for which $s(e) = t(e)$), nor multiple oriented edges (i.e. two or more edges with the same source and target: $s(e_1) = s(e_2)$, $t(e_1) = t(e_2)$).*

We will denote by $G_{n,m}$ the set of all admissible graphs with vertices from $V_{n,m}$.

Let us make few observations concerning the graphs in $G_{n,m}$: first of all, since multiple edges are prohibited, the set $G_{n,m}$ is finite, but its cardinality grows rapidly with the numbers n, m. In effect, there can be no more than $n + m - 1$ arrows that leave every vertex of

---

[63] Recall that *graph* is the term used for 1-dimensional simplicial complexes, i.e. graph is a collection of 1-simplices (*edges*) some of which are "glued" at their ends; a graph is called *oriented* if all its edges are oriented. An oriented edge is depicted by an arrow, connecting two vertices: the *source* and the *target* of this arrow. We will denote by V_Γ, E_Γ respectively the sets of vertices and edges of a graph Γ.

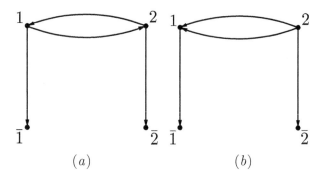

Figure 18.4: Admissible (*a*) and nonadmissible (*b*) graphs on the set $V = \{1,2\} \sqcup \{\bar{1},\bar{2}\}$.

the first type (one arrow for each remaining vertex in V_Γ), so in total there can be no more than $n(n + m - 1)$ arrows. Since every graph is uniquely determined by its edges we have

$$|G_{n,m}| = 2^{n(n+m-1)}.$$

The set of all arrows that leave from a point $v \in V_\Gamma$ (i.e. $e \in E_\Gamma$ such that $s(e) = v$) is called "the star of v" and denoted $star_\Gamma(v)$. Similarly, we will call the set of arrows that "hit" a vertex $v \in V_\Gamma$ "the costar of v": $costar_\Gamma(v) = \{e \in E_\Gamma \mid t(e) = v\}$. We will denote by $e_v^1, \ldots, e_v^{|star_\Gamma(v)|}$ the edges in $star_\Gamma(v)$ and by $f_v^1, \ldots, f_v^{|costar_\Gamma(v)|}$ the edges in $costar_\Gamma(v)$.

An admissible graph Γ is called *binary* if $|star_\Gamma(v)| = 2$ for all vertices $v \in V_\Gamma$ of the first type; we will denote by $G_{n,m}^2$ the set of all admissible binary graphs with n vertices of the first type, and m vertices of the second type. The number of elements in it is determined by all possible ways to choose edges for such graph and is given by the formula

$$|G_{n,m}^2| = \frac{1}{2}n(n + m - 1)(n + m - 2).$$

Indeed, for each one of n vertices of the first type we have $\binom{n+m-1}{2}$ ways to choose edges in its star. The set $G_{n,2}^2$ plays an important role in Kontsevich's construction.

18.2.2 Differential forms and weights

In order to introduce the weights in the formula (18.5), we need to do the integration over the spaces $\bar{C}_{n,m}$: these spaces are manifolds with corners and we can pull differential forms to these spaces from other manifolds. We begin with the space $\bar{C}_{2,0}$: this space is homotopy equivalent to a circle (see Fig. 18.2). In fact $\bar{C}_{2,0}$ is retractible to the circle $S^1 \cong \bar{C}_2$, which lies in its boundary.

Let us fix such retraction $\phi : \bar{C}_{2,0} \to S^1$, we will assume that in addition to being the identity map on the inner boundary (circle) of $\bar{C}_{2,0}$, the map ϕ also satisfies the condition that the upper half-circle $\bar{C}_{1,1}$ in the boundary of $\bar{C}_{2,0}$ is sent to one point. This map $\phi : \bar{C}_{2,0} \to \mathbb{R}/2\pi\mathbb{Z}$ can be interpreted as the *angle function*, where measure the angle in the circle from 0 to 2π in the counterclockwise direction beginning from the vertical line. There exist many different ways to define such angle maps; most part of the results that we cite here does not depend on the particular choice of the angle map ϕ, an example (one special choice) of such function ϕ is given by the angle between the Lobachevski line[64] that connects p_1 and p_2 and the vertical line in p_1; one can express this by the formula

$$\phi(p_1, p_2) = Arg\left(\frac{p_2 - p_1}{p_2 - \bar{p}_1}\right).$$

Let now Γ be an admissible graph in $G_{n,m}$, with $2n+m-2$ edges; we will once and for all identify the set of vertices V_Γ of this graph with the points $(p_1, \ldots, p_n; q_1, \ldots, q_m)$ of configurations in $Conf_{n,m}$; we will denote by V'_Γ, V''_Γ the subsets of vertices of the first and the second types respectively, i.e. V'_Γ corresponds to the points $(p_1, \ldots, p_n)$, and V''_Γ to $(q_1, \ldots, q_m)$.

For each edge $e \in E_\Gamma$ we define map $\phi_e : \bar{C}_{n,m} \to \mathbb{R}/2\pi\mathbb{Z}$ as the composition

$$\bar{C}_{n,m} \to \bar{C}_{2,0} \xrightarrow{\phi} \mathbb{R}/2\pi\mathbb{Z}.$$

Here the first arrow corresponds to the projection of $\bar{C}_{n,m}$ to $\bar{C}_{2,0}$ induced by the forgetful map that removes all points of the configuration except the endpoints of the edge e (the order of vertices

---

[64]Recall that in the Poincaré model of Lobachevski plane in the upper half-plane $\mathcal{H}$ straight lines are given by half-circles with center in $\mathbb{R}^1 \subset \bar{\mathcal{H}}$.

is prescribed by the direction of the edge). Observe that since the points of the second type are always targets of the oriented edges in E_Γ, image of the corresponding edges fall into the lower half-circle boundary of $\bar{C}_{2,0}$, and hence ϕ_e is nontrivial in this case.

We further use the standard orientation of $\mathcal{H}$, the induced (standard) orientation of its boundary and the order induced by the indices of the points p_i and q_j to introduce the orientation on the configuration space $Conf_{n,m} \subset (\mathbb{R}^1)^{\times m} \times (\mathbb{R}^2)^{\times n}$. Since the action of $G^{(1)}$ is orientation-preserving we obtain an orientation on $\bar{C}_{n,m}$. We shall denote by $\bar{C}_{n,m}^+$ the compactification of the connected component $C_{n,m}^+$, corresponding to the natural ordering of the points $q_i \in \mathbb{R}^1 : q_1 < q_2 < \cdots < q_m$. We also fix the order in the set E_Γ of edges of the graph by saying, that $e \prec e'$ if $s(e) < s(e')$ or $s(e) = s(e')$, $t(e) < t(e')$.

We finally define

$$W_\Gamma = \frac{1}{(2\pi)^{2n+m-2}} \prod_{v \in V'_\Gamma} \frac{1}{|star_\Gamma(v)|!} \int_{\bar{C}_{n,m}^+} \bigwedge_{e \in E_\Gamma} d\phi_e. \qquad (18.3)$$

Here the wedge product of $d\phi_e$ is taken in accordance with the order in the set E_Γ, defined above.

18.3 The formula

Let again $\Gamma \in G_{n,m}$ be an admissible graph with $2n + m - 2$ edges. We are going to describe the polydifferential operator $\mathcal{U}_\Gamma(P_1, \ldots, P_n)$ which we construct from Γ and the polyvector fields $P_1, \ldots, P_n$ on $\mathbb{R}^d$. To this end we fix the coordinates $(x^1, \ldots, x^d)$ on $\mathbb{R}^d$, and let $\partial_i = \frac{\partial}{\partial x^i}$. This gives the following formula for a k-vector field $P \in \mathcal{T}_{poly}^k(\mathbb{R}^d)$:

$$P = \sum_{1 \le i_1 < \cdots < i_k \le d} P^{i_1 \ldots i_k}(x) \partial_{i_1} \wedge \cdots \wedge \partial_{i_k}.$$

Let $v_1, \ldots, v_n$ be the vertices of the first type of the graph Γ and let $p_i = |star_\Gamma(v_i)|$. We are about to describe the value of the operator $\mathcal{U}_\Gamma(P_1, \ldots, P_n)$ on functions $f_1, \ldots, f_m \in C^\infty(\mathbb{R}^d)$. Take $P_i \in \mathcal{T}_{poly}^{p_i}(\mathbb{R}^d)$ and let $I : E_\Gamma \to \mathbf{d} = \{1, 2, \ldots, d\}$ be a map that associates to every edge of Γ an index from 1 to d (its label). Let for every vertex $v_i \in V'_\Gamma$ of the first type $(k_1, \ldots, k_{p_i}) = (I(e^1_{v_i}), \ldots, I(e^{p_i}_{v_i}))$ be the multiindex, determined by the labels on the edges in $star_\Gamma(v_i)$, taken in the natural order of the edges (see discussion before the formula (18.3)); then

$$\psi_{v_i} = \langle dx^{k_1} \otimes \cdots \otimes dx^{k_{p_i}}, P_i \rangle.$$

Here $\langle,\rangle$ is the natural pairing between the differential forms and polyvectors (see Section 4.3.3); in other words ψ_{v_i} is up to a sign the coefficient of the polyvector field P_i that stands at partial derivatives $\partial_{k_1} \wedge \cdots \wedge \partial_{k_{p_i}}$. Further, if $v_{\bar{j}} \in V''_\Gamma$ is a vertex of the second type, i.e. $v_{\bar{j}} = q_j$, we put $\psi_{v_{\bar{j}}} = f_j$.

Let eventually $(\ell_1, \ldots, \ell_{q_v}) = (I(f^1_v), \ldots, I(f^{q_v}_v))$ be the labels of the edges in the costar of a vertex $v \in V_\Gamma$ (in particular $q_v = |costar_\Gamma(v)|$). We put

$$\Phi_I(P_1, \ldots, P_n)(f_1, \ldots, f_m) = \prod_{v \in V_\Gamma} \left(\prod_{k=1}^{q_v} \partial_{\ell_k} \right) \psi_v.$$

In the end we put

$$\mathcal{U}_\Gamma(P_1, \ldots, P_n)(f_1, \ldots, f_m) = \sum_{I: E_\Gamma \to \mathbf{d}} \Phi_I(P_1, \ldots, P_n)(f_1, \ldots, f_m). \quad (18.4)$$

So eventually, we get the map $\mathcal{U}_n$:

$$\mathcal{U}_n(P_1, \ldots, P_n)(f_1, \ldots, f_m) = \sum_\Gamma W_\Gamma \mathcal{U}_\Gamma(P_1, \ldots, P_n)(f_1, \ldots, f_m). \quad (18.5)$$

Here the sum is taken over all admissible graphs $\Gamma \in G_{n,m}$ with prescribed sizes of the stars of the vertices $p_1, \ldots, p_n$. Then the following is true:

Theorem 18.2 (Kontsevich). *The collection of maps $\{\mathcal{U}_n\}$ given by the formula (18.5) determines an L_∞-morphism*

$$\mathcal{U}: \mathcal{T}^\cdot_{poly}(\mathbb{R}^d)[1][[\hbar]] \Rightarrow \mathcal{D}^\cdot_{poly}(\mathbb{R}^d)[1][[\hbar]],$$

called the Kontsevich's L_∞-morphism, such that $\mathcal{U}_1 = \chi$ is the Hochschild–Kostant–Rosenberg map (and hence $\mathcal{U}$ is a quasi-isomorphism of L_∞-algebras).

Now we have a way to describe the coefficients B_k of the $\star$-product, constructed by this approach. They are equal to the bidifferential operators $\mathcal{U}_k(\pi, \ldots, \pi)$ (here π appear exactly k times); in particular they can now be written in the form of the sums over admissible binary graphs:

$$B_k(f, g) = \sum_{\Gamma \in G^2_{k,2}} W_\Gamma \mathcal{U}_\Gamma(\pi, \ldots, \pi)(f, g). \qquad (18.6)$$

18.4 Proof of Theorem 18.2 (sketch)

We are going to give here only a brief explanation of the fact, why the map $\mathcal{U} = \{\mathcal{U}_n\}$ defined by the formula (18.5) is in effect an L_∞-morphism

$$\mathcal{U} = \{\mathcal{U}_n\} : \mathcal{T}^\cdot_{poly}(\mathbb{R}^d)[1][[\hbar]] \Rightarrow \mathcal{D}^\cdot_{poly}(\mathbb{R}^d)[1][[\hbar]],$$

for which $\mathcal{U}_1 = \chi$, is the Hochschild–Kostant–Rosenberg map. In this we follow the original reasoning from [Kon03] (many details of the proof were given only as small remarks in this paper and were later elaborated by other authors). For the interested readers we will give a detailed exposition of Tamarkin's approach to the same result, see Appendix B.

18.4.1 The term $\mathcal{U}_1$ is equal to the Hochschild–Kostant–Rosenberg map

We use the construction from Section 18.3 to determine $\mathcal{U}_1$. First of all, we observe that (up to a reordering of the points $q_1, \ldots, q_m$) there exists only one graph Γ in $G_{1,m}$, see Fig. 18.5. Thus for all $P \in \mathcal{T}^\cdot_{poly}(\mathbb{R}^d)$ we have $\mathcal{U}_1(P) = W_\Gamma \mathcal{U}_\Gamma(P)$. Now from the formula (18.4) we see that the polydifferential operator $\mathcal{U}_\Gamma(P)$ on functions $f_1, \ldots, f_m$ is given by

$$\mathcal{U}_\Gamma(P)(f_1, \ldots, f_m) = \sum_{I : E_\Gamma \to \mathbf{m}} \Phi_I(P)(f_1, \ldots, f_m).$$

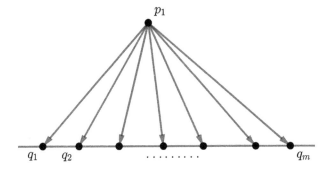

Figure 18.5: The graph in $G_{1,m}$.

The term $\Phi_I(P)(f_1, \ldots, f_m)$ is equal to 0 if $m > d$ and is equal to

$$\psi_{p_1} \prod_{k=1}^{m} \partial_{i_{I(1)}} f_1 \partial_{i_{I(2)}} f_2 \ldots \partial_{i_{I(m)}} f_m$$
$$= P^{i_{I(1)} i_{I(2)} \cdots i_{I(m)}} \partial_{i_{I(1)}} f_1 \partial_{i_{I(2)}} f_2 \ldots \partial_{i_{I(m)}} f_m$$

if $m \leq d$. Summing up for all labelings I gives

$$m! \sum_{1 \leq i_1 < \cdots < i_m \leq d} P^{i_1 \ldots i_m} \partial_{i_1} f_1 \ldots \partial_{i_m} f_m$$
$$= m! \langle df_1 \otimes \cdots \otimes df_m, P \rangle = (m!)^2 \chi(P)$$

where χ is the Hochschild–Kostant–Rosenberg map. We only need to compute the weight W_Γ to complete the proof. The weight, corresponding to this graph is

$$W_\Gamma = \frac{1}{(2\pi)^m} \frac{1}{m!} \int_{\bar{C}_{1,m}^+} \bigwedge_{e \in E_\Gamma} d\phi_e.$$

In order to calculate this integral, we can use the action of $G^{(1)}$ to fix the position of p_1 (for instance, put $p_1 = \sqrt{-1}$); then the projections $\bar{C}_{1,m} \to \bar{C}_{1,1}$ for all edges fall onto the lower boundary of "the eye" $\bar{C}_{2,0}$. Composing this projection with the angle map ϕ (see Section 18.2.2) gives us a bijective correspondence between points in $\bar{C}_{1,m}$ and the configuration of points on the segment $[0, 2\pi]$; the points in this configuration are ordered in a natural way, since we are only interested in the component $\bar{C}_{1,m}^+$. Thus this integral is equal to the volume of the m-dimensional simplex $0 \leq t_1 \leq t_2 \leq \cdots \leq t_m \leq 2\pi$, which is $\frac{(2\pi)^m}{m!}$.

18.4.2 The L_∞-morphism equation

To show that the map $\mathcal{U} = \{\mathcal{U}_n\}$ determined by Equation (18.5) is indeed an L_∞-morphism, we must check that it satisfies Equation (17.4). Let us outline here the main steps of this proof.

First of all since the differential in the algebra $\mathcal{T}_{poly}^\cdot(\mathbb{R}^d)[1][[\hbar]]$ is trivial, the left-hand side of Equation (17.4) will consist only of the term $d\mathcal{U}_n(P_1, \ldots, P_n)$, where $d = \delta$ is the Hochschild differential in $C_{diff}^\cdot(C^\infty(\mathbb{R}^d), C^\infty(\mathbb{R}^d)) = \mathcal{D}_{poly}^\cdot(\mathbb{R}^d)$. We now observe that the product of functions in $C^\infty(\mathbb{R}^d)$, which appears in the Hochschild differential is in effect equal to the operator $\mathcal{U}_\Gamma$ associated with the unique admissible graph Γ with 2 vertices of the second type and no vertices of the first type (and hence no edges). Hence the Hochschild differential of an operator $\mathcal{U}_\Gamma(P_1, \ldots P_n)$ can be obtained as the sum of the operators $\mathcal{U}_{\Gamma'}(P_1, \ldots, P_n)$ for suitable admissible graphs.

Similarly, the Schouten bracket of a p-vector field P and a q-vector field Q can be rewritten as the sum of the operators $W_{\Gamma_i} \mathcal{U}_{\Gamma_i}(P, Q)$ for suitable graphs $\Gamma_i \in G_{2,p+q-1}$ (see Exercise E.6.16). The number of edges in the graphs Γ_i, involved in the computation of the Schouten bracket $[P, Q]$ is equal to $p + q = 2 \cdot 2 + (p + q - 1) - 3$, and the same is true for the graphs Γ' that are used in the computation of the Hochschild differential of $\mathcal{U}_\Gamma(P_1, \ldots P_n)$. With a little extra work, which we omit, one can see that Gerstenhaber brackets of the operators $\mathcal{U}_{\Gamma_1}(P_1, \ldots, P_k)$ and $\mathcal{U}_{\Gamma_2}(P_{k+1}, \ldots, P_{k+l})$ is also expressible as a linear combination of the operators $\mathcal{U}_\Gamma(P_1, \ldots, P_{k+l})$ where $\Gamma \in G_{n,m}$, $n = k+l$ and the number of edges in Γ is equal to $2n + m - 3$.

It follows that we can rewrite Equation (17.4) as:

$$\sum_\Gamma \pm c_\Gamma \mathcal{U}_\Gamma(P_1, \ldots, P_n)(f_1, \ldots, f_m) = 0. \qquad (18.7)$$

Here the summation is over admissible graphs $\Gamma \in G_{n,m}$ with $2n + m - 3$ edges and the coefficients c_Γ appear as sums of the weights $W_{\Gamma'}$ and of their pairwise products, where Γ' are the admissible graphs which appear in the definition of the morphisms $\mathcal{U}_n$. In order to show that (18.7) holds, we need to check that $c_\Gamma = 0$ for all Γ.

Kontsevich then used the following idea: we can generalise the construction from Section 18.2.2 to define for every graph $\Gamma \in G_{n,m}$ the differential forms on $\bar{C}_{n,m}$ independently of the number of edges of this graph. Then we identify c_Γ with the integral of this form over the boundary of $\bar{C}_{n,m}$: this can be done since we have $\dim \bar{C}_{n,m} = 2n + m - 2$, so $\dim \partial \bar{C}_{n,m} = 2n + m - 3$ is precisely the number of edges of Γ, which defines c_Γ. Then by Stokes formula

$$c_\Gamma = \int_{\partial \bar{C}_{n,m}} \bigwedge_{e \in E_\Gamma} \phi_e = \int_{\bar{C}_{n,m}} d\left(\bigwedge_{e \in E_\Gamma} \phi_e\right) = 0,$$

since the form is closed.

Recall now that the codimension 1 strata in the boundary of $\bar{C}_{n,m}$ naturally come in two forms: $\partial_S \bar{C}_{n,m}$ and $\partial_{S,S'} \bar{C}_{n,m}$, see Section 18.1.3, Equations (18.1) and (18.2), so we have to identify c_Γ with the sum

$$\int_{\partial \bar{C}_{n,m}} \bigwedge_{e \in E_\Gamma} \phi_e = \sum_S \int_{\partial_S \bar{C}_{n,m}} \bigwedge_{e \in E_\Gamma} \phi_e + \sum_{S,S'} \int_{\partial_{S,S'} \bar{C}_{n,m}} \bigwedge_{e \in E_\Gamma} \phi_e.$$

Now, each and every term in the right-hand side of this formula Kontsevich identifies with an expression in the equation that determines the L_∞-morphism. For instance, let us consider $\partial_S \bar{C}_{n,m}$: let $|S| = 2$ and the vertices in S are connected by an edge. Then the subgraph, spanned by S can be identified with one of the graphs that appear in the expression for the Schouten bracket $[P, Q]$; the corresponding integral also coincides with the weight, that comes from the equation on L_∞-morphisms. In case, when $|S| > 2$ Kontsevich shows that the corresponding integral vanishes.[65] Similar identifications work for other cases; for instance the nontrivial portions of $\partial_{S,S'} \bar{C}_{n,m}$ correspond to the terms in Gerstenhaber bracket of two polydifferential operators. Thus the theorem is proved.

---

[65]This fact is in effect one of the most nontrivial remaining parts of the proof; it is based on the following statement:
Lemma: *Integral of the form $\bigwedge_\alpha d(Arg(f_\alpha))$, $\alpha = 1, \ldots, 2N$ where f_α are non-vanishing rational functions in an open set $U \subset M$ of a smooth N-dimensional complex algebraic variety M, over U is always equal to 0.*

Exercises 6: Higher homotopy algebras, Kontsevich's theorem

E.6.1 Prove that for every graded vector space V, its tensor algebra TV can be endowed with coproduct, making it cofree coalgebra, generated by V, i.e. that for every linear map $g : C \to V$ from a coalgebra C to V there exists a coalgebra homomorphism $\phi(g)$ commuting the Diagram (16.6).

E.6.2 Prove the uniqueness of cofree coalgebras (cocommutative and simple) up to an isomorphism of coalgebras.

E.6.3 Prove that any coderivation $M : C(V) \to C(V)$ (see Section 16.3.3) of a cofree coalgebra of V, i.e. $C(V) = TV$ (see Exercise E.6.1), generated by a vector space V is uniquely determined by its "Taylor series", i.e. by the maps $m_n : V^{\otimes n} \to V$ (and similarly for cocommutative coalgebras). Similarly, for graded cocommutative cofree coalgebra $CC(V) = \Lambda V$ every coderivation L is determined by the maps $\ell_n : \tilde{\Lambda}^n V \to V$ (here $\tilde{\Lambda}^n V$ is the same as $\Lambda^n V[1]$).

E.6.4 Show that the Taylor coefficients m_n of the coderivation M that determine an A_∞ algebra satisfy Equations (16.5) iff $M^2 = 0$.

E.6.5 Show that the equation $L^2 = 0$ for the structure map of an L_∞-algebra is equivalent to the following system of equations on its "Taylor coefficients":

$$\sum_{p+q=n+1} \sum_{\sigma \in \overline{Sh}_{p,q-1}} (-1)^{\epsilon(\sigma,|a_1|,\ldots,|a_n|)} \ell_q(\ell_p(a_{\sigma(1)},\ldots,a_{\sigma(p)}),$$

$$a_{\sigma(p+1)},\ldots,a_{\sigma(n)}) = 0.$$

Here $(-1)^{\epsilon(\sigma,|a_1|,\ldots,|a_n|)}$ is the sign determined by the Koszul rule for $V[1]$ (see the discussion after Definition 16.6), a_i are the homogenous elements of degrees $|a_i|$ in V and we use commas instead of the wedge sign again to make the formulas look more compact. Derive the equalities (16.7$_1$)–(16.7$_3$) with correct signs from this equations.

E.6.6 Let X be a smooth manifold, $n \geq 1$ and let $\omega \in \Omega^{n+1}(X)$ be a closed $n-1$-form on X. Consider the space

$$\mathrm{Ham}^{n-1}(X) = \{v + \varphi \in Vect(X) \oplus \Omega^{n-1}(X) \mid i_v\omega + d\varphi = 0\}.$$

($i_v\omega$ is the inner derivative of ω by the vector field v). Prove that the graded space

$$\mathscr{P}(X) = \bigoplus_{k=0}^{n-2} \Omega^k(X) \oplus \mathrm{Ham}^{n-1}(X)$$

bears the structure of an L_∞-algebra with the structure maps ℓ_p given by

$$\ell_1(\varphi) = d\varphi$$

for $\varphi \in \Omega^k(X)$, $k \leq n-2$ and d is de Rham differential (for $k = n-2$ we embed $d\varphi$ into the first summand of $\mathrm{Ham}^{n-1}(X)$), and

$$\ell_p(v_1 + \varphi_1, \ldots, v_p + \varphi_p) = -(-1)^{\binom{p+1}{2}} i_{v_1 \wedge \cdots \wedge v_p} \omega,$$

for all $v_i + \varphi_i \in \mathrm{Ham}^{n-1}(X)$ and $p = 2, \ldots, n+1$; all other maps ℓ_k are trivial. Here $i_{v_1 \wedge \cdots \wedge v_p}\omega$ denotes the convolution of ω with polyvector fields. This L_∞-algebra is called *Poisson bracket Lie n-algebra*.

E.6.7 Prove that the formula (17.3) determines a cochain differential on the space $\mathrm{Hom}(V, V')$ of linear maps between the chain complexes (graded by the homogenous degree of a map), i.e. that $d^2(\varphi) = 0$. Show that $H^0(\mathrm{Hom}(V, V'), d)$ can be identified with the space of cochain homotopy classes of cochain maps between the complexes.

E.6.8 (i) Let $(A, \{m_k\})$, $(A', \{m'_k\})$ be two A_∞ algebras, $F : V \Rightarrow V'$ be an A_∞-morphism between them (see Remark 17.3). Let $f_n : A^{\otimes n} \to A'$ be the "Taylor coefficients" of F; below we shall pick out f_1 and denote it by d, so we always assume that $n \geq 2$ in f_n. Prove that the condition that F is an A_∞-morphism is equivalent to the following system of equations:

$$df_n(a_1,\ldots,a_n)$$
$$= \sum_{p+q=n+1} \sum_{i=1}^{p} (-1)^{\epsilon_1} f_p(a_1,\ldots,a_{i-1}, m_q(a_i,\ldots$$
$$\ldots, a_{i+q-1}), a_{i+q},\ldots,a_n)$$
$$- \sum_{p=2} \sum_{q_1+\cdots+q_p=n} (-1)^{\epsilon_2} m_p(f_{q_1}(a_1,\ldots,a_{q_1}),\ldots$$
$$\ldots, f_{q_p}(a_{n-q_p+1},\ldots,a_n)).$$

Here the signs $(-1)^{\epsilon_1}$, $(-1)^{\epsilon_2}$ are given by the Koszul rule and the differential dF_n of a linear map between cochain complexes is given by the formula (17.3).

(ii) Find analogs of formulas (17.2_1), (17.2_2) for an A_∞ morphism. Show that every homomorphism of associative differential graded algebras induces an A_∞-morphism between them, but not vice-versa.

(iii) Let $(V,\{\ell_k\})$, $(V',\{\ell'_k\})$ be two L_∞-algebras, $F: V \Rightarrow V'$ be an L_∞-morphism between them. Find the equations, satisfied by ℓ_n, ℓ'_n similar to part (i) of this problem. Show that in case V, V' are given by DG Lie algebras $\mathfrak{g}^\cdot$, $\mathfrak{h}^\cdot$, this equations reduce to the system (17.4). Find similar equations for an A_∞-morphism between two DG algebras.

E.6.9 Prove that for every cochain map $f: C^\cdot \to D^\cdot$ of cochain complexes over a characteristic zero field that induces an isomorphism in cohomology, there exists a homotopy inverse cochain map $g: D^\cdot \to C^\cdot$. **Hint:** it is enough to consider the map $H^\cdot(C) \to C^\cdot$ that sends elements in cohomology into representing cochains; to prove this choose a basis in $C^\cdot$, that matches the decomposition of $C^\cdot$ into the image and the kernel of the differential.

E.6.10 Suppose that the DG Lie algebra homomorphism $f: \mathfrak{g}^\cdot \to \mathfrak{h}^\cdot$, cochain map $g_1: \mathfrak{h}^\cdot \to \mathfrak{g}^\cdot$ and cochain homotopy $h_1: \mathfrak{h}^\cdot \to$

$\mathfrak{h}^{-1}$ satisfy the standard conditions of *homology perturbation lemma*, i.e. the triple (f, g_1, h_1) satisfy the following relations:

$$g_1 f = \mathrm{id}_{\mathfrak{g}}, \quad f g_1 = \mathrm{id}_{\mathfrak{h}} + h_1 d + d h_1,$$
$$f h_1 = 0, \quad h_1 g_1 = 0, \quad h_1 h_1 = 0.$$

Find explicit formulas for the Taylor coefficients g_n of the L_∞-morphism G homotopy inverse to f and the cochain homotopy $H = \{h_n\}$ between $f \circ G$ and $\mathrm{id}_{\mathfrak{h}}$ (see Proposition 17.4) in terms of f, g_1 and h_1.

E.6.11 Let us call an element ϖ of degree 1 in an L_∞-algebra V a *Maurer-Cartan element*, if

$$\sum_{k=1}^{\infty} \frac{1}{k!} \ell_k (\underbrace{\varpi, \ldots, \varpi}_{k \text{ times}}) = 0.$$

Here and below ℓ_n are the "Taylor coefficients" of the L_∞ structure. Prove that for any L_∞-morphism $F = \{f_n\} : V \Rightarrow W$, the formula

$$F(\pi) = \sum_{k=1}^{\infty} \frac{1}{k!} f_k (\underbrace{\varpi, \ldots, \varpi}_{k \text{ times}})$$

determines a Maurer-Cartan element in W (we assume the convergence on the right-hand side).

E.6.12 Let ϖ be a Maurer-Cartan element in an L_∞-algebra V, and let $\epsilon \in V^0$. Show that if $\varpi_\epsilon(t) \in V$ satisfies the equation

$$\frac{d\varpi_\epsilon}{dt} = -\sum_{k=0}^{\infty} \frac{1}{k!} \ell_{k+1}(\epsilon, \underbrace{\varpi_\epsilon, \ldots, \varpi_\epsilon}_{k \text{ times}}),$$

then $\varpi_\epsilon(t) \in \mathfrak{MC}(V)$ for all t. Let us say, that the Maurer-Cartan elements ϖ_0 and ϖ_1 in V are equivalent, if one can connect them by a path in $\mathfrak{MC}(V)$ consisting of elements of the form $\varpi_\epsilon(t)$ (c.f. Definition 17.8). Show that the map $F : \mathfrak{MC}(V) \to \mathfrak{MC}(W)$ induced by an L_∞-morphism $F : V \Rightarrow W$ preserves this equivalence relation.

E.6.13 Let $\star$, $\star'$ be two $\star$-products on $C^\infty(M)[[\hbar]]$, and let B, $B' \in \mathfrak{MC}(\mathscr{D}_{poly}^\bullet(M)[1][[\hbar]])$ be the corresponding Maurer-Cartan elements. Prove, that $\star \sim \star'$ iff $B \sim B'$. **Hint:** let T be the formal power series differential operator, that determines the equivalence of $\star$ and $\star'$. Then $T \in \mathscr{D}_{poly}^0(M)[1][[\hbar]]$ determines the elementary equivalence $B \sim_T B'$ in the sense of Definition 17.8 and vice-versa.

E.6.14 Describe the spaces $\bar{C}_{0,3}$, $\bar{C}_{1,2}$, $\bar{C}_{2,1}$, $\bar{C}_{3,0}$ and $\bar{C}_3$.

E.6.15 Give a list of binary admissible graphs (see Definition 18.1) in $G_{n,2}^2$ for $n \leq 3$.

E.6.16 Let $\Gamma_{1,\sigma}, \Gamma_{2,\sigma'} \in G_{2,p+q-1}$, where σ, σ' run through all $(p-1, q)$ and $(p, q-1)$-shuffles respectively, be the admissible graphs with two vertices of the first type and $p + q - 1$ vertices of the second type. Let a and b be the vertices of the first type; then let
$$|star_{\Gamma_{1,\sigma}}(a)| = |star_{\Gamma_{2,\sigma'}}(a)| = p,$$
$$|star_{\Gamma_{1,\sigma}}(b)| = |star_{\Gamma_{2,\sigma'}}(b)| = q,$$
and
$$|costar_{\Gamma_{1,\sigma}}(a)| = |costar_{\Gamma_{2,\sigma'}}(b)| = 0,$$
$$|costar_{\Gamma_{2,\sigma'}}(a)| = |costar_{\Gamma_{1,\sigma}}(b)| = 1.$$
In other words, the only edge that connects a and b is directed from a to b in the first case, and from b to a in the second. The shuffles σ, σ' encode the order of the edges that connect a, b with the vertices of the second type. Show that for any polyvector fields $P \in \mathscr{T}_{poly}^p(\mathbb{R}^d)$, $Q \in \mathscr{T}_{poly}^q(\mathbb{R}^d)$ we have (up to a constant factor)
$$[P,Q] = \sum_\sigma \mathcal{U}_{\Gamma_{1,\sigma}}(P,Q) - (-1)^{(p-1)(q-1)} \sum_{\sigma'} \mathcal{U}_{\Gamma_{2,\sigma'}}(P,Q).$$

E.6.17 Use Kontsevich's graph calculus to compute the term B_2 in the formula for the $\star$-product.

19 Kontsevich's quantisation: Modifications and related questions

It is difficult to overemphasise the importance of Kontsevich's formality theorem: it is not just that it solved the deformation quantisation problem, it is rather the fact that the ideas and methods developed for its proof play an important role in many different branches of modern Mathematics. In particular, as we will see, the ideas of this proof are closely related with modern Physics (below we shall describe the relation of Kontsevich's formula with *Poisson sigma-model*, a version of Quantum Field Theory) so in addition to giving a solution of an old problem, it showed the importance of "physical" methods that had often been neglected by traditional Mathematics. Besides this, due to the fact that it was based on a the analysis of L_∞-morphisms it resuscitated the interest to higher homotopical structures and their role in various branches of Mathematics; in particular this achievement celebrated the renaissance of the operadic theory (see Appendix A): considered before that as a topological curiosity, these methods became progressively popular in various branches of Mathematics in 1990s. It is not surprising that the alternative proof of the formality theorem appeared from this direction.

In this section we shall briefly discuss some of the most important developments and results, related with Kontsevich's theorem. First of all we will briefly describe the "physical" approach to the Kontsevich's formula: it turns out that one can interpret the $\star$-product obtained by Kontsevich formality map in terms of the Feynmann diagram calculus for a certain continual integral. It is arguable, whether Kontsevich himself based his construction on this approach, but he was most probably aware of it and inspired by it. Besides this, we describe the way to pass from the *local* situation (Poisson structures on $\mathbb{R}^d$) to the *global* case: the generic Poisson manifold M with possibly nontrivial topology. This has been done by Kontsevich in his original paper [Kon03] in a rather schematic manner; we outline his reasoning, but we also describe the construction of Cattaneo, Felder and Tomassini that gives a solution of the same problem (see [CFT02]), which is close to Fedosov's construction. This construction can be further generalised to the case of Lie algebroids and other situations, where one has restrictions on the choice of vector fields, so we believe this construction is important for various applications, see for instance [Ca05].

19.1 "Physical" approach to the $\star$-product

Before we speak about how one can derive the Kontsevich $\star$-product from the physical principles, let us recall some basic ideas of the modern quantum Physics.

19.1.1 The path integral formalism

One of the central ideas in modern theoretical Physics is that the behaviour of various physical objects is governed by the "minimal action principle" (see Section 1.1.2). Loosely speaking every physical system is endowed with a function $\mathcal{L}(x, \dot{x}, t)$, the *Lagrangian* of the system, depending on various parameters, such as the time, velocity and possible positions of the points in the system; for every possible trajectory γ we consider the *action* of the system along this path: the integral $S[\gamma] = \int_\gamma \mathcal{L}(x, \dot{x}, t) dt$. Then the principle says that the evolution of the physical system should follow the trajectory γ_0 for which $S[\gamma_0] = \min_\gamma S(\gamma)$.

The development of the Quantum Mechanics brought new significance to this old idea. Namely, in 1948 departing from Paul Dirac's works Richard Feynman formulated the Quantum Mechanical principles in terms of the so-called "path integral": *for any two configurations P_0 and P_1 of a quantum system, the probability amplitude of passing from P_0 to P_1 when the time goes from 0 to T is up to a normalising factor equal to the integral*

$$\left\langle P_1 \left| \exp\left(-\frac{\sqrt{-1}}{\hbar} \hat{H} T\right) \right| P_0 \right\rangle = \int_{\mathscr{X}_{P_0}^{P_1}} e^{\frac{\sqrt{-1}}{\hbar} S[\gamma]} D\gamma. \tag{19.1}$$

Here $\hat{H}$ is the quantum Hamiltonian, corresponding to the classical Lagrangian $\mathcal{L}(x, \dot{x}, t)$ used to define the action $S[\gamma]$ and the integration is done over the space $\mathscr{X}_{P_0}^{P_1}$ of all possible trajectories γ of the quantum system, for which $\gamma(0) = P_0$, $\gamma(T) = P_1$; this includes not only trajectories, that are possible from the physical point of view, but all parametrised curves in the space, where our quantum system "lives", including the trajectories that "turn back in time direction". More generally, if we fix a moment in time $t \in [0, T]$ and ask, what is the expectation of the value of an observable F depending on our quantum system that goes from P_0 to P_1 at the moment t, then

$$\langle F(t) \rangle = \frac{\int_{\mathscr{X}_{P_0}^{P_1}} F(\gamma(t)) e^{\frac{\sqrt{-1}}{\hbar} S[\gamma]} D\gamma}{\int_{\mathscr{X}_{P_0}^{P_1}} e^{\frac{\sqrt{-1}}{\hbar} S[\gamma]} D\gamma}.$$

Of course, these integrals are not well-defined from the mathematical point of view: in effect, one can prove that there's no good measure on the infinite-dimensional space $\mathscr{X}_{P_0}^{P_1}$ of paths in a general situation, especially if we ask that this measure should be translation-invariant and satisfy other intuitively clear properties, similar to the properties of Lebesgues' measure in $\mathbb{R}^d$. However in the course of the 75 years that elapsed since Feynman's work was published, physicists found ways to ascribe certain unambiguous meaning to the integral (19.1). These methods are based on the Feynman diagram calculus; we are not going to explain them here as this will take too much time and space. An interested reader can find an accurate introduction to the Feynman diagram method, renormalisation etc., in various textbooks on Quantum Mechanics, see for instance [EMS04], [Ku18], etc.

Similar path integral methods were developed later in other physical theories, such as various versions of Quantum Field Theory. Nowadays these methods are central for the physical treatment of these theories. These methods are explained in many textbooks, see for instance [BS83], [Ra01], or [K93] (some more modern textbooks on the subject you can find in the list of references). In these theories the configuration of the system is represented by various fields, such as sections of vector bundles, connections on these bundles, metrics on manifolds etc. In that case the integration is often done over the space of all possible fields. One of such theories is the Poisson σ-model, which we are going to explain.

19.1.2 Poisson σ-model and the ⋆-product

Let (M, π) be a smooth manifold with a Poisson bivector π. We can assume that $M = \mathbb{R}^d$ with coordinates $(x^1, \ldots, x^d)$, or else we shall fix a local coordinates system $(x^1, \ldots, x^d)$ in M and work in this chart. Let $\pi = \pi^{ij} \frac{\partial}{\partial x^i} \otimes \frac{\partial}{\partial x^j}$ in these coordinates, so that $\pi^{ij} = -\pi^{ji}$. Let us denote by D^2 the unit disc in the plane:

$$D^2 = \{z \in \mathbb{C} \mid |z| \leq 1\}.$$

We denote by $u = (u^1, u^2)$ the coordinates in D^2; one can either regard (u^1, u^2) as the standard Euclidean coordinates (x, y) in the plane, or as the complex coordinates $(z, \bar{z})$.

The configuration space $\mathscr{X}(M)$ of the Poisson σ-model is given by the set of pairs (X, η), where $X : D^2 \to M$ is a smooth map and $\eta \in \Omega^1(D^2, X^*(T^*M))$ is a 1-form on the disc with values in the pull-back of the cotangent bundle on M. In local coordinates:

$$X(u) = (X^1(u), \ldots, X^d(u)),$$
$$\eta = (\eta_1(u), \ldots, \eta_d(u)),$$

where $\eta_k(u) = \eta_{k,\mu}(u) du^\mu$. For every pair (X, η) we consider the action functional:

$$S[X, \eta] = \int_{D^2} \left[\eta_i(u) \wedge dX^i(u) + \frac{1}{2} \pi^{ij}(X(u)) \eta_i(u) \wedge \eta_j(u) \right].$$

Then the "transition amplitude" of this theory is given by the integration of $e^{\frac{\sqrt{-1}}{\hbar} S[X,\eta]}$ over the space of pairs (X, η). Another important entities that are studied in Quantum Field Theory are the so-called "correlation functions" of various functionals on configuration space. For instance for any function f on M and any point p on D^2 we can compute

$$\langle f \rangle(p) = \int_{\mathscr{X}(M)} f(X(p)) e^{\frac{\sqrt{-1}}{\hbar} S[X,\eta]} DX D\eta.$$

It turns out (see [CF00], [CF01]) that one can define the ⋆-product on $C^\infty(M)[[\hbar]]$ with the help of a similar path integral: let $f, g \in C^\infty(M)$. Then we put

$$(f \star g)(x) = \int_{\mathscr{X}_x(M)} f(X(0)) g(X(1)) e^{\frac{\sqrt{-1}}{\hbar} S[X,\eta]} DX D\eta. \qquad (19.2)$$

In this formula we fix three points 0, 1 and ∞ on the boundary S^1 of D^2. For instance, we identify D^2 with the upper half-plane $\bar{\mathcal{H}} = \{z \in \mathbb{C} \mid \Im z \geq 0\}$, and the points 0, 1 and ∞ lie on the boundary $\mathbb{R}^1$ of $\bar{\mathcal{H}}$; then we integrate over the subspace $\mathscr{X}'_x(M)$ of all pairs $(X, \eta) \in \mathscr{X}(M)$ with $X(\infty) = x$ and $\eta(\xi) = 0$ on $\partial D^2 = S^1$, where ξ is the vector field tangent to S^1.

19.1.3 Associativity of the $\star$-product

Of course, Equation (19.2) is rather an intuitive justification for the search of $\star$-product in $C^\infty(M)$ than a well-defined formula; in order to make it more explicit, one needs to use perturbative approach to path integrals, renormalisation and other methods of theoretical Physics that allow one to replace the right-hand side of this formula by a formal power series. This was done by Cattaneo and Felder: see [CF01], where they also showed that the perturbative series associated with the integral (19.2) coincides with Kontsevich's formula. In order to explain this result we would have to give an introduction into various physical theories such as the BRST complex, renormalisation group etc.; this excursion into Theoretical Physics would take too much time and space, so we omit it. Instead, let us sketch the proof of associativity of the operation $f \star g$ we have introduced: it turns out that this fact follows straightforwardly from the elementary intuitively clear operations with integrals.

To this end we let $f, g, h \in C^\infty(M)$ be three arbitrary functions. We need to show that

$$f \star (g \star h) = (f \star g) \star h,$$

where $\star$ is the operation given by (19.2). To this end we compute:

$(f \star (g \star h))(x)$

$$= \int_{\mathscr{X}'_x(M)} f(X(0))(g \star h)(X(1)) e^{\frac{\sqrt{-1}}{\hbar} S[X, \eta]} DX D\eta$$

$$= \int_{\mathscr{X}'_x(M)} f(X(0)) \left(\int_{\mathscr{Y}'_{X(1)}(M)} g(Y(0)) h(Y(1)) e^{\frac{\sqrt{-1}}{\hbar} S[Y, \zeta]} DY D\zeta \right) e^{\frac{\sqrt{-1}}{\hbar} S[X, \eta]} DX D\eta$$

where for a given $X \in \mathscr{X}'_x(M)$ we denote by $\mathscr{Y}'_{X(1)}(M)$ the space of pairs (Y, ζ) similar to (X, η), i.e. $Y : D^2 \to M$ is a smooth map for which $Y(\infty) = X(1)$ and $\zeta \in \Omega^1(D^2, Y^*(T^*M))$ is a 1-form with $\zeta(\xi) = 0$ for $\xi \in TS^1$ (here $S^1 = \partial D^2$).

The product of the same functions taken in the same order, but when we multiply first f and g and then take the product with h is given by a similar integral (where we use the same notation as above):

$$((f \star g) \star h)(x)$$
$$= \int_{\mathscr{Y}_x'} (f \star g)(Y(0)) h(Y(1)) e^{\frac{\sqrt{-1}}{\hbar} S[Y,\zeta]} DY D\zeta$$
$$= \int_{\mathscr{Y}_x'} \left(\int_{\mathscr{X}_{Y(0)}'} f(X(0)) g(X(1)) e^{\frac{\sqrt{-1}}{\hbar} S[X,\eta]} DX D\eta \right) h(Y(1)) e^{\frac{\sqrt{-1}}{\hbar} S[Y,\zeta]} DY D\zeta.$$

In both cases, deforming the configurations on the top of Fig. 19.1 to the circles in the lower row thereof, we can regard the resulting formula as a path integral over the space of maps $Z : D^2 \to M$ and 1-forms $\theta \in \Omega^1(D^2, Z^*(T^*M))$ with functions f, g and h "inserted"

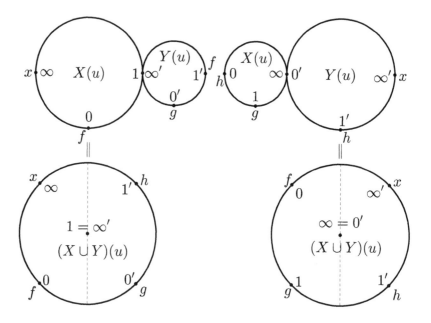

Figure 19.1: Associativity of the $\star$-product from the formula (19.2).

in the points on the boundary:

$$(f \star (g \star h))(x) = \int_{\mathscr{L}_x'} f(Z(0))g(Z(0'))h(Z(1'))e^{\frac{\sqrt{-1}}{\hbar}S[Z,\theta]} DZD\theta,$$

$$((f \star g) \star h)(x) = \int_{\mathscr{L}_x''} f(Z(0))g(Z(1))h(Z(1'))e^{\frac{\sqrt{-1}}{\hbar}S[Z,\theta]} DZD\theta,$$

where $\mathscr{L}_x'$ denotes the maps with $Z(\infty) = x$, and $\mathscr{L}_x''$ refers to the maps with $Z(\infty') = x$. Clearly both integrals give the same result, as the integration domains are related by a reparametrisation.

Remark 19.1. In fact, one can define Kontsevich's formality L_∞-morphism $\mathcal{U} = \{\mathcal{U}_n\}$ in terms of the similar path integrals too. However, even if we would like just to give an accurate exposition of this construction we would need to speak about the BV and BRST formalisms, superfields etc., introducing which would take lots of time and efforts, so we refer the interested reader to the original works of Cattaneo and Felder [CF00], [CF01] and to numerous introductory texts and monographs, see for instance [Sza00].

19.2 Globalisation of the $\star$-product

As we have already mentioned, Kontsevich's Theorem 17.12 deals with the Poisson structures on Euclidean spaces. It follows in particular, that for every coordinate chart in a Poisson manifold M it gives an associative $\star$-product on the algebra of smooth functions on the coordinate neighbourhoods $U \subseteq M$. However we still need to "paste together" the constructions on coordinate charts to obtain the $\star$-product on the space of globally defined functions on the manifold; more generally one can try to extend the formality of the Lie algebra $\mathscr{D}_{poly}^{\cdot}(\mathbb{R}^d)[1][[\hbar]]$ to the statement about the formality of $\mathscr{D}_{poly}^{\cdot}(M)[1][[\hbar]]$, the algebra of global polydifferential operators on a smooth manifold M (shifted and enhanced by the formal variable $\hbar$). Of course, this will mean that every (formal) Poisson structure on M gives rise to an associative $\star$-product. It is this global formality statement that we are going to discuss now.

The process of passing from a result valid on coordinate charts to the statement for a whole manifold is promptly called "globalisation". In what follows we shall describe two ways to look at this problem:

one, due to Kontsevich himself, and another one later formulated by various people, first of all by Dolgushev ([Do05]), Cattaneo, Felder and Tomassini ([CFT02]) and Calaque ([Ca05]).

19.2.1 Few words about jets and stuff

In his original paper [Kon03] (published online in 1997) Kontsevich gave a brief outline of a method that allows one to derive the formality of the algebra of polydifferential operators on an arbitrary manifold from the formality of $\mathscr{T}_{poly}^{\cdot}(\mathbb{R}^d)[1][[\hbar]]$. This reasoning[66] was based on some facts about the structure of the infinite jet spaces on manifolds, which is the subject of what is often called *Formal Geometry*. As usual, there are plenty of textbooks and introductions to this theory, of which we can recommend to the interested reader the book [Sa89]; however, to make our text as self-contained as possible, we will briefly recall basic notions from this field.

So let us begin with the definitions. First of all, we shall use the following notation: $\mathbb{R}^d_{formal}$ is abbreviation for $\mathbb{R}[[y^1, \ldots, y^d]]$, i.e. the algebra of formal power series in d variables; similarly $\mathscr{T}_{poly}^{\cdot}(\mathbb{R}^d_{formal})$, $\mathscr{D}_{poly}^{\cdot}(\mathbb{R}^d_{formal})$ will denote the polyvector fields, polydifferential operators etc. with formal coefficients. Further:

Definition 19.2. *Let M be a smooth d-manifold, then for any point $p \in M$, any integer $r \geq 0$ and any two functions f, g defined on a coordinate chart with coordinates $(x^1, \ldots, x^d)$ around p, we shall say, that f and g have the same r-jet in p if for every multiindex $I = (i_1, \ldots, i_d)$ with $|I| \leq r$ (where $|I| = \sum_{k=1}^{d} i_k$) we have*

$$\frac{\partial^{|I|} f}{\partial x^I}\Big|_p = \frac{\partial^{|I|} g}{\partial x^I}\Big|_p. \qquad (19.3)$$

Since the change of coordinates acts in the same way on both sides of Equation (19.3) and involves the partial derivatives of degrees less or equal to r, it is clear that the property of having the same r-jet does not depend on the choice of coordinates, so we do not need to mention the coordinate system any more. Moreover, equality (19.3) is an equivalence relation, and both sides of it are linear over the ground

---
[66] And most of the others proofs of the global formality, known to the author.

field, so we obtain a well-defined vector space of equivalence classes of functions. This space is called the *space of r-jets of smooth functions in p* and is denoted $J_p^r(M)$. The element of $J_p^r(M)$, corresponding to a function $f \in C^\infty(M)$ is denoted by $j_p^r(f)$; it is called *the r-th order jet of the function f at the point p*.

Clearly $J_p^r(M) \cong \bigoplus_{k=0}^r S^k(\mathbb{R}^d)$ as vector spaces, the isomorphism being established by sending $j_p^r(f)$ to the collection of the Taylor coefficients of f at p up to degree r in a fixed coordinate system. Then the change of coordinates gives a linear transformation of $J_p^r(M)$, so we obtain a vector bundle, called the *r-jet bundle over M*: it is equal to the union of the jet spaces over M, topologised with the help of local coordinates in M, similarly to the tangent and cotangent bundles of a smooth manifold:

$$J^r(M) = \bigcup_{p \in M} J_p^r(M).$$

In effect $J^1(M) \cong \varepsilon_1(M) \oplus T^*M$, where $\varepsilon_1(M) = M \times \mathbb{R}^1$ is the trivial bundle. More generally, there are natural projections

$$0 \leftarrow J^0(M) = \varepsilon_1(M) \leftarrow J^1(M) \leftarrow J^2(M) \leftarrow \cdots \leftarrow J^r(M) \leftarrow \cdots$$

commuting with the projection onto M (corresponding to the "forgetful map", that erases the Taylor coefficients in top degrees), so we can pass to projective limit and define:

$$J^\infty(M) = \varprojlim_r J^r(M).$$

Then $J^\infty(M)$ is an infinite-dimensional vector bundle over M with fibre $\mathbb{R}_{formal}^d$; it is called the *infinite jet-bundle over M*. This construction can be easily generalised to any vector bundle $E \to M$: one can use the local trivialisations in E to define the jet spaces $J_p^r(E)$, consisting of the *jets of local sections* of E at $p \in M$ and the jet bundles $J^r(E)$ for $r = 0, 1, 2, \ldots, \infty$.

Still more generally we can consider a similar equivalence relation for the local (i.e. determined in an open neighbourhood of the origin $0 \in \mathbb{R}^d$) diffeomorphisms $F : \mathbb{R}^d \to M$ with $F(0) = p$, namely: we say that $j_p^r(F) = j_p^r(G)$ if for all multiindices I with $|I| \leq r$

$$\frac{\partial^{|I|}F}{\partial x^I}\Big|_0 = \frac{\partial^{|I|}G}{\partial x^I}\Big|_0,$$

where we regard F, G as the vector-valued functions (with respect to some local coordinates in M, same for F and G). Again $j_p^r(F) = j_p^r(G)$ is an equivalence relation for all $r \geq 0$, and we denote by $J_p^r(\mathbb{R}^d; M)$ the set of equivalence classes. These sets for all r have a natural topology, induced from the group of diffeomorphisms; this topology is respected by the natural projections and we put

$$J_p^\infty(\mathbb{R}^d; M) = \varprojlim J_p^r(\mathbb{R}^d; M)$$

as before. One can regard $J_p^r(\mathbb{R}^d; M)$, $r = 0, 1, 2, \ldots, \infty$ as the spaces of infinitesimal germs (up to degree r) of local coordinate systems at p. The spaces $J_p^r(\mathbb{R}^d; M)$, $r = 0, 1, \ldots, \infty$ have a group structure (induced by the composition of diffeomorphisms), so that $J^r(\mathbb{R}^d; M) = \bigcup_p J_p^r(\mathbb{R}^d; M)$ is a principal bundle over M; for instance $J^1(\mathbb{R}^d; M)$ is the frame bundle associated with the tangent bundle of M.

We denote $J^\infty(\mathbb{R}^d; M)$ by M^{coord}: one can regard it as the bundle of all formal coordinate systems (infinite germs of coordinate systems) on M. This is a principal bundle over M with the structure group $\text{Diffeo}^\infty(\mathbb{R}^d)$ of infinitesimal diffeomorphisms of $\mathbb{R}^d$ preserving the origin, this group corresponds to the Lie algebra $\widetilde{\mathcal{W}_d}$ of formal vector fields on $\mathbb{R}^d$, vanishing at the origin.[67] This group acts in a natural way on the space $\mathbb{R}^d_{formal}$ (recall, that $\mathbb{R}^d_{formal} = \mathbb{R}[[y^1, \ldots, y^d]]$, algebra of formal power series in d variables), so one can form the associated vector bundle; then we have:

$$M^{coord} \times_{\text{Diffeo}^\infty(\mathbb{R}^d)} \mathbb{R}^d_{formal} \cong J^\infty(M).$$

As a matter of fact, the $\widetilde{\mathcal{W}_d}$ corresponds to the *vertical* subspace in $T_{j_p^\infty(F)} M^{coord}$ (i.e. the kernel of the differential of projection $M^{coord} \to M$); the whole space $T_{j_p^\infty(F)} M^{coord}$ can be naturally identified with the Lie algebra $\mathcal{W}_d$ of vector fields on $\mathbb{R}^d$ with coefficients in formal power series.

---

[67] We use a slightly sloppy language here, since a thorough introduction to formal geometry would take too much time and efforts. A reader familiar with the corresponding terminology should think of $\text{Diffeo}^\infty(\mathbb{R}^d)$ as of a pro-Lie group.

Also observe that the group of linear transformations $GL_d(\mathbb{R})$ is a subgroup of $\text{Diffeo}^\infty(\mathbb{R}^d)$, so we can consider the quotient fibration $M^{aff} = M^{coord}/GL_d(\mathbb{R})$. In his proof of the global version of formality Kontsevich uses the following fact about M^{aff}:

Proposition 19.3. *The fibres of the bundle $M^{aff} \to M$ are contractible.*

The consequence of this fact is that there always exists a global section of the bundle $M^{aff} \to M$, moreover, all such sections are homotopic to each other.

Another important property of the bundle $J^\infty(M)$, and more generally, $J^\infty(E)$ for any vector bundle $E \to M$, is the existence of a canonical flat connection on them. To get an idea of this connection, it is enough to consider the local basis in a generic fibre of the bundle $J^\infty(E)$: if e_α, $\alpha = 1, \ldots, n$ is a (local) basis of sections in E at a point p, and $x^1, \ldots, x^d$ are local coordinates around p, then the space $J_p^\infty(E)$ is spanned by the elements $e_{\alpha;(i_1 i_2 \ldots i_N)}$ where $N \geq 0$ and $i_k = 1, \ldots, d$ and

$$e_{\alpha;(i_1 i_2 \ldots i_N)} = e_{\alpha;(i_{\sigma(1)} i_{\sigma(2)} \ldots i_{\sigma(N)})}$$

for any permutation $\sigma \in S_N$. In fact:

$$e_{\alpha;(i_1 i_2 \ldots i_N)} = j_p^\infty \left(\frac{\partial^N e_\alpha}{\partial x^{i_1} \ldots \partial x^{i_N}} \right).$$

Then we can define the connection ∇^0 on $J^\infty(E)$ simply by the formula:

$$\nabla^0_{\frac{\partial}{\partial x^j}} \left(e_{\alpha;(i_1 i_2 \ldots i_N)} \right) = e_{\alpha;(i_1 i_2 \ldots i_N j)}.$$

Then

$$\nabla^0_{\frac{\partial}{\partial x^j}} \nabla^0_{\frac{\partial}{\partial x^k}} = \nabla^0_{\frac{\partial}{\partial x^k}} \nabla^0_{\frac{\partial}{\partial x^j}} \text{ for all } j, k,$$

so ∇^0 is a flat connection; a similar construction gives flat connection on the principal bundle $M^{coord} \to M$. Below we shall often refer to it as the *Maurer–Cartan connection*; in some modern texts similar connections are also called *Grothendieck connections*. Just as a usual connection on principal bundles, Maurer–Cartan connection on

M^{coord} determines a differential 1-form on M^{coord} with values in $\widetilde{\mathcal{W}}_d$. One can in effect extend it to a form ω_{MC} on M^{coord} with values in $\mathcal{W}_d$: for any $\xi \in T_{j_p^\infty(F)}M^{coord}$ we put

$$\omega_{MC}(\xi) = j_0^\infty \left(dF_p^{-1} \left(\frac{dG_t}{dt}\Big|_{t=0} \right) \right), \qquad (19.4)$$

where G_t is a one parameter family of infinitesimal diffeomorphisms $G_t : \mathbb{R}^d \to M$, such that G_0 sends 0 to $p \in M$ and $\frac{dG_t}{dt}\big|_{t=0} = \xi$. Observe that ω_{MC} is an isomorphism from $T_{j_p^\infty(F)}M^{coord}$ to $\mathcal{W}_d$; its pointwise inverse is a "lift" map from $\mathcal{W}_d$ to the vector fields on M^{coord} which is a homomorphism of Lie algebras,[68] since ∇^0 is flat.

We eventually can define the bundles $J^\infty(M; \mathcal{T}_{poly}^{\cdot})$ and $J^\infty(M; \mathcal{D}_{poly}^{\cdot})$ of infinite jets of polyvector fields and polydifferential operators on M; these bundles inherit flat connections from the previously discussed construction and also bear the differential Lie algebra structure at the same time. They play important role in our constructions.

19.2.2 Kontsevich's globalisation method (via the formal geometry)

Kontsevich begins with the following observation: the L_∞-morphism $\mathcal{U} = \{\mathcal{U}_n\}$, constructed in Section 18 (Proposition 18.2) is not unique. So it is natural to assume that $\mathcal{U}$ satisfies some properties, which do not hold for a generic L_∞-morphism of this sort. This is indeed the case, in particular the following is true:

Proposition 19.4. *The map*

$$\mathcal{U} : \mathcal{T}_{poly}^{\cdot}(\mathbb{R}^d)[1][[\hbar]] \to \mathcal{D}_{poly}^{\cdot}(\mathbb{R}^d)[1][[\hbar]],$$

which we constructed in previous section, satisfies the following additional conditions:

(P₁) *The morphism $\mathcal{U}$ is naturally extended to an L_∞-morphism between $\mathcal{T}_{poly}^{\cdot}(\mathbb{R}^d_{formal})$ and $\mathcal{D}_{poly}^{\cdot}(\mathbb{R}^d_{formal})$ (with due shifts);*

---

[68] In effect, ω_{MC} is just the canonical identification of the infinitesimal vector fields on M with $\widetilde{\mathcal{W}}_d$ induced by the infinitesimal coordinate systems on M.

(**P₂**) Let $\xi \in \mathcal{W}_d$ be a formal vector field; we can identify it with an element $m_T(\xi) \in \mathcal{T}_{poly}^{\cdot}(\mathbb{R}_{formal}^d)$ and an element $m_D(\xi) \in \mathcal{D}_{poly}^{\cdot}(\mathbb{R}_{formal}^d)$; then

$$\mathcal{U}_1(m_T(\xi)) = m_D(\xi);$$

(**P₃**) $\mathcal{U}$ is $GL_d(\mathbb{R})$ equivariant (here we let $GL_d(\mathbb{R})$ act on $\mathbb{R}_{formal}^d$ in a natural way);

(**P₄**) For any $k \geq 2$ and any $\xi_1, \ldots, \xi_k \in \mathcal{W}_d$ we have

$$\mathcal{U}_k(m_T(\xi_1), m_T(\xi_2), \ldots, m_T(\xi_k)) = 0.$$

(**P₅**) Consider the natural inclusion $\mathfrak{gl}_d \subset \mathcal{W}_d$; then for any $k \geq 2$, $\xi \in \mathfrak{gl}_d$ and any $\eta_2, \ldots, \eta_k \in \mathcal{T}_{poly}^{\cdot}(\mathbb{R}_{formal}^d)$ we have

$$\mathcal{U}_k(m_T(\xi), \eta_2, \ldots, \eta_k) = 0.$$

In effect properties (**P₁**)–(**P₃**) are evident from the construction, and the remaining two properties can be proved by accurate unraveling of the definitions and explicit computation of the integrals.

Now we can use the properties of $\mathcal{U}$ to prove the globalisation result. Let us sketch the main steps of the proof: first we observe that for every formal coordinate system on M centred at some point $p \in M$ (i.e. for every point $j_p^\infty(\varphi) \in M^{coord}$) we can identify the jets of polyvector fields and polydifferential operators at p with the spaces $\mathcal{T} = \mathcal{T}_{poly}^{\cdot}(\mathbb{R}_{formal}^d)$ and $\mathcal{D} = \mathcal{D}_{poly}^{\cdot}(\mathbb{R}_{formal}^d)$ respectively; applying $\mathcal{U}$ from Section 18 at every point $j_p^\infty(\varphi)$ we obtain a family of L_∞-morphisms

$$\mathcal{U}^{coord} : \mathcal{T} \times M^{coord} \Rightarrow \mathcal{D} \times M^{coord},$$

commuting with the action of $GL_d(\mathbb{R})$ (this follows from properties (**P₃**) – (**P₅**) and the existence of the flat connection ∇^0).[69] Hence we can use the map $\mathcal{U}$ to induce a family of morphisms

$$\mathcal{U}^{aff} : \mathcal{T} \times M^{aff} \Rightarrow \mathcal{D} \times M^{aff}.$$

---

[69] Here it is convenient to interpret L_∞-algebra structures and L_∞-morphisms as formal square-zero vector fields on formal manifolds (Q-manifolds) and maps between them.

On the other hand, since the fibre of the bundle $M^{aff} \to M$ is contractible, one can find a section $s^{aff} : M \to M^{aff}$; for instance one can choose a symmetric connection ∇ on M and put $s^{aff}(p)$ to be the jet at $p \in M$ of the normal coordinate system associated with ∇. Using this section we can pull the families of morphisms $\mathcal{U}^{aff}$ down to M to obtain a family of L_∞-morphisms between the pull-backs $\mathcal{T}_{s^{aff}} \to M$ and $\mathcal{D}_{s^{aff}} \to M$ of $\mathcal{T} \times M^{aff}$ and $\mathcal{D} \times M^{aff}$ respectively (here $\mathcal{T}_{s^{aff}} \to M$ and $\mathcal{D}_{s^{aff}}$ are flat families of differential Lie algebras over M; in effect they can be identified with the bundles $J^\infty(M; \mathcal{T}_{poly})$ and $J^\infty(M; \mathcal{D}_{poly})$ independently of the choice of the section s^{aff}). It remains to be observed that the formal completions of the spaces of global sections of $\mathcal{T}_{s^{aff}}$ and $\mathcal{D}_{s^{aff}}$ can be identified with the differential Lie algebras $\mathcal{T}_{poly}(M)[1][[\hbar]]$ and $\mathcal{D}_{poly}(M)[1][[\hbar]]$ respectively, so the formality map between them is given by restriction of the corresponding morphism $\mathcal{U}_{s^{aff}} : \mathcal{T}_{s^{aff}} \Rightarrow \mathcal{D}_{s^{aff}}$. Summing up, we obtain the following result:

Proposition 19.5. *For every smooth manifold M there exists an L_∞-quasi-isomorphism*

$$\widehat{\mathcal{U}} : \mathcal{T}_{poly}(M)[1][[\hbar]] \Rightarrow \mathcal{D}_{poly}(M)[1][[\hbar]],$$

extending the Hochschild-Kostant-Rosenberg map on M, i.e. $\chi_M = \widehat{\mathcal{U}}_1$.

19.2.3 Fedosov-type constructions and the globalisation

As one could see connections on jet bundles and similar objects play crucial role in the process of globalisation of the formality morphism. So one is tempted to assume that this construction is somehow related with the Fedosov's construction of flat (abelian) connections. And this is indeed so: the global formality morphism can be obtained as a process of "pasting together" the local (in effect, infinitesimal) Kontsevich's morphisms with the help of a suitable flat connection, just like Fedosov $\star$-product can be regarded as the result of the similar process applied to local (infinitesimal) Moyal $\star$-products.

There are few different ways to describe this process. In particular one can concentrate just on the $\star$-product i.e. construct the global ⋆-product on a Poisson manifold M from the locally defined Kontsevich's ⋆-products, or go further and obtain a flat family of formality morphisms, that establishes the quasi-isomorphism between $\mathscr{T}_{poly}^{\cdot}(M)[1][[\hbar]]$ and $\mathscr{D}_{poly}^{\cdot}(M)[1][[\hbar]]$, similar to the map $\widehat{\mathcal{U}}$ (see Proposition 19.5). In this section we shall describe the former approach due to Cattaneo, Felder and Tomassini [CFT02]; an interested reader, who would like to know more about the globalisation of formality morphisms, can refer to the papers of Dolgushev, Calaque, Bursztin and others, see our list of references.

We begin with slightly modifying the construction of $\mathbb{R}^d_{formal}$ by introducing the formal variable $\hbar$ into it: we set

$$\widetilde{\mathbb{R}}^d_{formal} = \mathbb{R}[[y^1, \ldots, y^d]][[\hbar]].$$

Any Poisson bivector π on M induces a Poisson structure in all formal coordinate neighbourhoods of every point in M (i.e. at every point of M^{coord}), thus using the Kontsevich's construction we obtain a ⋆-product on the space of sections of the trivial bundle $M^{coord} \times \widetilde{\mathbb{R}}^d_{formal}$: for every point $j_p^\infty(F) \in M^{coord}$ we use the local (infinitesimal) coordinate system F at p to write down the infinite jet of π at p as a bivector $\hat{\pi}$ in $\mathbb{R}^d_{formal}$ (see formula (19.5)) and use Kontsevich's construction for the ⋆-product.

Further we can consider the infinite-dimensional vector bundle $\widetilde{E}$ over M^{aff}:

$$\widetilde{E} = M^{coord} \times_{GL_d(\mathbb{R})} \widetilde{\mathbb{R}}^d_{formal} \to M^{aff},$$

which we can pull down to an infinite-dimensional bundle $E \to M$ with the help of a global section $s^{aff} : M \to M^{aff}$; Kontsevich's construction being $GL_d(\mathbb{R})$-equivariant (see property (**P₃**), Proposition 19.4), so is Kontsevich's ⋆-product, hence we obtain a ⋆-product in the space of global sections of $\widetilde{E}$ and also in the space of sections of E.

Now the construction of global ⋆-product will go on in the manner analogous to the Fedosov's ideas: we want to find a flat (abelian) Fedosov connection D on the vector bundle E. To this end we need to begin with a connection ∇ on E, which satisfies the Leibniz identity with respect to the point-wise ⋆-product in E and modify it by adding the commutator with a suitable section $\gamma \in \Omega^1(M, E)$. The main difficulty here is the necessity to choose the initial connection ∇: unlike in the classical case, considered by Fedosov himself, it is not enough to take a symplectic connection on M (here we can regard it as a connection ∇' on TM for which $\nabla'(\pi) = 0$[70]), since this does not guarantee the compatibility of connection and the $\star$-product, as it involves derivatives of the coefficients of π.

It turns out however that Kontsevich's construction provides us with a canonical way to produce such connections from any flat connection on M^{coord}, for instance we can begin with the Maurer–Cartan connection ∇^0, more accurately with connection form ω_{MC}. Let for every $\hat{\xi} \in \mathcal{W}_d$ and every bivector $\hat{\pi}_0$ on $\mathbb{R}^d_{formal}$

$$A_{\hat{\pi}_0}(\hat{\xi}) = \sum_{k=0}^{\infty} \frac{\hbar^k}{k!} \mathcal{U}_{k+1}(\hat{\xi}, \underbrace{\hat{\pi}_0, \ldots, \hat{\pi}_0}_{k \text{ times}}).$$

Let π now be a Poisson bivector on M; recall that $\hat{\pi}$ denotes the function on M^{coord} with values in $\mathcal{T}^2_{poly}(\mathbb{R}^d_{formal})$, induced by π, i.e.

$$\hat{\pi}(j_p^\infty(F)) = j_0^\infty(\pi(x_F^1, \ldots, x_F^d)), \tag{19.5}$$

i.e. $\hat{\pi}$ is given by the infinite jets at the origin of $\mathbb{R}^d$ of π expressed in terms of the (infinitesimal) local coordinates system $F \in M^{coord}$. We put further

$$\hat{A}_{\hat{\pi}}(\xi) = A_{\hat{\pi}}(\omega_{MC}(\xi)) \text{ for any } \xi \in T_{j_p^\infty(F)} M^{coord}.$$

One can regard $\hat{A}_{\hat{\pi}}$ as a differential 1-form on M^{coord} with coefficients in (usual) differential operators on $\mathbb{R}^d_{formal}$.

In order to define a connection on E we begin with the following operator on the trivial bundle $M^{coord} \times \mathbb{R}^d_{formal}$: for any function $f: M^{coord} \to \mathbb{R}^d_{formal}$ we put

$$\tilde{\nabla}(f) = df + \hat{A}_{\hat{\pi}}(f). \tag{19.6}$$

---

[70] One can naturally extend connections from the tangent bundle to the infinite jet bundle, and M^{coord} is the corresponding principal bundle.

Here for any vector $\xi \in T_{j_p^\infty(F)} M^{coord}$ we let $\hat{A}_{\hat{\pi}}(\xi)$ act on the value $s(j_p^\infty(F))$ of the section s by the corresponding differential operator.

We leave it as an exercise to the reader to show that the operator $\tilde{\nabla}$ satisfies the Leibniz rule with respect to the fibre-wise $\star$-product (see Exercise E.7.1). Now one can induce a $\star$-product compatible connection on E from the connection $\tilde{\nabla}$ on trivial bundle $M^{coord} \times \mathbb{R}^d_{formal}$; to this end we first use its restriction to the bundle $\tilde{E}$ over M^{aff} (which is possible, since Kontsevich's construction is $GL_d(\mathbb{R})$-equivariant) and then we pull it down to a connection ∇ on E along a section $s^{aff}: M \to M^{aff}$. An alternative point of view is to consider a family of local sections $s_\alpha: M \supseteq U_\alpha \to M^{coord}$ which differ by a family of linear transformations. Then locally (after the pullback along s_α) we can identify any section of E with a $\mathbb{R}^d_{formal}$-valued function on M and we have:

$$\nabla(f) = df + s_\alpha^*(\hat{A}_{\hat{\pi}})(f). \tag{19.7}$$

These formulas match on the intersections of the open sets $U_\alpha \cap U_\beta$, since $s_\alpha = g_{\alpha\beta} s_\beta$, where $g_{\alpha\beta}: U_{\alpha\beta} \to GL_d(\mathbb{R})$ and the maps $\mathcal{U}_k$ are $GL_d(\mathbb{R})$-equivariant.

In this way we obtain a connection ∇ on E, that satisfies the Leibniz rule with respect to the $\star$-product; however, this connection is not flat; let F_∇ be its curvature form (see Exercise E.7.2). In order to obtain a flat connection D we can now imitate the construction of Fedosov connection from Sections 13.3.2–13.3.5: we use iterations to solve the equation on flatness of the connection

$$D(f) = \nabla(f) + [\gamma, f],$$

where γ is a 1-form on M with values in E and $[,]$ denotes the commutator with respect to the $\star$-product in E. Namely: $D^2 = 0$ iff

$$F_\nabla + \nabla \gamma + \frac{1}{2}[\gamma, \gamma] \in \Omega^2(M)$$

where we identify $\Omega^*(M)$ with the differential forms on M with values in $\mathbb{R} \subset \mathbb{R}^d_{formal}$, which is central with respect to the $\star$-product (see the cited paper [CFT02] for details). One can also show that the space of flat sections E_D of the bundle E is naturally isomorphic to the space of functions on M, which gives the desired $\star$-product on $C^\infty(M)$. In both cases this follows from the fact that the "classical part" of our ∇ (which is just the Maurer–Cartan connection) has trivial cohomology in degree 2 and its cohomology in degree 0 is isomorphic to $C^\infty(M)$.

Remark 19.6. In effect, as we have already mentioned, the globalised version of formality morphism can also be obtained from a similar construction: we consider the bundles $\mathcal{T}(M)$, $\mathcal{D}(M)$ over M with fibres equal to the polyvector fields and polydifferential operators on $\mathbb{R}^d_{formal}$ (these bundles are induced from trivial bundles of polyvector fields and polydifferential operators over M^{coord} in a way similar to the construction of the bundle E) and use fibrewise Kontsevich's map $\mathcal{U}$ between them, which gives a fibrewise quasi-isomorphism. After this we can find flat Fedosov-type connections on these bundles such that the formality morphism $\mathcal{U}$ commutes with these connections, and hence sends flat sections to flat sections. It only remains to identify the space of flat sections of $\mathcal{T}(M)$ and $\mathcal{D}(M)$ with polyvector fields and polydifferential operators on M.

The main difference however between this case and Fedosov's construction, involving the $\star$-product is the fact that flat connections on $\mathcal{T}(M)$, $\mathcal{D}(M)$ are essentially unique: one cannot modify them by choosing various central elements, like in the case of usual Fedosov connection. However, if the Poisson structure π on M is induced by a symplectic form ω, the $\star$-products, obtained in this way, are classified by the same space $\frac{\omega}{\hbar} + H^2_{dR}(M)[[\hbar]]$ as in Fedosov's construction: in effect, one can identify the equivalence classes of formal Poisson structures, beginning with ω^{-1} on M with this space; moreover, the class of a formal Poisson structure is identified with Fedosov's class of the $\star$-product, associated with it, see [BDW12].

20 Applications of Kontsevich's quantisation: Duflo's isomorphism

Kontsevich's construction can be applied in various situations; one is tempted to believe that it allows to establish relation between the properties of the initial Poisson structure and the corresponding quantised algebra. In many cases this is so, although the relations of these two objects in general remain quite mysterious.

In this section we will describe one particular construction, where Kontsevich's quantisation allows to find a simpler way to prove a classical result: Duflo's formula, which gives an isomorphism between the algebra of Ad_G-invariant functions on the (dual space of) a Lie algebra $\mathfrak{g}$ (here G is the corresponding Lie group) and the center of the universal enveloping algebra of $\mathfrak{g}$.[71] We will follow the exposition by Kontsevich sketched in his original paper [Kon03], already cited in previous sections (it was originally published online in 1997, see https://arxiv.org/pdf/q-alg/9709040.pdf), which was later elaborated by Calaque, van den Bergh, see [CvdB10], where similar constructions appear in the context of (the sheaves of sections of) Lie algebroids rather than Lie algebras.

20.1 The tangent map

One of the many advantages of Kontsevich's construction is that it gives a very straightforward and simple answer to the following question: what is the relation between the center of the Poisson Lie algebra $(C^\infty(M), \{,\})$ (i.e. the algebra $C^\infty(M)$, regarded as Lie algebra with respect to the Poisson bracket, induced by a Poisson bivector π) and the quantised algebra $(C^\infty(M)[[\hbar]], \star)$? Here we say that an element $f \in C^\infty(M)$ is central in Poisson Lie algebra if $\{f, g\} = 0$ for all $g \in C^\infty(M)$. Such functions f are often called *Casimir functions* or just *Casimirs*; these functions are characterised by the condition that their Hamiltonian fields vanish identically, see Section 6.2.3. On the other hand, the center of an associative algebra $(C^\infty(M)[[\hbar]], \star)$ is just the set of elements F, which commute with everything with

---

[71] Here and below $\mathfrak{g}$ will denote any real or complex Lie algebra; we will use $\Bbbk$ to denote the ground field of $\mathfrak{g}$, i.e. $\Bbbk = \mathbb{R}$ or $\mathbb{C}$.

respect to the $\star$-product. Clearly, in both cases center is a subalgebra in the corresponding associative algebra.

From the very start it is clear that these two centres are very closely related. Indeed, let $F = f_0 + \hbar f_1 + \hbar^2 f_2 + \ldots$ be a central element in $(C^\infty(M)[[\hbar]], \star)$. Consider arbitrary $g \in C^\infty(M) \subset C^\infty(M)[[\hbar]]$. Then

$$0 = F \star g - g \star F = \hbar \{f_0, g\} + o(\hbar),$$

and so F is central only if f_0 is central. One can ask, if the opposite is also true, i.e. if one can restore the central elements of the quantised algebra from the center of the Poisson Lie algebra. It turns out that this is indeed so, if we allow the formal power series in $\hbar$ on the Poisson side. More accurately, the following is true:

Proposition 20.1. *Let $(C^\infty(M)[[\hbar]], \{,\})$ be the Poisson algebra (where we extend $\{,\}$ to an $\hbar$-linear bracket on $C^\infty(M)[[\hbar]]$), then the centre of this algebra is naturally isomorphic (as $\mathbb{C}[[\hbar]]$-module) to the centre of the quantised algebra $(C^\infty(M)[[\hbar]], \star)$, so that for every f in the Poisson centre, the corresponding element in the center of quantised algebra is equal to f modulo $\hbar$.*

Proof. First of all, observe that the centre of $(C^\infty(M)[[\hbar]], \{,\})$ is equal to $\mathcal{Z}_\pi(C^\infty(M))[[\hbar]]$, where $\mathcal{Z}_\pi(C^\infty(M))$ is the Poisson centre of $C^\infty(M)$. Let us now construct a $\mathbb{C}[[\hbar]]$-linear map that sends elements of $\mathcal{Z}_\pi(C^\infty(M))$ to the centre of the quantised algebra. To this end let $\mathcal{U} = \{\mathcal{U}_n\}$ be an L_∞-morphism of the differential Lie algebras

$$\mathcal{U} : \mathcal{T}_{poly}(M)[1][[\hbar]] \Rightarrow \mathcal{D}_{poly}(M)[1][[\hbar]]$$

(for instance, $\mathcal{U}$ is Kontsevich's quasi-isomorphism); consider the map $\mathcal{U}_1^\pi : C^\infty(M) \to C^\infty(M)[[\hbar]]$, given by the formula

$$\mathcal{U}_1^\pi(f) = \sum_{k=0}^\infty \frac{\hbar^k}{k!} \mathcal{U}_{k+1}(f, \underbrace{\pi, \ldots, \pi}_{k \text{ times}}). \tag{20.1}$$

By a slight abuse of terminology, we shall often call $\mathcal{U}_1^\pi$ *tangent map* below (more accurately one should speak about the tangent L_∞-morphism, see Remark 20.2). Recall now that for any two elements

$f, g \in C^\infty(M)[[\hbar]]$, their $\star$-product in Kontsevich's construction is given by the formula

$$f \star g = fg + B(f, g)$$

where $B \in \mathscr{D}^2_{poly}(M)[[\hbar]]$ is a Maurer–Cartan element; hence the commutator can be expressed as follows

$$f \star g - g \star f = B(f, g) - B(g, f) = [B, f](g),$$

where $[B, f]$ denotes the Gerstenhaber bracket of B and f. Further recall, that the Maurer–Cartan element B in this construction is given by the formula

$$B = \sum_{k=1}^{\infty} \frac{\hbar^k}{k!} \mathcal{U}_k(\underbrace{\pi, \ldots, \pi}_{k \text{ times}}).$$

Now, for any $f \in \mathcal{Z}_\pi(C^\infty(M))$ we have, in full analogy with the proof of Proposition 17.5 (here δ denotes the Hochschild differential):

$$\delta(\mathcal{U}_1^\pi(f)) = \sum_{k=0}^{\infty} \frac{\hbar^k}{k!} \delta(\mathcal{U}_{k+1}(f, \pi, \ldots, \pi))$$

$$= [\text{since the differential in } \mathcal{T}_{poly}^{\cdot}(M) \text{ is trivial}]$$

$$= \sum_{k=0}^{\infty} \frac{\hbar^k}{(k-1)!} \mathcal{U}_{k+1}([f, \pi], \pi, \ldots, \pi)$$

$$+ \sum_{k=0}^{\infty} \frac{\hbar^k}{k!} \binom{k}{2} \mathcal{U}_{k+1}([\pi, \pi], f, \pi, \ldots, \pi)$$

$$+ \sum_{k=0}^{\infty} \frac{\hbar^k}{k!} \sum_{p+q=k} \binom{k}{p} [\mathcal{U}_{p+1}(f, \pi, \ldots, \pi), \mathcal{U}_q(\pi, \ldots, \pi)]$$

$$= \begin{bmatrix} \text{since } \pi \text{ is Poisson, its Schouten brackets vanish:} \\ [\pi, \pi] = 0; \text{ since } f \text{ is Poisson-central, its Hamil-} \\ \text{tonian field vanishes: } [f, \pi] = -X_f = 0 \end{bmatrix}$$

$$= \sum_{k=0}^{\infty} \sum_{p+q=k} \frac{\hbar^k}{p!q!} [\mathcal{U}_{p+1}(f, \pi, \ldots, \pi), \mathcal{U}_q(\pi, \ldots, \pi)].$$

Now $\mathcal{U}_1^\pi(f) \in C^\infty(M)[[\hbar]]$, and Hochschild differential δ vanishes on $C^\infty(M)$, since this algebra is commutative, hence we have

$$[B, \mathcal{U}_1^\pi(f)] = \sum_{p,q} \frac{\hbar^{p+q}}{p!q!} [\mathcal{U}_p(\pi,\ldots,\pi), \mathcal{U}_{q+1}(f,\pi,\ldots,\pi)] = 0.$$

Clearly, the map $\mathcal{U}_1^\pi$ extends to a $\mathbb{C}[[\hbar]]$-module map $\mathcal{Z}_\pi(C^\infty(M)[[\hbar]]) \to \mathcal{Z}(C^\infty(M)[[\hbar]], \star)$. In order to obtain the inverse map, we consider any element $F \in \mathcal{Z}(C^\infty(M)[[\hbar]], \star)$; as we already mentioned above $F = f_0 + \hbar f_1 + \hbar^2 f_2 + \ldots$, where $f_0 \in \mathcal{Z}_\pi(C^\infty(M))$. Now consider $F' = F - \mathcal{U}_1^\pi(f_0)$; clearly

$$F' \in \mathcal{Z}(C^\infty(M)[[\hbar]], \star), \text{ and } F' = 0 \mod \hbar.$$

If $F' = \hbar f_1' + \hbar^2 f_2' + \ldots$, then as before we have $f_1' \in \mathcal{Z}_\pi(C^\infty(M))$ and we can consider

$$F'' = F' - \mathcal{U}_1^\pi(\hbar f_1') \in \mathcal{Z}(C^\infty(M)[[\hbar]], \star), \text{ and } F'' = 0 \mod \hbar^2.$$

Going on in this manner, we obtain the elements $f_k^{(k)} \in \mathcal{Z}_\pi(C^\infty(M))$ (with $f_0 = f_0^{(0)}$, $f_1' = f_1^{(1)}$ etc.) such that

$$F = \mathcal{U}_1^\pi(\widehat{F}), \text{ for } \widehat{F} = \sum_{k=0}^\infty \hbar^k f_k^{(k)} \in \mathcal{Z}_\pi(C^\infty(M)[[\hbar]]).$$

Then the correspondence $F \mapsto \widehat{F}$ is the $\mathbb{C}[[\hbar]]$-linear morphism inverse to $\mathcal{U}_1^\pi$. □

Remark 20.2. In effect (c.f. Exercise E.7.3) Maurer–Cartan elements in L_∞-algebras give rise to the deformations of the corresponding L_∞-structures. Applying this procedure to the algebras $\mathcal{T}_{poly}(M)[1][[\hbar]]$ and $\mathcal{D}_{poly}(M)[1][[\hbar]]$ that we use here and the Maurer–Cartan elements π and B (Poisson bivector and the deformation series), we get new L_∞-structures on polyvector fields and polydifferential operators. In particular, the differentials in the deformed algebras are given by the following formulas: for any $P \in \mathcal{T}_{poly}(M)$ and any $\varphi \in \mathcal{D}_{poly}(M)$ we put

$$d_\pi(P) = [\pi, P], \ d_B(\varphi) = b\varphi + [B, \varphi], \tag{20.2}$$

where the brackets on the right denote the Schouten and Gerstenhaber Lie brackets; in particular d_π is just the differential in the Lichnerowicz's Poisson cohomology complex, see Section 6.2.3. Then (see Exercise E.7.4) one can show that $\mathcal{U}_1^\pi$ is the first Taylor coefficient of an L_∞-morphism $\mathcal{U}^\pi$ between the L_∞-algebras $\mathcal{T}_{poly}^{\cdot}(M)[1][[\hbar]]$ and $\mathcal{D}_{poly}^{\cdot}(M)[1][[\hbar]]$. These deformed DG algebras are called *the tangent L_∞-algebras* and the map between them is called *the tangent L_∞-morphism* by Kontsevich; this terminology is due to his interpretation of the L_∞-algebras as "fibres" in a formal Q-manifold.

20.2 Centre of a universal enveloping algebra

As we see, centres of the classical and quantised algebras are closely related; in fact, they are isomorphic as modules over $\mathbb{C}[[\hbar]]$. This result is not very surprising, and has in many situations been known much earlier. As an example, let us consider quantisation of the symmetric algebra $S(\mathfrak{g})$: here $\mathfrak{g}$ is a Lie algebra and we regard $S(\mathfrak{g})$ as the algebra of polynomial functions on the dual space $\mathfrak{g}^*$ (see Section 5, especially 5.1 and 5.3). In this case one can regard the universal enveloping algebra of $\mathfrak{g}$ as a quantisation of $S(\mathfrak{g})$ (see Section 5.3).[72] Then the statement of Proposition 20.1 is equivalent to the following well-known result:

Proposition 20.3. *The symmetrisation map* $\sigma : S(\mathfrak{g}) \to U\mathfrak{g}$, *see Equation* (5.8'), *gives by restriction a linear isomorphism of the centres* $\sigma : \mathcal{Z}_\pi(G(\mathfrak{g})) \to \mathcal{Z}(U\mathfrak{g})$.

Proof. In order to prove this statement, let us recall a result from the classical Lie theory: let G be the 1-connected Lie group with Lie algebra $\mathfrak{g}$. This group acts by conjugations on itself:

$$Ad : G \to \text{End}(G), \ g \mapsto Ad_g, \text{ where } Ad_g(x) = gxg^{-1}, \ \forall g, x \in G.$$

This action restricts to the adjoint action $Ad : G \to \text{End}(\mathfrak{g})$ of the group G on its Lie algebra $\mathfrak{g}$ by Lie algebra homomorphisms. It follows that Ad extends to the actions of G on algebras $S(\mathfrak{g})$ and $U\mathfrak{g}$

---

[72] Below we shall show that Kontsevich's construction gives essentially the same result.

by algebraic homomorphisms:
$$Ad: G \to \mathrm{End}(S(\mathfrak{g})), \quad Ad: G \to \mathrm{End}(U\mathfrak{g}).$$

A straightforward inspection of definitions shows that the symmetrisation map σ intertwines these two actions, i.e.
$$\sigma(Ad_g(f)) = Ad_g(\sigma(f)), \quad \forall g \in G, \; f \in S(\mathfrak{g}).$$

In particular, σ sends Ad_G-invariant elements in $S(\mathfrak{g})$ to Ad_G-invariant elements in $U\mathfrak{g}$. Now the statement of the proposition follows from the well-known result: *an element $f \in S(\mathfrak{g})$ is Poisson-central iff it is invariant with respect to the Ad_G-action; similarly $\hat{f} \in U\mathfrak{g}$ is central iff it is Ad_G-invariant.* We will prove this here for the sake of completeness.

To this end we observe that $f \in S(\mathfrak{g})$ is Poisson-central, iff it Poisson-commutes with all linear functions $X \in \mathfrak{g}$ on $\mathfrak{g}^*$. On the other hand, if $g = \exp(tX)$, then by definitions for any $Y \in \mathfrak{g}$
$$\frac{d}{dt}(Ad_{\exp(tX)})(Y)|_{t=0} = [X,Y] = \{X,Y\},$$

and so for any $f \in S(\mathfrak{g})$ we have
$$\frac{d}{dt}(Ad_{\exp(tX)})(f)|_{t=0} = \{X,f\}.$$

Hence, every Ad_G-invariant $f \in S(\mathfrak{g})$ is Poisson-central, and since elements $\exp X$, $X \in \mathfrak{g}$ generate the group G, the opposite is also true. Similarly, since $U\mathfrak{g}$ is generated by $\mathfrak{g}$, an element $\hat{f}$ is central iff $X \cdot \hat{f} - \hat{f} \cdot X = 0$ for all $X \in \mathfrak{g}$. On the other hand for any $X, Y \in \mathfrak{g} \subseteq U\mathfrak{g}$ we have
$$X \cdot Y - Y \cdot X = [X,Y] \in U\mathfrak{g},$$

and hence for all $\hat{f} \in U\mathfrak{g}$
$$\frac{d}{dt}(Ad_{\exp(tX)})(\hat{f})|_{t=0} = X \cdot \hat{f} - \hat{f} \cdot X,$$

which completes the proof. □

Below we will usually denote the subalgebras of G-invariant elements in $S(\mathfrak{g})$ and $U\mathfrak{g}$ by $S(\mathfrak{g})^G$ and $U\mathfrak{g}^G$, respectively.

20.3 Duflo's isomorphism: The original construction

We saw above that the symmetrisation map gives a linear isomorphism $\sigma : S(\mathfrak{g})^G \to U\mathfrak{g}^G$; however, this map is not a homomorphism of algebras: $\sigma(fg) \neq \sigma(f) \cdot \sigma(g)$. On the other hand it is known that both algebras are finitely generated free commutative algebras, so if we choose free generators $f_1, \ldots, f_p$ of the algebra algebras $S(\mathfrak{g})^G$ and if $\hat{f}_1, \ldots, \hat{f}_p$ such that $\hat{f}_i = \sigma(f_i)$, $i = 1, \ldots, p$ are free generators of $U\mathfrak{g}^G$, we would be able to define an isomorphism of algebras simply by sending f_k to $\hat{f}_k$. In many cases (for example for semisimple algebras) this can be done, see for instance Exercises E.7.5, E.7.6 and E.7.7. However, this method is inevitably an *ad hoc* one, and depends a great deal on the choice of the Lie algebra and of the generators in both centres. So one might wonder, if something can be done in a generic case? More accurately: *is it possible to modify the symmetrisation map σ by adding some correction terms depending on f to $\sigma(f)$ in a universal explicitly determined way, so that the new map $\mathfrak{S}$ will restrict to an isomorphism of the commutative algebras $\mathfrak{S} : \mathcal{Z}_\pi(S(\mathfrak{g})) \to \mathcal{Z}(U\mathfrak{g})$?*

This question was first answered by Duflo in his papers [Du70], [Du77]: it turns out that the answer is positive, moreover, one can find an explicit formula for the series of correction terms that turns σ into a homomorphism of algebras. Let us briefly explain this construction. To this end we consider the formal power series

$$S(x) = \frac{\sinh \frac{x}{2}}{\frac{x}{2}} = 1 + \frac{1}{3!} \cdot \frac{x^2}{2^3} + \frac{1}{5!} \cdot \frac{x^4}{2^5} + \cdots.$$

Let $X \in \mathfrak{g}$ be any element; consider the matrix $ad_X \in \text{End}(\mathfrak{g})$ and put

$$j(X) = \left(\det(S(ad_X))\right)^{\frac{1}{2}}.$$

We can regard $ad : X \to \text{End}(\mathfrak{g})$ as a linear matrix-valued function on $\mathfrak{g}$, i.e. as an element of the tensor product $\mathfrak{g}^* \otimes \text{End}(\mathfrak{g})$. This allows us regard $j(X)$ as a formal power series function of $X \in \mathfrak{g}$, that is as the element in the completion $\widehat{S}(\mathfrak{g}^*)$ of the symmetric algebra $S(\mathfrak{g}^*)$. Using the natural pairing $\mathfrak{g} \otimes \mathfrak{g}^* \to \Bbbk$ we can identify elements of $\mathfrak{g}^*$ with partial derivatives on the algebra $S(\mathfrak{g})$ and hence the elements of $S(\mathfrak{g}^*)$ can be regarded as differential operators with constant coefficients on $S(\mathfrak{g})$. Now we can regard $j(X)$ as a differential operator on $S(\mathfrak{g})$ (of infinite order); it is easy to see now that for any $f \in S(\mathfrak{g})$ and any $g \in G$ we have

$$j(X)(Ad_g(f)) = Ad_g(j(X)(f)),$$

hence $j(X)$ sends the Poisson center $S(\mathfrak{g})^G$ of $S(\mathfrak{g})$ to itself. In effect, the following is true:

Theorem 20.4 (Duflo's formula). *The map $\sigma_j : S(\mathfrak{g}) \to U\mathfrak{g}$, equal to the composition of the action of $j(X)$ and the symmetrisation map σ:*

$$\sigma_j(f) = \sigma(j(X)(f)),$$

restricts to an isomorphism of commutative algebras $\sigma_j : S(\mathfrak{g})^G \xrightarrow{\cong} U\mathfrak{g}^G$.

The original proof of this formula due to Duflo worked for semisimple Lie algebras, it was based on the study of their characters and made use of the *Kirillov's orbits integration method*; we are not going to reproduce it here. Instead we shall show that the map $\mathcal{U}_1^\pi$ that we introduced above, see Equation (20.1), does in effect induce an isomorphism of algebras, when we apply it to the Poisson centre of $S(\mathfrak{g})$. Moreover, we will show that σ_j does in fact coincide with $\mathcal{U}_1^\pi$.

20.4 Kontsevich's quantisation of $S(\mathfrak{g})$

Before we can start proving the relation between Kontsevich's tangent map $\mathcal{U}_1^\pi$ and Duflo's isomorphism, the first thing we need to show, is the relation of Kontsevich quantisation of $S(\mathfrak{g})$ and the universal enveloping algebra $U\mathfrak{g}$. So we begin with the following fact, first observed by Kontsevich in the original paper [Kon03]:

Proposition 20.5. *Kontsevich's quantisation $(S(\mathfrak{g})[[\hbar]], \star)$ of the algebra $S(\mathfrak{g})$ equipped with Kirillov–Kostant–Souriau bivector π (Definition 5.1) can be restricted to $S(\mathfrak{g})$, where it induces an associative product; then $S(\mathfrak{g})$ equipped with this new product is isomorphic to the universal enveloping algebra $U\mathfrak{g}$ of $\mathfrak{g}$.*

In effect, we already saw in Section 9.5 that the deformation quantisation of $\mathfrak{g}^*$ equipped with the canonical Kirillov–Kostant–Souriau Poisson bracket is essentially unique, i.e. that all $\star$-products, associated with this structure, are equivalent. So, all possible quantisations of $S(\mathfrak{g})$ should be isomorphic, in particular, they all are equivalent to the PBW-quantisation, see Section 5.3, whence the result. However, we feel it proper to prove this fact independently for Kontsevich $\star$-product; this can be regarded as yet another proof of the Poincaré–Birkhoff–Witt theorem.

Proof. First of all we observe, that the total degrees of the differential operators B_k, that appear in Kontsevich's formula grow indefinitely with k, when we apply this construction to $\mathfrak{g}^*$ with Kirillov–Kostant–Souriau bivector π. To this end recall that

$$B_k = \mathcal{U}_k(\underbrace{\pi, \ldots, \pi}_{k \text{ times}}),$$

where $\mathcal{U}_k(\pi, \ldots, \pi)$ is expressible as a weighted sum of operators $\mathcal{U}_\Gamma(\pi, \ldots, \pi)$, indexed by admissible binary graphs $\Gamma \in G_{n,2}^2$, see Equation (18.6). But since the coefficients of the bivector π are linear with respect to the natural coordinates on $\mathfrak{g}^*$ (see Definition 5.1), we conclude that of all the admissible binary graphs in $G_{n,2}^2$ in the formula (18.6) the ones that give nonzero contribution to the sum has the following property: *the number of arrows that enter any vertex $p_1, \ldots, p_k$ (vertex of the first type) in this graph cannot be larger, than 1*, because otherwise the formula will contain a factor, equal to the second partial derivative of a coefficient of π, which is identically equal to 0. Since the total number of edges in Γ is equal to $2k$ and no more than k of these arrows end up in the vertices of the first type, we conclude that the total degree of B_k is greater or equal to k.

It follows immediately from the previous observation that for any two polynomials $f, g \in S(\mathfrak{g})$, their product $f \star g$ belongs to the algebra $S(\mathfrak{g})[\hbar]$, i.e. the infinite sum $\sum_k \hbar^k B_k(f, g)$ in $f \star g$ contains only a finite number of nonzero terms and each term is in fact a polynomial function. This means that we can in fact set $\hbar = 1$ in the formula for the $\star$-product; let us denote the algebra $S(\mathfrak{g})$ with this product by $B(\mathfrak{g})$. Thus we have an associative product on $S(\mathfrak{g})$; it is easy to see that for any $X, Y \in \mathfrak{g} \subset S(\mathfrak{g})$ we have

$$X \star Y - Y \star X = [X, Y],$$

and hence due to the universal property of $U\mathfrak{g}$ we have a homomorphism $U\mathfrak{g} \to B(\mathfrak{g})$. This homomorphism is clearly epimorphic, since the algebra $B(\mathfrak{g})$ is generated by $\mathfrak{g}$ (to see this it is enough to remark that the top degree element of $f \star g$ coincides with fg for any $f, g \in S(\mathfrak{g})$). And hence it follows from the simple part of Poincaré–Birkhoff–Witt theorem (i.e. from the fact that $U\mathfrak{g}$ is spanned by the elements in the image of σ as linear space), that the natural map $U\mathfrak{g} \to B(\mathfrak{g})$ we just defined is an isomorphism. □

Remark 20.6. Kontsevich denotes the isomorphism $U\mathfrak{g} \xrightarrow{\cong} (S(\mathfrak{g}), \star) = B(\mathfrak{g})$ by I_{alg}. Clearly, the map I_{alg} induces an isomorphism of the centres and commutes with the adjoint action of G.

In order to proceed, let us observe that the map $\mathcal{U}_1^\pi$, where we use Kontsevich's L_∞-morphism $\mathcal{U}$, defined in Section 18 is equal to the sum of differential operators $\hbar^k D_k$ acting on the functions:

$$D_k(f) = \frac{1}{k!} \mathcal{U}_{k+1}(f, \underbrace{\pi, \ldots, \pi}_{k \text{ times}}).$$

Just as in the case of the $\star$-product, one can show that since the bivector π has linear coefficients, the degrees of the operators D_k grow indefinitely as k goes to infinity, so the map $\mathcal{U}_1^\pi$ restricts to a map $I_T : S(\mathfrak{g}) \to B(\mathfrak{g})$ (once again we use Kontsevich's notation), taking its composition with I_{alg}^{-1} we obtain a map $S(\mathfrak{g}) \to U\mathfrak{g}$. Since both factors in this composition commute with the adjoint action of G, the map $I_{alg}^{-1} \circ I_T$ sends Ad_G-invariant part of the Poisson algebra $S(\mathfrak{g})$ to $U\mathfrak{g}^G$; it is this map that gives an isomorphism of the centres. To show this it is sufficient to show that $\mathcal{U}_1^\pi$ is a homomorphism on centres, which we do in the next section.

20.5 Properties of the map $\mathcal{U}_1^\pi$

As we have earlier mentioned, the L_∞-morphism, constructed by Kontsevich is not the unique L_∞-quasi-isomorphism between the algebras of polyvector fields and polydifferential operators on an affine space $\mathbb{R}^d$. In effect, there exist infinitely many such morphisms; they are classified with the help of the *Grothendieck-Teichmüller group* action and *Drinfeld associators* (we briefly touch this subject in the appendix). All these morphisms induce $\mathbb{C}[[\hbar]]$-linear isomorphisms between the centres of the Poisson and the quantum algebra; however not all of them induce an isomorphism of the centres as commutative algebras (given by the Duflo's or some other formula).

The important property of Kontsevich's construction, defined in Section 18, which sets it apart from other possible L_∞-morphisms, is that the map $\mathcal{U}_1^\pi$ associated with it, does in effect give not just an isomorphism of the centres, and not just the L_∞-quasi-isomorphism of the deformed L_∞-algebras (see Remark 20.2 and Exercises E.7.3 and E.7.4); in addition to this as Kontsevich observed in the paper [Kon03] *it intertwines, up to a chain homotopy, the $\wedge$-product of polyvectors on $\mathbb{R}^d$ with the $\cup$-product of polydifferential operators on $\mathbb{R}^d$ viewed as the Hochschild complex of the deformed algebra.*[73]

As in many other situations, Kontsevich in his paper gave a rather brief although comprehensive explanation of this fact. Some details were later provided by other authors. In particular, working in this direction Shoikhet, Calaque, Rossi [Sh03], [CR11] and others generalised the construction of Kontsevich to the formality theorem result for Hochschild chains: i.e. instead of the cohomology complex (polydifferential operators) of functions on a manifold they consider the Hochschild homology complex of the same algebra. In these lecture notes we will follow the original exposition of [Kon03]: Kontsevich's heuristic approach is well-suited for the purposes of the first acquaintance with the subject, which we pursue here. So we recall that the operators $\mathcal{U}_n(P_1, \ldots, P_n)$ associated with polyvector fields $P_1, \ldots, P_n$ are equal to the weighted sums of the operators $\mathcal{U}_\Gamma(P_1, \ldots, P_k)$ indexed by the suitable admissible graphs Γ from $G_{n,m}$; the vertices of the first type in Γ correspond to the entries $P_1, \ldots, P_n$. For instance, if $n = 2$ and we take the only (up to permutations of the vertices of the second type) graph $\Gamma \in G_{2,m}$ with no edges connecting p_1 and p_2 (where p_1, p_2 are the vertices of the first type), then $\mathcal{U}_\Gamma(P_1, P_2)$ is equal to $\chi(P_1 \wedge P_2)$ up to a constant

---

[73] Later it was shown that Kontsevich's map is in effect a quasi-isomorphism of $\mathcal{G}_\infty$-algebras: we will explain this concept in appendix.

factor. Moreover, one can modify the construction by taking appropriate weight W_Γ (see Exercise E.7.8), so that the this factor is taken into consideration.

Modifying this observation one can describe in the terms of graphs the element of the form $\mathcal{U}_1^\pi(P_1 \wedge P_2)$: it turns out that this expression is equal to the sum of the operators, corresponding to the graphs, in which the vertices associated with P_1, P_2 are infinitesimally close and not connected by an edge, see Fig. 20.1. Here the thin circle around the points p_1, p_2 is used to show that they are infinitesimally close, i.e. that in fact we deal with a boundary component of the corresponding space $\bar{C}_{n,m}$; one can show that there should be no other points in this circle, if the corresponding operator appears with nontrivial weight in Kontsevich's formula; all the other vertices of the first type in this graph should have degree 2, and the number of vertices of the second type is equal to the degree of $P_1 \wedge P_2$.[74] On the other hand, if we are to describe the element $\mathcal{U}_1^\pi(P_1) \cup \mathcal{U}_1^\pi(P_2)$ (where the $\cup$-product is taken with respect to the $\star$-product of $C^\infty(\mathbb{R}^d)[[\hbar]]$), then we should consider the graphs of the form shown at Fig. 20.2: these graphs are in the preimage of the point

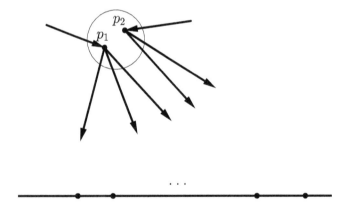

Figure 20.1: The graphs, representing the elements $\mathcal{U}_1^\pi(P_1 \wedge P_2)$.

---

[74]Here we mutely assume that the grading of polyvector fields and polydifferential operators is determined in the usual way; Kontsevich in effect considers more general situation where both polyvectors and polydifferential operators take values in a graded Lie algebra.

Duflo's isomorphism 323

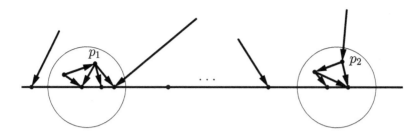

Figure 20.2: The graphs, representing the elements $\mathcal{U}_1^\pi(P_1) \cup \mathcal{U}_1^\pi(P_2)$.

$\bar{C}_{0,2}^+ \subset \bar{C}_{2,0}$, see Fig. 18.2, under the natural projection $\bar{C}_{n,m} \to \bar{C}_{2,0}$. These graphs are characterised by the fact that the points p_1 and p_2 lie infinitesimally close to the boundary straight line $\mathbb{R}^1 \subset \bar{\mathcal{H}}$, the corresponding collapsing clusters of points are inside the thin circles in the figure, and there are no arrows exiting these two encircled regions.

We eventually conclude by observing that one can always homotopy the graph from Fig. 20.1 to a graph from Fig. 20.2 simply by letting p_1 and p_2 move in appropriate direction. In algebraic terms this gives a (co)chain homotopy between the two products. In other words, we have a map $H : (\mathcal{T}_{poly}^\cdot(\mathbb{R}^d)[1][[\hbar]])^{\otimes 2} \to \mathcal{D}_{poly}^\cdot(\mathbb{R}^d)[1][[\hbar]]$ of degree -1 for which

$$\mathcal{U}_1^\pi(P_1 \wedge P_2) - \mathcal{U}_1^\pi(P_1) \cup \mathcal{U}_1^\pi(P_2)$$
$$= H(d_\pi(P_1), P_2) \pm H(P_1, d_\pi(P_2)) \pm d_B(H(P_1, P_2)), \qquad (20.3)$$

where P_1, P_2 are some polyvector fields on $\mathbb{R}^d$ and d_π, d_B are the differentials in the tangent algebras (see formula (20.2)). In particular, when $P_1 = f$, $P_2 = g$ are Casimir functions, then the first two terms on the right-hand side of the formula (20.3) vanish; on the other hand $H(f,g) = 0$ since the degree of H is -1, and we get:

$$\mathcal{U}_1^\pi(fg) = \mathcal{U}_1^\pi(f) \cup \mathcal{U}_1^\pi(g).$$

It remains to observe that the $\cup$-product in this case is just $\mathcal{U}_1^\pi(f) \star \mathcal{U}_1^\pi(g)$.

20.6 Comparison of Kontsevich's and Duflo's isomorphisms

We have just shown that the map $I_{alg}^{-1} \circ I_T : S(\mathfrak{g}) \to U\mathfrak{g}$ induces an isomorphism between the centres of the Poisson algebra on the left and the universal enveloping algebra on the right; here I_T is the restriction to $S(\mathfrak{g})$ of the Kontsevich's tangent morphism $\mathcal{U}_1^\pi$. Let us now outline the method to show that this map is in fact equal to the Duflo's isomorphism.

20.6.1 The graphs and the "wheels"

Below we shall need a more detailed analysis of the graphs that appear in Kontsevich's construction: it turns out that in case of the standard Poisson structure on $\mathfrak{g}^*$ these graphs have a very peculiar properties; more accurately it turns out that the properties of the coefficients mean that the graphs in the definition of $\mathcal{U}_1^\pi$ should have a very special form, which we are about to describe.

Let us recall that the bivector π that determines the Poisson structure on $\mathfrak{g}^*$ is linear with respect to any affine coordinate system on $\mathfrak{g}^* \cong \mathbb{R}^d$ (i.e. with respect to any coordinate system that arises from a choice of basis in $\mathfrak{g}^*$). Hence the corresponding vertices of the first type (i.e. the vertices into which we substitute the bivector) in any graph that gives a nonzero contribution to the formula, should have two arrows originating in them and no more than one arrow abutting into them; otherwise the corresponding operator $\mathcal{U}_\Gamma$ would vanish. Taking this into account, consider the terms $\mathcal{U}_\Gamma(f, \pi, \ldots, \pi)$ that appear in the formula for $\mathcal{U}_1^\pi(f)$: observe that since as a result we should obtain a 0-differential operator, i.e. a function, there should be no vertices of the second type in Γ.

Let $n + 1$ be the total number of vertices in Γ, so $\Gamma \in G_{n+1,0}$. In this graph there should be one vertex with empty *star* (i.e. with no arrows leaving it), this is the vertex where f will stand; we will denote it by p_0. All the other vertices of Γ (all these vertices are of the first type) will have two arrows leaving them and no more than one arrow abutting into them; let us denote these vertices by $p_1, \ldots, p_n$. Remark that the total number of arrows in Γ is $2n$. We claim that exactly one of the pair of arrows leaving each of the vertices $p_1, \ldots, p_n$

should go into the vertex p_0. Indeed, if they both go to p_0, then $\mathcal{U}_\Gamma(f,\pi,\ldots,\pi) = 0$ due to the skew symmetry of π. On the other hand if both arrows that exit one of the vertices $p_1,\ldots,p_n$ end up in some of the vertices from the same set, then for the remaining $2n-2$ arrows that leave these vertices there remain only $n-1$ spots, where they can reach: $n-2$ vertices of the set $p_1,\ldots,p_n$ and p_0. Since each of the vertices $p_1,\ldots,p_n$ can receive no more than 1 arrow, this means that at least n arrows leaving $n-1$ vertices should end in p_0. By Dirichlet's principle this means that there should be a vertex, from which both arrows go to p_0, and hence the corresponding operator vanishes again.

We conclude that one of the arrows leaving each of the vertices $p_1,\ldots,p_n$ should go to p_0 and the other one goes to one of the vertices $p_1,\ldots,p_n$. Taking the chains of the vertices p_i, connected by the arrows we end up by concluding that the graph Γ is in effect a collection of "wheels" centred at the vertex p_0, see Fig. 20.3, where one of the wheels is drawn with solid arrows, and the other one in

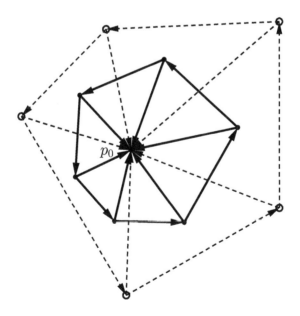

Figure 20.3: "Wheels" in the formula for $\mathcal{U}_1^\pi(f)$.

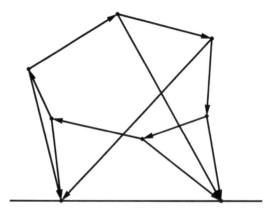

Figure 20.4: "Axle-less wheel" that appear in the formula for Kontsevich's $\star$-product.

dashed lines. I.e. a *wheel with n spokes* is an oriented graph with one vertex in the "center" (axis of the wheel) and an oriented n-gon, surrounding it; every vertex of an n-gon has exactly two arrows leaving it: one of these arrows goes to the center and the other one connects it with the next vertex on the "rim" of the wheel.

Remark 20.7. Similar "wheels" are ubiquitous in the studies of Kontsevich's product on the dual space of Lie algebras; graphs of this kind (see Fig. 20.4) appear naturally in the formula for the $\star$-product in this case. In effect, every graph that enters this formula is equal to a union of graphs that are either of this form, or are formed from a binary tree whose leaves are glued to the two vertices of the second type on the boundary line. Moreover, one can show that although in many situations "wheels" are redundant (see Exercise E.7.9), one cannot avoid them in the universal formula, i.e. in the formula valid for the Poisson bivectors of generic form, see [Di15].

20.6.2 The exponent

Let us eventually show that the structure of the formula for $\mathcal{U}_1^\pi(f)$ is similar to the Duflo's formula (in fact these formulas are identical, but here we will not give an accurate proof of this fact, only a sketch).

To this end we consider the result of application of a "wheel" to a function $f \in S(\mathfrak{g})$. Namely, the following is true:

Proposition 20.8. *The contribution of the n-spoked wheel graph to $\mathcal{U}_1^\pi(f)$ is equal to the action on f of the differential operator, induced by the polynomial $Tr(ad_X^n)$ (see Section 20.3 for the definitions).*

Proof. Let $X = x^\alpha e_\alpha$ for some basis $e_1, \ldots, e_N$ of $\mathfrak{g}$; let $C_{\alpha\beta}^\gamma$ be the corresponding structure constants:

$$[e_\alpha, e_\beta] = C_{\alpha\beta}^\gamma e_\gamma.$$

Then we have the following formula:

$$Tr(ad_X^n) = \sum_{\bar\alpha, \bar\beta} x^{\alpha_1} x^{\alpha_2} \ldots x^{\alpha_n} C_{\alpha_1 \beta_2}^{\beta_1} C_{\alpha_2 \beta_3}^{\beta_2} \ldots C_{\alpha_n \beta_1}^{\beta_n},$$

where $\bar\alpha, \bar\beta$ are the multi-indices

$$\bar\alpha = (\alpha_1, \alpha_2, \ldots, \alpha_n), \quad \bar\beta = (\beta_1, \beta_2, \ldots, \beta_n)$$

for $\alpha_i, \beta_j = 1, \ldots, N$. So the operator induced by this polynomial on f is

$$Tr(ad_X^n)(f) = \sum_{\bar\alpha, \bar\beta} C_{\alpha_1 \beta_2}^{\beta_1} C_{\alpha_2 \beta_3}^{\beta_2} \ldots C_{\alpha_n \beta_1}^{\beta_n} \partial_{\alpha_1} \partial_{\alpha_2} \ldots \partial_{\alpha_n}(f).$$

On the other hand a straightforward inspection of definitions shows that the operator, induced by an n-spoked wheel has the same form. $\square$

Let us make a pair of observations: first, *the wheels with odd number of spokes do not contribute to the final formula* since the corresponding weighted sums cancel out. This follows from the skew symmetry of the weights corresponding to these pictures with respect to the map $z \mapsto -\bar z$, $\mathcal{H} \to \mathcal{H}$. Second, *the graph consisting of several wheels corresponds to the action of the composition of the corresponding operators $Tr(ad_X^n)$ on f.*

Taking into consideration the combinatoric structure of the weights that appear in Kontsevich's formula, we conclude that the following claim holds: *the map I_T is associated with the formal power series*

$$S_1(X) = \exp\left(\sum_{k=0}^{\infty} a_{2k} Tr\left(ad_X^{2k}\right)\right) \qquad (20.4)$$

in the same way Duflo's homomorphism is associated with $j(X)$. Here a_{2k} are certain real coefficients. In previous notation, we have

$$I_T(f) = \widehat{\exp}\left(\sum_{k=0}^{\infty} a_{2k} Tr\left(ad_X^{2k}\right)\right)(f),$$

where $\widehat{\cdot}$ denotes the differential operator associated with a power series; recall that I_T denotes the "algebraic part" of $\mathcal{U}_1^T$. Kontsevich explains why this formula is sufficient to demonstrate that the map $I_{alg}^{-1} \circ I_T$ is equal to the Duflo isomorphism: in effect by an accurate inspection of definitions in Section 20.3 one can show that Duflo's isomorphism is given by composition of the symmetrisation map with the action of the differential operator associated with the following formal power series:

$$S_2(X) = \exp\left(\sum_{k=0}^{\infty} \frac{B_{2k}}{4k(2k)!} Tr\left(ad_X^{2k}\right)\right); \qquad (20.5)$$

here B_{2k} are the even Bernoulli numbers; so we need to show that

$$a_{2k} = \frac{B_{2k}}{4k(2k)!}.$$

To prove this, assume that the power series S_1 and S_2 do not coincide. In this case we get a universal automorphism of the algebra $S(\mathfrak{g})^G$, given by the same power series for all Lie algebras. Applying this automorphism to $\mathfrak{gl}_n$ with n large enough we see that this can only happen when the automorphism is equal to the identity and so $S_1(X) = S_2(X)$. A more detailed proof of this statement was given later by van den Bergh and Calaque, see [CvdB10] and references in this paper; their proof was partly based on explicit computation of the coefficients a_{2k}. In particular, they showed that the formula (20.4) coincides with the Duflo isomorphism (i.e. that a_{2k} is equal to $\frac{B_{2k}}{4k(2k)!}$) by direct computations.

Exercises 7: Kontsevich's quantisation: Properties and applications

E.7.1 Check, that the operator $\tilde{\nabla}$ in Equation (19.6) satisfies the Leibniz rule with respect to the fibrewise $\star$-product in $M^{coord} \times \mathbb{R}^d_{formal}$. **Hint:** recall that the $\star$-product in the fibre $\mathbb{R}^d_{formal}$ of $M^{coord} \times \mathbb{R}^d_{formal}$ over a point $x = j_p^\infty(F) \in M^{coord}$ is given by the Maurer-Cartan element

$$B = \sum_{k=1}^{\infty} \frac{\hbar^k}{k!} \mathcal{U}_k(\underbrace{\hat{\pi}(x), \ldots, \hat{\pi}(x)}_{k \text{ times}}).$$

E.7.2 Prove that locally the curvature of connection ∇ (see formula (19.7)) is equal to the commutator with the 2-forms $s_\alpha^* F_{\hat{\pi}}$ on $U_\alpha \subseteq M$ (here s_α are the local sections) with values in $\mathbb{R}^d_{formal}$, where $F_{\hat{\pi}}$ is a $\mathbb{R}^d_{formal}$-valued 2-form on M^{coord} given by

$$F_{\hat{\pi}}(\xi, \eta) = \sum_{k=0}^{\infty} \frac{\hbar^k}{k!} \mathcal{U}_{k+2}(\omega_{MC}(\xi), \omega_{MC}(\eta), \underbrace{\hat{\pi}(x), \ldots, \hat{\pi}(x)}_{k \text{ times}})$$

for any two vectors $\xi, \eta \in T_{j_p^\infty(F)} M^{coord}$. Also check that $s_\alpha^* F_{\hat{\pi}} = s_\beta^* F_{\hat{\pi}}$ on the intersection of U_α and U_β, so that there exists a well-defined E-valued 2-form F_∇ on M such that $\nabla^2(f) = [F_\nabla, f]$.

E.7.3 Let $(V, \{\ell_n\})$ be an L_∞-algebra with structure maps ℓ_n, $n \geq 1$ and $\varpi \in \mathfrak{MC}(V)$ a Maurer-Cartan element in V. Put:[75]

$$\ell_n^\varpi(v_1, \ldots, v_n) - \sum_{k=0}^{\infty} \frac{1}{k!} \ell_{k+n}(v_1, \ldots, v_n, \underbrace{\varpi, \ldots, \varpi}_{k \text{ times}}).$$

Show that $(V, \{\ell_n^\varpi\})$ is again an L_∞-algebra with respect to the structure maps ℓ_n^ϖ, $n \geq 1$.

---

[75] Recall that we assume that all the infinite sums appearing in this and other similar formulas are either finite, or converge with respect to some algebraic topology, for example with respect to the topology of formal power series: this is the case of the formulas in Exercises E.7.1 and E.7.2.

E.7.4 Let $\mathcal{F} : (V, \{\ell_n\}) \Rightarrow (W, \{\tilde{\ell}_n\})$ be an L_∞-map of L_∞-algebras. Let $\varpi \in \mathfrak{MC}(V)$ and $\mathcal{F}(\varpi) \in \mathfrak{MC}(W)$ its image under $\mathcal{F}$ (see Proposition 17.5). Show that the formulas

$$\mathcal{F}_n^\varpi(v_1,\ldots,v_n) = \sum_{k=0}^\infty \frac{1}{k!} \mathcal{F}_{k+n}(v_1,\ldots,v_n,\underbrace{\varpi,\ldots,\varpi}_{k \text{ times}})$$

determine an L_∞-map

$$\mathcal{F}^\varpi : (V, \{\ell_n^\varpi\}) \Rightarrow (W, \{\tilde{\ell}_n^{\mathcal{F}(\varpi)}\}).$$

Moreover, if $\mathcal{F}$ is a quasi-isomorphism (i.e. if $\mathcal{F}_1$ induces an isomorphism in cohomology), then so is $\mathcal{F}^\varpi$.

E.7.5 Let $\mathfrak{gl}_n$ be the Lie algebra of matrices (over $\mathbb{C}$ or $\mathbb{R}$) with respect to the matrix commutator, and let $e_{ij} \in \mathfrak{gl}_n$ be the canonical generators (matrix units) of $\mathfrak{gl}_n$; we shall use the same notation for e_{ij} regarded as elements of $S(\mathfrak{gl}_n)$. Let L be the matrix $L = (e_{ij}) \in \mathrm{Mat}_n(S(\mathfrak{gl}_n))$. Prove that the subalgebra $S(\mathfrak{gl}_n)^{GL_n}$ of Ad_{GL_n}-invariant polynomial functions on $\mathfrak{gl}_n^*$ is freely generated by the functions

(i) $\sigma_k(L) \in S(\mathfrak{gl}_n)$, $k = 0,\ldots,n$, equal to the coefficients of the characteristic polynomial of L:

$$\det(L - \lambda \mathbb{1}) = \sum_{k=0}^n (-1)^k \lambda^k \sigma_{n-k}(L).$$

(ii) $t_k(L)$, $k = 0,\ldots,n$, equal to the traces of powers of L:

$$t_k(L) = Tr(L^k) = \sum_{\alpha_1,\ldots,\alpha_k=1}^n e_{\alpha_1\alpha_2} e_{\alpha_2\alpha_3} \cdots e_{\alpha_k\alpha_1}.$$

Here we put $\sigma_0(L) = t_0(L) = 1$ by definition.

E.7.6 Use the results of previous exercise (but do not use Duflo's theorem) to show that the elements $\hat{\sigma}_k(L) = \sigma(\sigma_k(L))$ (where $\sigma : S(\mathfrak{g}) \to U\mathfrak{g}$ is the symmetrisation map) are free generators of the algebra $U\mathfrak{gl}_n^{GL_n}$.

E.7.7 Let $\hat{e}_{ij} \in U\mathfrak{gl}_n$ be the generators, consider the matrix $\hat{L} = (\hat{e}_{ij}) \in \mathrm{Mat}_n(U\mathfrak{gl}_n)$. Prove that the elements

$$\hat{t}_k = Tr(\hat{L}^k) = \sum_{\alpha_1,\ldots,\alpha_k=1}^n \hat{e}_{\alpha_1\alpha_2}\hat{e}_{\alpha_2\alpha_3}\cdots\hat{e}_{\alpha_k\alpha_1}$$

are free generators of the centre of $U\mathfrak{gl}_n$.

E.7.8 Show that the operator $\mathcal{U}_\Gamma(P_1, P_2)$, associated with the graph Γ with no internal edges (i.e. no edges, connecting the vertices of the first type) is up to the coefficient equal to $\chi(P_1 \wedge P_2)$, where χ is the Hochschild-Kostant-Rosenberg map. Let now p_1 and p_2 collapse to a single point on the boundary of $\bar{C}_{2,m}$, and put W_Γ to be the normalised integral of the form $\bigwedge_{e \in E_\Gamma} d\phi_e$ over the fibre of the forgetful map $\bar{C}_{2,m} \to C_2 \subset \bar{C}_{2,0}$, which sends the graph to the pair of infinitesimally close points p_1, p_2 (see the formula (18.3)). Check that the weight W_Γ is such that $W_\Gamma \mathcal{U}_\Gamma(P_1, P_2) = \chi(P_1 \wedge P_2)$.

E.7.9 Let $\mathfrak{g}$ be a nilpotent Lie algebra; prove that the Kontsevich's formula for $\star$-product on $\mathfrak{g}^*$ in this case does not contain "wheels", see Fig. 20.4.

E.7.10 Find the image of the generators $\sigma_k(L)$ and $t_k(L)$ of $S(\mathfrak{gl}_n)^{GL_n}$ for $k \leq 3$ under Duflo's isomorphism $S(\mathfrak{gl}_n)^{GL_n} \cong U\mathfrak{gl}_n^{GL_n}$.

A Operads: History and definitions

The purpose of this and the following appendix is to explain the renowned Tamarkin's proof of Kontsevich's formality theorem, announced by Dmitry Tamarkin in the paper "Another proof of M. Kontsevich formality theorem" published in the online arxiv: https://arxiv.org/abs/math/9803025 (we couldn't find any trace of this text being published in a refereed journal ever since). A little later a streamlined version of this reasoning was published by Vladimir Hinich, see [Hi03]. It is based on a clever application of the operad theory: it turns out that the homotopy Gerstenhaber algebras operad has an important property, that reduces the proof of Kontsevich's formality theorem to the computation of certain cohomology spaces, easily done "by hand" (but on expense of the difficult theorem by Kazhdan and Etingof, which describes the quantisations of Lie bialgebras, see Theorem B.12). In order to give meaning to this statement, we begin with a brief outline of the operads theory, just sufficient for the understanding of Tamarkin's reasoning. The reader, who would like to get a deeper insight into the operads and their role in modern Mathematics and Mathematical Physics is kindly referred to more specialised texts, see [Sm01], or [KM95]; a more algebraic approach can be found in the appendix to Loday's book on cyclic homology [Lo98] and in the fundamental paper by Ginzburg and Kapranov [GK95]; if you look for some insight on the role of operads in modern Physics you can look into the book [MSS02]. We should at once warn the reader however, that in our treatment of operads we pay very little attention to the existing variation of definitions, categorification of the properties, relation with homotopy theory, graph complexes and other modern developments of the theory. We also give preference to the clearness and visual interpretation of constructions over rigorousness of definitions and detailed proofs.

A.1 Topological preliminaries (a bit of history)

Just as the theory of strong homotopy algebra structures, see Section 16, operad theory originated from the study of the loop spaces. Recall (see Definition 16.1), that for a pointed space $(X, *)$ we define its based loop space $\Omega_* X$ as

$$\Omega_* X = \{\gamma : [0,1] \to X \mid \gamma(0) = \gamma(1) = *\}.$$

The same construction can be further applied to $\Omega_* X$ endowed with the base point given by the path ω_0, where $\omega_0(t) \equiv *$. This allows one to iterate the construction, giving iterated loop spaces $\Omega_*^2 X$, $\Omega_*^3 X, \ldots$ etc. As we mentioned earlier, below we will usually omit the asterisks in the subscripts, writing ΩX, $\Omega^2 X$ etc. instead of $\Omega_* X$, $\Omega_*^2 X$, etc.

Stasheff's Theorem 16.3 characterises the based loop space of a cell complex X in terms of the A_∞-space structure. Of course an iterated loop space satisfies the same criterion, however not every loop space is equivalent to an iterated loop space. For example, it is well-known that the concatenation of loops in case of $\Omega_*^2 X$, $\Omega_*^3 X$ etc. is a homotopy-commutative operation (see Fig. A.1[76]), which is not at all the case for the usual (noniterated) loops.

So the A_∞-space structure, which detects the loop spaces is not sufficient to distinguish the iterated loop spaces. It turned out that the adequate structures appear if one takes into consideration all possible operations on loop spaces that are produced from the loop compositions in a broad sense. The appropriate construction was due to John Peter May [May72] (at the same period an alternative, but equivalent treatment was suggested by Michael Boardman and Rainer Vogt [BV73]).

In order to explain May's construction, consider the *n-cube spaces*: by definition these are the spaces $E_n(k)$, $k \geq 1$ consisting of all possible ordered configurations of k nonoverlapping right-angled n-dimensional parallelepipeds in the unit n-cube, with their edges parallel to the edges of the cube. Since every such parallelepiped is uniquely determined by the positions of its "first" and "last" vertices (e.g. the south-western and the north-eastern vertices in the case of

---

[76]This figure shows the homotopy, connecting the composition $f * g$ of 2-dimensional loops f and g with $g*f$; the homotopy is achieved by first "shrinking" the domains of f and g inside the square (shaded area in the bottom squares is mapped to the base point, as well as the boundaries of all the domains in the picture), then using the established freedom to change the positions of these domains and finally "blowing up" them back to normal.

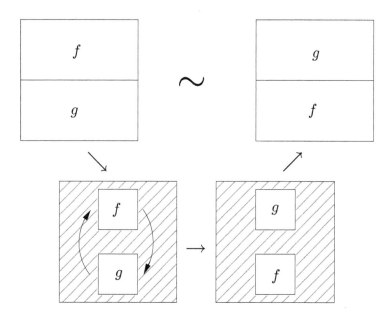

Figure A.1: Homotopy commutativity of the higher loop spaces.

rectangles), we can topologise $E_n(k)$ as a subspace of the $\mathbb{R}^{2kn}$: if

$$\{(a_1, b_1), \ldots, (a_k, b_k),\ a_i, b_j \in \mathbb{R}^n\} \in E_n(k)$$

is such a configuration, where $a_i = (a_i^1, \ldots, a_i^n)$, $b_j = (b_j^1, \ldots, b_j^n)$ and $0 \leq a_i^m \leq b_i^m \leq 1$ for all i, m, then the i-th parallelepiped in the configuration is determined by the inequalities:

$$P_i = \{(x^1, \ldots, x^n) \in \mathbb{R}^n \mid a_i^m \leq x^m \leq b_i^m,\ m = 1, \ldots, n\}.$$

Of course, not every k-tuple of pairs (a_i, b_i) will determine a collection of parallelepipeds. The condition on the configuration of (a_i, b_i) we need, would look like the noncompatibility of inequalities: for all i, j there doesn't exist $x = (x^1, \ldots, x^n) \in \mathbb{R}^n$ for which the systems of inequalities

$$a_i^m \leq x^m \leq b_i^m,\ m = 1, \ldots, n \text{ and } a_j^m \leq x^m \leq b_j^m,\ m = 1, \ldots, n$$

hold simultaneously. Remark that the spaces $E_n(k)$ are homotopy equivalent to the configuration spaces $Config_n(k)$ of k distinct points in $\mathbb{R}^n$: the map $E_n(k) \to Config_n(k)$ can be obtained by sending each parallelepiped to its center. Also observe that the group S_k acts naturally on $E_n(k)$ by exchanging the order of the parallelepipeds.

There are two important operations on the small cubes spaces, that play crucial role in the May's criterion of the iterated loop spaces and eventually in the operad theory. First, given an element α in $E_n(k)$ and a collection of elements $\beta_1 \in E_n(\ell_1), \ldots, \beta_k \in E_n(\ell_k)$ we can construct an element $\gamma(\alpha; \beta_1, \ldots, \beta_k) \in E_n(K)$, where $K = \ell_1 + \cdots + \ell_k$. The construction is very natural: the parallelepipeds $(a_j^{\ell_i}, b_j^{\ell_i})$, $j = 1, \ldots, \ell_i$ from β_i are sent to the parallelepipeds $(u_i(a_j^{\ell_i}), u_i(b_j^{\ell_i}))$, where u_i is the unique affine map that identifies the unit n-cube with the parallelepiped (a_i, b_i) from α preserving orientations of all the edges; it is clear that this construction intertwines the natural action of the permutation group S_k on $E_n(k)$ with its action on the product $E_n(\ell_1) \times \cdots \times E_n(\ell_k)$ by permutations of the factors.

Speaking more rigorously we can say that γ is a system of maps:

$$\gamma: E_n(k) \times_{S_k} (E_n(\ell_1) \times \cdots \times E_n(\ell_k)) \to E_n(\ell_1 + \cdots + \ell_k), \quad (A.1)$$

for all k and all $\ell_1, \ldots, \ell_k$. For the most of time we will not use separate notations for the individual maps from this system; in effect one can treat them as a unique map from a suitable space. If necessary γ_k will denote the map γ with $E_n(k)$ on the left.

The second map relates the spaces $E_n(k)$ with the iterated loops $\Omega^n X$: given an element $\alpha \in E_n(k)$ and k elements $f_1, \ldots, f_k \in \Omega^n X$ we can define an iterated loop $m(\alpha; f_1, \ldots, f_k) \in \Omega^n X$ (a k-fold composition of loops, induced by α): $m(\alpha; f_1, \ldots, f_k)$ is just the map that is equal to the composition $f \circ u_i^{-1}$ on $u_i(I^n)$ (recall that u_i is the unique affine isomorphism between the unit n-cube and (a_i, b_i) parallelepiped in α, which we used earlier) for all $i = 1, \ldots, k$. The rest of the cube is mapped by $m(\alpha; f_1, \ldots, f_k)$ to the basepoint, see Fig. A.2, which represents the composition $m(\alpha; f_1, \ldots, f_4)$ of an element α from $E_2(4)$ with four double loops. One more time we are not going to distinguish the maps m for different k. Clearly, this map is invariant with respect to the S_k action as before, so we can write it as

$$m: E_n(k) \times_{S_k} (\Omega^n X)^{\times k} \to \Omega^n X. \quad (A.2)$$

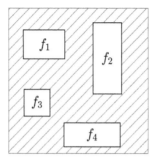

Figure A.2: The action of the small cubes on loop spaces: the shaded area goes to the base point.

An important property of these two maps is their compatibility: for all possible $k, \ell_1, \ldots, \ell_k$ the following diagram is commutative:

$$\begin{array}{ccc}
E_n(k) \times (E_n(\ell_1) \times \cdots \times E_n(\ell_k)) \times (\Omega^n X)^{\times K} & \xrightarrow{\gamma \times 1_{\Omega^n X}^{\times K}} & E_n(K) \times (\Omega^n X)^{\times K} \\
\downarrow & & \\
E_n(k) \times ((E_n(\ell_1) \times (\Omega^n X)^{\times \ell_1}) \times \cdots \times (E_n(\ell_k) \times (\Omega^n X)^{\times \ell_k})) & & m \\
{\scriptstyle 1_{E_n(k)} \times m^{\times k}} \downarrow & & \\
E_n(k) \times (\Omega^n X)^{\times k} & \xrightarrow{m} & \Omega^n X
\end{array}$$

(A.3)

Here we used the notation $K = \ell_1 + \cdots + \ell_k$ for the sake of brevity; the upper vertical arrow on the left is just the rearrangement of the factors in a Cartesian product. The equivariance properties of these maps with respect to the action of permutation groups can be traced here to the equivariance of the whole diagram with respect to the action of the subgroup of "block" permutations in S_K, generated by S_k and $S_{\ell_1}, \ldots, S_{\ell_k}$: one first permutes the elements inside the blocks of lengths $\ell_1, \ldots, \ell_k$ and then permutes the blocks as a whole.

It is easy to see that a similar diagram with $(\Omega^n X)^{\times K}$ replaced by $E_n(s_1) \times \cdots \times E_n(s_K)$ and with γ instead of m in proper places will also be commutative, so we shall have the following "associativity condition" (if we suppress the permutation of the Cartesian factors from our notation):

$$\gamma_K \left(\gamma_k \times \prod_{i=1}^{K} \mathbb{1}_{E_n(s_i)} \right) = \gamma_k(\gamma_{\ell_1} \times \cdots \times \gamma_{\ell_k}). \qquad (A.4)$$

Systems of topological spaces $\mathscr{E} = \{E(k)\}$ equipped with the action of permutation groups S_k and maps $\gamma = \gamma_\mathscr{E}$ similar to (A.1), that satisfy the associativity condition (A.4) were called *operads* or *topological operads* by May. If a topological space T admits the system of maps

$$m = m_T : E(k) \times_{S_k} T^{\times k} \to T$$

(c.f. (A.2)) that satisfies the condition (A.3), then it is said that $\mathscr{E}$ *acts on* T, or that T *is an* $\mathscr{E}$-*algebra*. Thus we can say that *the spaces* $\{E_n(k)\}_{k \geq 1}$ *make up an operad, called the* **little** *n*-**cubes operad**, *denoted by* $\mathscr{E}_n$.[77] *Any n-loop space* $\Omega^n X$ *is an algebra over* $\mathscr{E}_n$.

It turns out that this seemingly trivial observation is a clue to the solution of our main question, the characterisation of iterated loop spaces. Namely, in his seminal paper, J.P. May proved the inverse statement:

Theorem A.1 (May, 1974). *If a cell space T is homotopy equivalent to an n-loop space $\Omega^n X$, then it can be endowed with a structure of an $\mathscr{E}_n$-algebra. Moreover, the property of being an $\mathscr{E}_n$-algebra characterises n-loop spaces up to a homotopy equivalence: T is homotopy equivalent to some space $\Omega^n X$ iff it is an $\mathscr{E}_n$-algebra.*

This result matches well with the results of Section 16.2: in effect one can show that being A_∞-space is equivalent to being a $\mathscr{E}_1$-algebra. More generally if one can replace the spaces $E_n(k)$ by homotopy-equivalent cell spaces $F'_n(k)$ that form an operad on their own, so that the composition maps will survive (up to homotopy) in the process, then the result of Theorem A.1 will hold for $\mathscr{E}'_n(k)$ too.[78] Theorem A.1 has many interesting and important topological corollaries and generalisations; in particular one can let n in the original

---

[77] Instead of parallelepipeds, one can consider small balls inside the unit n-dimensional disc and repeat all these constructions; the result will be called **little disc operad**; it is clearly homotopy equivalent to $\mathscr{E}_n$.

[78] In particular, little discs (see footnote 77) work just as well, as little cubes.

question go to the infinity, i.e. to speak about *infinite loop spaces*; May's criterion works well in that case as well, although the operad $\mathscr{E}_\infty$ that detects such spaces is not constructed in as explicit way as $\mathscr{E}_n$ for finite n.

A.2 Algebraic operads

One can say that since their inception in 1974 operads were an important (albeit not central) subject of studies in Topology. However, operads were totally neglected in the most other branches of Mathematics; it was a general opinion that this theory is unnecessarily difficult and can be of some use only in Algebraic Topology, where similar abstract constructions are quite widespread.

However this common point of view changed rapidly, when in the end of the 1980s and the beginning of the 1990s strong homotopy algebras and operads made their way into Mathematical Physics: it turned out that many difficult relations that came from the Perturbative Field Theories can be organised in a short and concise way if one regards them as the equalities satisfied by the structure maps in a strong homotopy algebra. As it was known from Topology, homotopy algebra structures usually appear from the action of suitable operads (just as the A_∞-space structure on a loop space is related with the $\mathscr{E}_1$-algebra structure). Hence physicists' interest towards the operads was quite natural, genuine and earnest, and this fact brought forward the so-called "Renaissance of the Operads" in the 1990s. One can say that the operads' theory was rediscovered, or rather reinvented in that time, as many modern definitions and concepts (such as the Koszul property, etc.) were eventually coined in the 1990s.

In this section we finally give an explicit definition of operads, phrased algebraically (this approach in particular, allows one give the dual definition of co-operads and comodules over them); we also outline the major results and theorems from the operad theory, in particular we speak about the resolutions of operads and other features of operads in the context of Homological Algebra. After this we shall talk about one specific kind of operads that play an important role in Tamarkin's proof: we mean quadratic operads, in particular, Koszul operads. Our exposition of this subject is therefore primarily based on the seminal Ginzburg and Kapranov's paper [GK95], where the concepts of quadratic operads and Koszul property were introduced. We also use the wonderful paper by Hinich [Hi03], which is a primary source for our exposition in this and the next sections. Let us once again warn the reader that in this book we almost completely neglect the vast field of research and terminology that has grown around the operads theory since the beginning of the century; an interested reader should consult original papers, some of which we cite in the bibliography.

A.2.1 The general operads theory: Definitions and examples

In what follows we will restrict our attention to the operad structures that are defined on vector spaces. So let us fix $\Bbbk$ to be the ground field, $\operatorname{char} \Bbbk = 0$ (usually $\Bbbk = \mathbb{R}$ or $\mathbb{C}$). Below (unless stated otherwise) all vector spaces will be considered over $\Bbbk$, all morphisms will be $\Bbbk$-linear and all tensor products are taken over $\Bbbk$. One can say that we work in the strict monoidal category $\underline{Vect_\Bbbk}$ of $\Bbbk$-vector spaces.[79]

Further, we shall say that *an $\mathbb{S}$-object in $\underline{Vect_\Bbbk}$* or *$\mathbb{S}$-vector space* for short is the collection $\mathcal{X} = \{X(n)\}_{n \geq 1}$ of vector spaces $X(n)$, such that for every $n \geq 0$, the space $X(n)$ is equipped with the structure of an S_n-module. The class of $\mathbb{S}$-vector spaces and action-preserving linear maps between their elements has a natural structure of category, which we shall denote $\underline{\mathbb{S}Vect_\Bbbk}$; since one can regard every vector space as S_1-module, $\underline{Vect_\Bbbk}$ can be regarded as a subcategory of $\underline{\mathbb{S}Vect_\Bbbk}$.

---

[79] If you are not familiar with the terminology of category theory, you can ignore this and the following similar remarks altogether. However, the general idea is pretty simple: strict monoidal category is a category equipped with an analog of tensor product, satisfying the usual coherence conditions **except the commutativity of the tensor product**. Most part of the theory in this section is applicable in a word-for-word manner to any monoidal category, e.g. to the category $\underline{Top}$ of topological spaces equipped with cartesian product.

It turns out that $\underline{\mathbb{S}Vect_\Bbbk}$ is a strict monoidal category, the "tensor product" bifunctor being given by the formula[80]

$$(\mathcal{X}\circ\mathcal{Y})(n) = \bigoplus_{k, m_1+\cdots+m_k=n} [X(k)\otimes Y(m_1)\otimes\cdots\otimes Y(m_k)]\otimes_{G_{k,(m_1,\ldots,m_k)}} \Bbbk[S_n] \qquad (A.5)$$

for any $\underline{\mathbb{S}Vect_\Bbbk}$-spaces $\mathcal{X}$, $\mathcal{Y}$. Here $G_{k,(m_1,\ldots,m_k)}$ denotes the group of "block permutations" induced by S_k and $S_{m_1},\ldots,S_{m_k}$; this is the group of permutations of $m_1+\cdots+m_k = n$ elements, consisting of the permutations, which first change the order inside the blocks of lengths $m_1,\ldots,m_k$ of $(1,\ldots,n)$, and then permute these blocks "as a whole" (reader familiar with the notion of crossed product of groups should recognise it immediately). The group ring $\Bbbk[S_n]$ on the right induces the S_n-action on the resulting space. Observe that this monoidal structure is associative (and in the future we shall omit the parenthesis in iterated ∘-products) but it is not commutative, even if we allow nontrivial isomorphisms.

Given an $\mathbb{S}$-space $\mathcal{X}$, we can define the "Schur functor" $\mathbb{S}(\mathcal{X}) : \underline{Vect_\Bbbk} \to \underline{Vect_\Bbbk}$: for any vector space V we put

$$\mathbb{S}(\mathcal{X})(V) = \bigoplus_n X(n) \otimes_{S_n} (V^{\otimes n}). \qquad (A.6)$$

Let us eventually give the definition of operads in $\underline{Vect_\Bbbk}$:

Definition A.2. *An $\mathbb{S}$-space $\mathscr{O}$ is called an **operad**, more accurately operad in monoidal category $\underline{Vect_\Bbbk}$, if it is equipped with an $\mathbb{S}$-linear map $\gamma : \mathscr{O} \circ \mathscr{O} \to \mathscr{O}$, satisfying the "associativity condition"*

$$\gamma(\gamma \circ \mathbb{1}) = \gamma(\mathbb{1} \circ \gamma). \qquad (A.4')$$

Morphism of operads *is a morphism of $\mathbb{S}$-spaces that commutes with the structure morphisms. The operad is called **unital** if there is an element $e \in \mathscr{O}(1)$, such that*

$$\gamma(\mathbb{1}; a) = a \text{ and } \gamma(a; \underbrace{1, \ldots, 1}_{k \text{ times}}) = a$$

for all $a \in \mathscr{O}(k)$. Below we shall always assume that our operads are unital.

---

[80] From now on the symbol ∘ will be reserved for this operation; the composition of maps will be denoted simply by concatenation of symbols.

Example A.1.

A.1.1 With every vector space V one can associate the operad of its endomorphisms: $\mathscr{E}_V = \{\mathrm{Hom}(V^{\otimes n}, V)\}_{n \geq 1}$. The composition γ of $\mathscr{E}_V$ is given just by substituting the values of homomorphisms from $\mathscr{E}_V(\ell_i)$ into the homomorphism from $\mathscr{E}_V(k)$, and the groups S_n act by permuting the entries of the homomorphisms.

A.1.2 For every $\mathbb{S}$-vector space $\mathcal{V}$ one can define the *free operad* $\mathbb{T}(\mathcal{V})$ by the formula

$$\mathbb{T}(\mathcal{V}) = \bigoplus_k \underbrace{\mathcal{V} \circ \cdots \circ \mathcal{V}}_{k \text{ times}}(n)$$

with the γ map induced from the tautological identifications. The notation is chosen so as to resemble the tensor algebra of a vector space. In particular, just like in tensor algebras, the identification $\mathcal{V} = \mathcal{V}^{\circ 1}$ (i.e. $\mathcal{V}$ raised to the first power with respect to the monoidal structure $\circ$) is a canonical inclusion $u: \mathcal{V} \subseteq \mathbb{T}(\mathcal{V})$ as $\mathbb{S}$-spaces. Observe that just as in tensor algebra case, a 0-fold $\circ$-product of $\mathcal{V}$ inside $\mathbb{T}(\mathcal{V})$ can be identified with the $\mathbb{S}$-space $\underline{*}$, $*(1) = \mathbb{k}$, $*(k) = 0$, $k \neq 1$.

There is a slightly different interpretation of this construction, which is useful in many situations. Namely, let us regard the space $V(k)$ of $\mathcal{V}$ as the linear span of the set of k-adic operations on an abstract object. Let us choose a basis $e_1(k), \ldots, e_p(k)$ in $V(k)$; we can represent every element $e_i(k)$ as a planar graph-tree with one vertex and $k+1$ edges. One of these edges is regarded as an "output" of the operation $e_i(k)$, and k remaining edges are "inputs" of this operation, see Fig. A.3, (i). We label the inputs by $1, 2, \ldots, k$ and use this labelling to construct larger trees from the blocks in $\mathcal{V}$: we paste the blocks from $V(\ell_1), \ldots, V(\ell_k)$ into the inputs of the graphs, representing elements of $V(k)$ and iterate this process few times so that as a result we get a tree with n leaves similar to the one shown at Fig. A.3, (ii), where we omit the labels of its vertices. This tree represents an element in $\mathbb{T}(\mathcal{V})(n)$, more

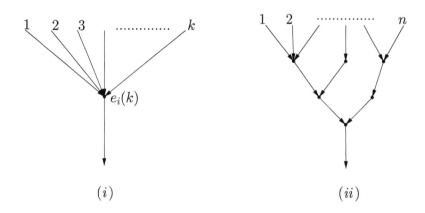

Figure A.3: The graph representation of operads and free operads.

accurately $\mathbb{T}(\mathcal{V})(n)$ is the linear span of such trees modulo the action of permutation group, which is induced from the combination of S_k and $S_{\ell_1}, \ldots, S_{\ell_k}$ action on $V(k), V(\ell_1), \ldots, V(\ell_k)$. This point of view on operads has its benefits and below we will often use it.

It is important to keep in mind that *free operads*, just like free algebras satisfy the universal lifting property: for every $\mathbb{S}$-space $\mathcal{V}$ and every morphism of $\mathbb{S}$-spaces $\phi : \mathcal{V} \to \mathscr{O}$, where $\mathscr{O}$ is an operad, there exists a unique morphism of operads $\mathbb{T}(\phi) : \mathbb{T}(\mathcal{V}) \to \mathscr{O}$, which makes the following diagram commutative

$$\begin{array}{ccc} \mathcal{V} & \xrightarrow{u} & \mathbb{T}(\mathcal{V}) \\ & \searrow{\phi} & \downarrow{\mathbb{T}(\phi)} \\ & & \mathscr{O} \end{array} \qquad (A.7)$$

Definition A.3. *A vector space V is said to bear the structure of **algebra over an operad** $\mathscr{O}$ or simply $\mathscr{O}$-algebra, when it is endowed with a morphism $m = m_V : \mathscr{O} \to \mathscr{E}_V$; if V, W are two algebras over $\mathscr{O}$, then a linear map $f : V \to W$ is a **morphism of $\mathscr{O}$-algebras**,* *if the following diagram is commutative for all n*

$$\begin{array}{ccc} \mathcal{O}(n) & \xrightarrow{m_V} & \mathrm{Hom}(V^{\otimes n}, V) \\ \| & & \downarrow{f_*} \\ & & \mathrm{Hom}(V^{\otimes n}, W) \\ \| & & \uparrow{f^*} \\ \mathcal{O}(n) & \xrightarrow{m_W} & \mathrm{Hom}(W^{\otimes n}, W) \end{array} \quad (A.8)$$

where f_*, f^* are the natural morphisms of the Hom-spaces, induced by f.

Remark A.4.

A.4.1 Definition A.3 can be rewritten as follows: one says, that V is an algebra over $\mathcal{O}$ if there is given a morphism $m_V : \mathbb{S}(\mathcal{O})(V) \to V$ that satisfies the following associativity condition
$$\begin{aligned} m_V(\mathbb{S}(\gamma)(\mathbb{1}_V)) &= m_V(\mathbb{S}(\mathbb{1}_{\mathcal{O}})(m_V)) : \\ \mathbb{S}(\mathcal{O} \circ \mathcal{O})(V) &= \mathbb{S}(\mathcal{O})(\mathbb{S}(\mathcal{O})(V)) \to V, \end{aligned} \quad (A.3')$$

where we used the identification $\mathbb{S}(\mathcal{X})(\mathbb{S}(\mathcal{Y})(V)) = \mathbb{S}(\mathcal{X} \circ \mathcal{Y})(V)$ for all $\mathbb{S}$-spaces $\mathcal{X}$, $\mathcal{Y}$ and any vector space V. The definition of morphisms between $\mathcal{O}$-algebras then amounts to the commutativity of a diagram, similar to (A.8), which we omit. If we "expand" the Schur functor as the sum of tensor products $\mathcal{O}(n) \otimes_{S_n} (V^{\otimes n})$, we will see that the structure of $\mathcal{O}$-algebra is given by a collection of maps
$$m_{V n} : \mathcal{O}(n) \otimes_{S_n} (V^{\otimes n}) \to V, \quad (A.3'')$$
satisfying the "associativity" condition with respect to the structure map γ of $\mathcal{O}$.

A.4.2 For every operad $\mathcal{O}$ and every vector space V, the Schur functor $\mathbb{S}(\mathcal{O})(V)$ is in effect an algebra over $\mathcal{O}$, the morphism m being given by the composition
$$\mathbb{S}(\mathcal{O})(\mathbb{S}(\mathcal{O})(V)) \xrightarrow{\cong} \mathbb{S}(\mathcal{O} \circ \mathcal{O})(V) \xrightarrow{\mathbb{S}(\gamma)(\mathbb{1}_V)} \mathbb{S}(\mathcal{O})(V).$$

We will call it the *free algebra over* $\mathcal{O}$ and denote by $\mathbb{F}_{\mathcal{O}}(V)$. Just as the usual free algebra $\mathbb{F}_{\mathcal{O}}(V)$ satisfies the universal property: for any $\mathcal{O}$-algebra X and any linear map $f : V \to X$ there exists a unique morphism of $\mathcal{O}$-algebras $\mathbb{F}_{\mathcal{O}}(V) \to X$ that extends f.

A.4.3 Also observe that if $f : \mathscr{A} \to \mathscr{B}$ is a morphism of operads, then every $\mathscr{B}$-algebra bears a natural structure of $\mathscr{A}$-algebra: in order to define the structure map one can just use the composition

$$\mathbb{S}(\mathscr{A})(V) \xrightarrow{\mathbb{S}(f)(\mathbf{1}_V)} \mathbb{S}(\mathscr{B})(V) \xrightarrow{m} V.$$

The opposite is also true, at least if $\mathscr{B}(1) \cong \Bbbk$. Indeed, since by assumption every $\mathscr{B}$-algebra is an $\mathscr{A}$-algebra, we can apply it to the free $\mathscr{B}$-algebra $\mathbb{F}_{\mathscr{B}}(\Bbbk)$ of $\Bbbk$. In this case $\mathbb{F}_{\mathscr{B}}(\Bbbk) \cong \bigoplus_n \mathscr{B}(n)$ and the structure map of $\mathscr{B}$-algebra coincides with $\gamma_{\mathscr{B}}$. On the other hand the structure of $\mathscr{A}$-algebra gives a map $\mathscr{A} \circ \mathscr{B} \to \mathscr{B}$ of $\mathbb{S}$-spaces. When restricted to $\mathscr{A}(n) \otimes (\mathscr{B}(1)^{\otimes n}) \cong \mathscr{A}(n)$ we obtain the morphism of operads.

Remark A.5. The same definitions can be applied word-for-word for the operads in other strict monoidal categories; in particular, $\mathscr{E}_n$ is an example of operad in topological category, and $\Omega^n X$ is algebra over $\mathscr{E}_n$ in the strict sense of these words.

Example A.2. Let us give three important examples of operads, that will often appear later:

A.2.1 Let A_n be the free noncommutative associative unital algebra on n variables; one can regard it as the tensor algebra $T^{\otimes}\mathbb{R}^n$ but we will not use this observation here. Let $\mathtt{ASS}(n)$ be the subspace of A_n spanned by the monomials of degree n that contain all n variables; then $\dim \mathtt{ASS}(n) = n!$ and we can identify this space with the group algebra $\Bbbk[S_n]$ by allowing S_n act on $\mathtt{ASS}(n)$ by permutations of variables. We can now introduce the composition

$$\gamma : \mathtt{ASS}(k) \otimes \mathtt{ASS}(\ell_1) \otimes \cdots \otimes \mathtt{ASS}(\ell_k) \to \mathtt{ASS}(\ell_1 + \cdots + \ell_k)$$

by the following rule: in every monomial $x_{i_1} \ldots x_{i_k}$ from $\mathtt{ASS}(k)$ we replace every variable x_p, $p = 1, \ldots, k$ by the given monomial $y_{j_1} \ldots y_{j_{\ell_p}}$ from $\mathrm{\underline{ASS}}(\ell_p)$ and then identify the variables y_s from all these monomials with $x_1, \ldots, x_{\ell_1 + \cdots + \ell_k}$ from $\mathrm{\underline{ASS}}(\ell_1 + \cdots + \ell_k)$. Clearly, this map is associative (as prescribed by Equation (A.4′)) and respects the group action. Thus we obtain an operad $\mathrm{\underline{ASS}}$. Algebras over this operad are precisely the associative algebras over $\Bbbk$; hence the name of this operad is *the operad of associative algebras*. To see this, remark that $\mathrm{\underline{ASS}}(1) = \Bbbk$ and whenever V is an algebra, then we have the structure map

$$\mathrm{\underline{ASS}}(2) \otimes V \otimes V \to V;$$

since $S_2 = \mathbb{Z}/2\mathbb{Z}$ has only two elements, this gives us two maps $m_0 : V \otimes V \to V$ and $m_1 : V \otimes V \to V$ such that $m_1(x, y) = m_0(y, x)$. Putting $m_0(x, y) = x \cdot y \in V$ we get the product on V, which is associative since by the condition (A.3′) on the action we have

$$m(\gamma \otimes \mathbb{1}) = m(\mathbb{1} \otimes m) : \mathrm{\underline{ASS}}(2) \otimes (\mathrm{\underline{ASS}}(2) \otimes \mathrm{\underline{ASS}}(1)) \otimes V^{\otimes 3} \to V$$
$$m(\gamma \otimes \mathbb{1}) = m(\mathbb{1} \otimes m) : \mathrm{\underline{ASS}}(2) \otimes (\mathrm{\underline{ASS}}(1) \otimes \mathrm{\underline{ASS}}(2)) \otimes V^{\otimes 3} \to V;$$

where the right maps of these identities when restricted to m_0 elements in $\mathrm{\underline{ASS}}(2)$ send $x \otimes y \otimes z \in V^{\otimes 3}$ to $(x \cdot y) \cdot z$ and $x \cdot (y \cdot z)$ respectively, while the left maps of these formulas differ just by the action of permutation group, and hence coincide. The opposite is also true: every associative $\Bbbk$-algebra is an algebra over the operad $\mathrm{\underline{ASS}}$, the structure map being given by the n-fold product of the elements in the prescribed order.

A.2.2 Similarly, let $\mathrm{\underline{COM}}$ be the operad such that $\mathrm{\underline{COM}}(n) = \Bbbk$ for all n with the trivial S_n-action and trivial composition law. Then (we recommend that you prove it as an exercise) algebras over $\mathrm{\underline{COM}}$ are precisely the commutative associative $\Bbbk$-algebras. Thus $\mathrm{\underline{COM}}$ is called *the operad of commutative algebras* or simply *the operad of commutative algebras*.

A.2.3 Let now L_n be the free Lie algebra with n generators $x_1, \ldots, x_n$; as vector space L_n is the linear span of expressions

$$[x_{i_1}, [x_{i_2}, [x_{i_3}, \ldots, [x_{i_{p-1}}, x_{i_p}] \ldots]]], \tag{A.9}$$

where
$$x_{i_k} \in \{x_1, \ldots, x_n\}, \text{ for all natural } p \geq 1.$$

We also impose the conditions of anti-symmetricity and the Jacobi identity for the brackets. We put LIE(n) to be the subspace of L_n, spanned by the expressions (A.9) which contain every variable $x_1, \ldots, x_n$ exactly once. The group S_n acts on LIE(n) by permutations of the variables[81] and the operad structure is given by substituting the expressions of the form (A.9) instead of the variables $x_1, \ldots, x_n$. Thus we obtain the operad **LIE**. One can show (again a good exercise) that V is a **LIE**-algebra iff V has the structure of a $\Bbbk$-Lie algebra. Hence the name of **LIE** is *the operad of Lie algebras*.

A.2.2 Homological algebra of operads: Co-operads and resolutions

As we have mentioned earlier, operads can be defined for any (small) strict monoidal category $\underline{A}$. To this end we first introduce the notion of $\mathbb{S}$-objects in this category as the collections $\{C(n)\}$, $C(n) \in Ob(\underline{A})$, such that $C(n)$ is acted upon by the group S_n; let $\mathbb{S}\underline{A}$ denote the category of $\mathbb{S}$-objects in $\underline{A}$. Next, we use the same formula (A.5) to define the $\circ$-product in $\mathbb{S}\underline{A}$ (clearly, one should use the inherent monoidal structure in $\underline{A}$ instead of the tensor product in this formula now). Then the definition of operads in $\underline{A}$ reproduces in a word for word manner the Definition A.2.

In particular one can now define operads in the category $\underline{GrVect_{\Bbbk}}$ of graded vector spaces, or more generally the category $\underline{Ch_{\Bbbk}}$ of chain complexes of $\Bbbk$-vector spaces; in what follows we shall assume that the complexes are $\mathbb{Z}$-graded, this will allow to neglect the difference between chain and cochain settings: every cochain complex $D^{\cdot}$ turns into a chain complex cD. by setting $cD_n = D^{-n}$.

Working with chain complexes allows one introduce into the operads theory many concepts from the Homological algebra, which we are going to do now. Since every vector space can be regarded as a

---

[81] In particular LIE(2) = $\Bbbk$ as a linear space, but the action of S_2 on it is nontrivial; in fact it is given by the sign of the transposition.

chain complex concentrated at degree 0 we have a natural embedding (fully faithful functor) of categories $\underline{Vect_\Bbbk} \to \underline{Ch_\Bbbk}$, in particular every $\underline{Vect_\Bbbk}$-operad is a $\underline{Ch_\Bbbk}$-operad.

Before we proceed, let us briefly recall that $C[n]$ denotes the chain complex C, shifted n times with appropriately changed sign of the differential. Also one should keep in mind that for any two chain complexes A, B we denote by $\operatorname{Hom}(A, B)$ the complex of homogenous linear homomorphisms from A to B, i.e.

$$\operatorname{Hom}(A, B) = \bigoplus_{n \in \mathbb{Z}} \operatorname{Hom}_n(A, B),$$

$$\operatorname{Hom}_n(A, B) = \{f : A \to B \mid f(A_p) \subseteq B_{p+n}\}.$$

The differential is given by commutators with the differentials in A and B. In particular, we obtain the dual chain complex $A^\dagger = \operatorname{Hom}(A, \Bbbk)$; its grading is negative, if the grading of A is positive. The tensor product in $\underline{Ch_\Bbbk}$ is given by the formula

$$(A \otimes B)_n = \bigoplus_{p+q=n} A_p \otimes B_q, \ d(a \otimes b) = d(a) \otimes b + (-1)^{|a|} a \otimes d(b).$$

Observe that the direct sum on the right can be infinite if A and B are not bounded.

When we consider $\mathbb{S}$-objects in $\underline{Ch_\Bbbk}$, we should keep in mind that the S_n-action does in effect depend on the grading. In particular, when shift is applied to an $\mathbb{S}$-chain complex $\mathcal{C}$, the S_n action on $\mathcal{C}.(n)$ should be modified: we put

$$(\mathcal{C}[1]).(n) = \mathcal{C}_{.+1}(n) \otimes \hat{\Bbbk}, \tag{A.10}$$

as the representation of S_n, where $\hat{\Bbbk}$ is the 1-dimensional representation of S_n in $\Bbbk$ by the signs of permutations. Similarly, to define the functors $\operatorname{Hom}(\mathcal{A}, \mathcal{B})$, $\mathcal{A} \otimes \mathcal{B}$ and $\mathcal{A}^\dagger$ for $\mathbb{S}$-chain complexes, one should keep the track of representations of S_n at every stage.

Just like in usual homological algebra one can now speak about resolutions of the operads, homotopy equivalence of operads etc.; the definitions of these objects more or less imitate the definitions of the usual homological algebra. For instance: *free resolution of an operad $\mathcal{O}$ is a free chain complex operad $\mathbb{T}$ with a morphism of operads*

$\mathbb{T} \to \mathscr{O}$ such that on the level of cohomology this morphism turns into an isomorphism.

The most important question now is to prove the existence of resolutions of this sort, and also check, if one can find somewhat smaller resolutions of certain operads. It turns out that the resolution of an operad can be constructed in a manner similar to the well-known *bar-cobar resolution* of an algebra.[82] In order to sketch this construction, we say that *co-operad* is an $\mathbb{S}$-space $\mathscr{C} = \{C(n)\}$ equipped with a *coproduct* map $\bar{\gamma} : \mathscr{C} \to \mathscr{C} \circ \mathscr{C}$ that satisfies the coassociativity condition, dual to (A.4'):

$$(\bar{\gamma} \circ \mathbb{1})\bar{\gamma} = (\mathbb{1} \circ \bar{\gamma})\bar{\gamma}. \qquad (A.11)$$

As before all cooperads that we consider here will have counit, which can be regarded as a morphism into a trivial cooperad $\bar{*}$, where $*(1) = \Bbbk$, $*(k) = 0$, $k \neq 1$.

Remark A.6. If $\mathscr{C}$ is a cooperad (in $\underline{Vect_\Bbbk}$, $\underline{GrVect_\Bbbk}$ or $\underline{Ch_\Bbbk}$), then its dual $\mathscr{C}^\dagger$ (i.e. $\mathscr{C}^\dagger = \{C^\dagger(n)\}$ as $\mathbb{S}$-space) is automatically an operad, the structure map being induced by duality. The opposite, however is not necessarily true, since the range of the map dual to a product might be equal to a completion of the tensor product.

Similarly one defines *coalgebra over a cooperad* $\mathscr{C}$ as a vector space (or a chain complex) W with morphism $\bar{m}_W : W \to \mathbb{S}(\mathscr{C})(W)$, for which the coassociativity condition, dual to (A.3') holds:

$$\begin{aligned} \mathbb{S}(\bar{\gamma})(\mathbb{1}_W)\bar{m}_W &= \mathbb{S}(\mathbb{1}_\mathscr{C})(\bar{m}_W)\bar{m}_W : \\ W &\to \mathbb{S}(\mathscr{C} \circ \mathscr{C})(W) = \mathbb{S}(\mathscr{C})(\mathbb{S}(\mathscr{C}(W))). \end{aligned} \qquad (A.12)$$

The definitions of free operads and free algebras over operads are readily dualised to give the definitions of *free cooperad* $\mathbb{T}^*(\mathcal{C})$, cogenerated by an $\mathbb{S}$-space $\mathcal{C}$

$$\mathbb{T}^*(\mathcal{C}) = \bigoplus_k \underbrace{\mathcal{C} \circ \cdots \circ \mathcal{C}}_{k \text{ times}} \qquad (A.13)$$

(with the coproduct given by interpreting one of the $\circ$ in this formula as "external") and *free coalgebra over a cooperad* $\mathscr{C}$, cogenerated

---
[82] A reader, unfamiliar with this construction, can look it up in [Sm01].

by W, denoted by $\mathbb{F}_{\mathscr{C}}^*(W)$. In this case one puts:

$$\mathbb{F}_{\mathscr{C}}^*(W) = \bigoplus_n (C(n) \otimes W^{\otimes n})^{S_n}. \tag{A.14}$$

Here S_n in the superscript denotes the S_n-invariant part of the space; however *over a characteristic zero field there is no difference between invariants and coinvariants of a finite group action; in particular* $(C(n) \otimes W^{\otimes n})^{S_n} \cong C(n) \otimes_{S_n} (W^{\otimes n})$, *if* char $\mathbb{k} = 0$.

Let now $\mathscr{O}$ be an operad in the category of chain complexes. Then the free co-operad $\mathbb{T}^*(\mathscr{O}[-1])$ bears (in addition to the inherent differential in $\mathscr{O}$) a natural differential, compatible with the coproduct. It is induced by the operad structure map γ of $\mathscr{O}$ (the shift is necessary to make sure that this differential has proper degree). Now the free operad $\mathbb{T}(\mathbb{T}^*(\mathscr{O}[-1])[1])$ (pay attention to the shift again) inherits the differentials from $\mathbb{T}^*(\mathscr{O}[-1])$ and also gets one from the cooperad structure of $\mathbb{T}^*(\mathscr{O}[-1])$. It is the free operad $\mathbb{T}(\mathbb{T}^*(\mathscr{O}[-1])[1])$, equipped with the sum of all these differentials that one calls *the bar-cobar resolution of the operad $\mathscr{O}$*. This is justified by the fact that under natural conditions $\mathbb{T}(\mathbb{T}^*(\mathscr{O}[-1])[1])$ has a functorial operadic morphism into $\mathscr{O}$ which induces an isomorphism in homology.

This construction is rather universal, but the size of the operad $\mathbb{T}(\mathbb{T}^*(\mathscr{O}[-1])[1])$ is extremely large, so one has quite small control over the structure of algebras over it, which is essential for Tamarkin's proof. Hence in order to cut down the size of resolutions we are going to use the notions of Koszul operads, introduced by Ginzburg and Kapranov, see [GK95].

A.2.3 Quadratic operads

In order to describe this construction, we will need a couple more definitions:

Definition A.7. *Let $\mathscr{O}$ be an operad (in $\underline{\text{Vect}_{\mathbb{k}}}$ or $\underline{\text{Ch}_{\mathbb{k}}}$), then an $\mathbb{S}$-subspace $\mathcal{I} \subseteq \mathscr{O}$ is called **ideal in** $\mathscr{O}$, if the composition $\gamma(a; b_1, \ldots, b_k) \in \mathcal{I}$ as soon as one of the elements a or $b_1, \ldots, b_k$ is in $\mathcal{I}$.*

It is easy to see that the quotient operad construction is applicable in this case, giving the quotient operad $\mathscr{O}/\mathcal{I}$ and that the analog of the homomorphism theorem holds for operads (one can say that operadic ideals play the role of ring ideals). In particular, one can speak about the quotient operads of free operads: $\mathcal{O} = \mathbb{T}/\mathcal{I}$. Also just like in the case of rings one can speak about ideals, generated by an $\mathbb{S}$-subspace $\mathcal{R}$ in $\mathcal{O}$: we just put $\mathcal{I} = \langle \mathcal{R} \rangle$ to be the union of all possible iterative compositions in $\mathcal{O}$, where at least one of the elements is from $\mathcal{R}$.

We can now give the following definition:

Definition A.8. *Let $\mathcal{V}$ be an $\mathbb{S}$-vector space (graded or not graded) such that $V(n) = 0$ if $n \neq 2$ and $V(2) = V$; let $\mathcal{R}$ be an $\mathbb{S}$-subspace in $\mathbb{T}(\mathcal{V})$, such that $R(k) = 0$, if $k \neq 3$ and $R(3) = R$. Then we shall say that $\mathcal{O} = \mathbb{T}(\mathcal{V})/\langle \mathcal{R} \rangle$ is* **quadratic operad with generators** V **and (space of) relations** R.

Vice-versa, if $\mathcal{O}$ is an operad, we define an $\mathbb{S}$-space $\mathcal{V}$ so that $V(2) = \mathcal{O}(2)$, $V(n) = 0$, $n \neq 2$. Then by the universal property of free operads we have a canonical morphism of operads $p : \mathbb{T}(\mathcal{V}) \to \mathcal{O}$. Suppose that

(i) the morphism p is surjective;

(ii) $\ker p$ is an ideal in $\mathbb{T}(\mathcal{V})$, generated by an $\mathbb{S}$-subspace $\mathcal{R}$, such that $R(3) = R \subseteq \mathbb{T}(\mathcal{V})(3)$ and $R(k) = 0$, $k \neq 3$.

In this case we will say that $\mathcal{O}$ **is quadratic operad with generators** $V = \mathcal{O}(2)$ **and (space of) relations** R.

Below we shall often denote the $\mathbb{S}$-spaces $\mathcal{V}$ and $\mathcal{R}$ constructed from V and R simply by V and R, i.e. we will write $\mathcal{O} = \mathbb{T}(V)/\langle R \rangle$ etc., when this notation is nonambiguous.

Remark A.9. A reader, familiar with the terminology of category theory, can compare this definition with the following: let $*$ be the trivial operad, i.e. $*(k) = 0$ for all $k \neq 1$, $*(1) = \Bbbk$ with the only possible composition law. Then *the operad $\mathcal{O}$ is quadratic with generators $\mathcal{V}$ and relations $\mathcal{R}$ concentrated in degrees 2 and 3 respectively, iff the following diagram is a pushout (a cocartesian square) in the category*

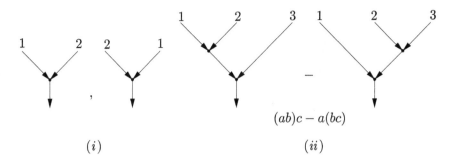

Figure A.4: Generators (*i*) and relations (*ii*) of the operad **ASS**.

of operads:
$$\begin{array}{ccc} \mathbb{T}(\mathcal{R}) & \longrightarrow & \mathbb{T}(\mathcal{V}) \\ \downarrow & & \downarrow p \\ * & \longrightarrow & \mathcal{O} \end{array} \qquad (A.15)$$

Here the left vertical arrow is a canonical projection onto the trivial operad, upper horizontal arrow is the natural morphism of free operad $\mathbb{T}(\mathcal{R})$, induced by inclusion $\mathcal{R} \subseteq \mathbb{T}(\mathcal{V})$ and the lower horizontal arrow is the morphism, induced by the inclusion of the unit in $\mathcal{O}$.

Example A.3. The operads **ASS**, **COM** and **LIE** (see Example A.2) are quadratic. Indeed, take for instance **ASS**: take $\mathcal{V} = \Bbbk^2$ with generators, depicted by two graphs, see Fig. A.4, (*i*), and let the space $R \subset \mathbb{T}(\mathcal{V})$ of relations be generated by the difference of graphs depicted at Fig. A.4, (*ii*) (up to the action of S_3). Here we use the description of free operads in terms of graphs, see Example A.1.2; also observe that we suppress the unit of the free operad in our notation, and just draw the "long branches" in the graphs at Fig. A.4, (*ii*). It is easy to see that the resulting operad is equal to **ASS**. We recommend the reader to find the corresponding presentations for the operads **COM** and **LIE**.

Dually to the definitions above we can define *coideals* in cooperads, as well as *quadratic cooperads*. We recommend the reader to work through these definition on his or her own. To put it short, similarly to Remark A.9, we say that *cooperad $\mathscr{C}$ is quadratic with* **cogenerators** $\mathcal{V} = \{V(n)\}$ where $V(2) = V = \mathcal{O}(2)$, $V(n) = 0$, $n \neq 2$ and

correlations $\mathcal{Q} = \{Q(n)\}$ *with*

$$Q(3) = Q = \mathbb{T}^*(\mathcal{V})(3)/R,\ Q(n) = 0,\ n \neq 3,$$

where R is a S_3-invariant subspace, iff the following diagram is a pullback (a cartesian square) in the category of cooperads:

$$\begin{array}{ccc} \mathscr{C} & \xrightarrow{p} & \mathbb{T}^*(\mathcal{V}) \\ \downarrow & & \downarrow \\ \overline{*} & \longrightarrow & \mathbb{T}^*(\mathcal{Q}) \end{array} \qquad (\text{A}.15')$$

As earlier, the left vertical arrow is just the operadic counit, and the lower horizontal arrow is the inclusion of the unit cooperad. Also, as before we will usually abbreviate our notation simply to V and Q (or V and R, if $Q = \mathbb{T}^*(V)(3)/R$) instead of $\mathcal{V}$ and $\mathcal{Q}$, (or $\mathcal{R}$) when this doesn't cause confusion.

A.2.4 Quadratic duals and Koszul property

Warning! In this section and further we change the point of view on the differentials in chain complexes, so that now our differentials have degree $+1$ and not -1 as before. We expect that the reader can easily make all the necessary amendments.

One of the main constructions, associated with quadratic operads and cooperads is the construction of *quadratic dual operad/cooperad*. From now on we work in the category of graded vector spaces and chain complexes.

Definition A.10.

(*i*) *Let $\mathscr{O}$ be a quadratic operad, let $V = \mathscr{O}(2)$ and $R \subseteq \mathbb{T}(V)(3)$ be its generators and relations. Then the **quadratic dual cooperad of** $\mathscr{O}$ is the quadratic cooperad $\mathscr{O}^{\perp}$ with cogenerators $V[1]$ (shifted space of generators of $\mathscr{O}$) and correlation space $R[2]$.*

(*ii*) *Similarly, if $\mathscr{C}$ is a quadratic cooperad with cogenerators $V = \mathscr{C}(2)$ and correlations $Q = \mathbb{T}^*(V)(3)/R$, $R \subseteq \mathbb{T}^*(V)(3)$, then the **quadratic dual operad of** $\mathscr{C}$ is the quadratic operad $\mathscr{C}^{\perp}$ with generators $V[-1]$ and relations $R[-2]$.*

Observe that in both cases the resulting dual (co)operads are naturally graded, even if the original quadratic (co)operads were not graded.

Remark A.11. In many sources one speaks about *quadratic dual operad* of a quadratic operad; one can obtain such operad by dualising the cooperad $\mathscr{O}^\perp$: we put $\widehat{\mathscr{O}} = (\mathscr{O}^\perp[-1])^\dagger$. In this case $\widehat{\mathscr{O}}$ is a quadratic operad with generators $\widehat{V} = V^\dagger$ and relations $\widehat{R} = R^\perp \subseteq \mathbb{T}(V^\dagger)(2) \subseteq \mathbb{T}(V)(2)^\dagger$. Here $L^\perp$ denotes the annulator of a subspace $L \subset W$ in $W^\dagger$. However, this construction cannot be moved to the case of quadratic dual operad of a quadratic cooperad: this is due to the fact, that dualising operad to get a cooperad is not always possible, see Remark A.6.

On the other hand, in many cases quadratic dual operad $\widehat{\mathscr{O}}$ of an operad $\mathscr{O}$ can be identified with another important operad. For instance consider <u>ASS</u>: in this case $\dim R = 3$, so $\dim \widehat{R} = 6 - 3 = 3$, and we see that $\widehat{R}$ is identified with R, when we identify $\widehat{V}$ with V using a basis. We conclude that <u>ASS</u> coincides with (or rather is isomorphic to) $\widehat{\underline{\text{ASS}}}$,

$$\underline{\text{ASS}} = \widehat{\underline{\text{ASS}}}, \text{ or } \underline{\text{ASS}}^\dagger = \underline{\text{ASS}}[-1]. \quad (A.16)$$

Similarly one can show (a good exercise) that

$$\underline{\text{LIE}} = \widehat{\underline{\text{COM}}}, \text{ or } \underline{\text{LIE}}^\dagger = \underline{\text{COM}}[-1], \quad (A.17)$$

$$\underline{\text{COM}} = \widehat{\underline{\text{LIE}}}, \text{ or } \underline{\text{COM}}^\dagger = \underline{\text{LIE}}[-1]. \quad (A.18)$$

Definition A.12. Let $\mathscr{O}$ be a quadratic operad; we shall say that a (graded) vector space X bears the structure of an $\mathscr{O}_\infty$-**algebra**, if the cofree coalgebra, cogenerated by X over the quadratic dual cooperad $\mathscr{O}^\perp$, i.e. $\mathbb{F}^*_{\mathscr{O}^\perp}(X)$, can be equipped with a codifferential[83] $Q: \mathbb{F}^*_{\mathscr{O}^\perp}(X) \to \mathbb{F}^*_{\mathscr{O}^\perp}(X)$ of degree 1 such that $Q^2 = 0$.

Just as in the case of A_∞ and L_∞ algebras, any such $\mathscr{O}_\infty$-structure Q is uniquely determined by its "Taylor coefficients"; in this case it

---

[83] Here we use the word codifferential in the sense of coalgebra over cooperad: one extends Q to a natural linear endomorphism of $\mathbb{S}(\mathscr{O}^\perp)(\mathbb{F}^*_{\mathscr{O}^\perp}(X))$; then the condition that Q is codifferential means that Q commutes with the structure map $\bar{m}_X: \mathbb{F}^*_{\mathscr{O}^\perp}(X) \to \mathbb{S}(\mathscr{O}^\perp)(\mathbb{F}^*_{\mathscr{O}^\perp}(X))$ of the free coalgebra.

means the restriction of Q to the direct factors in the definition of $\mathbb{F}^*_{\mathcal{O}^\perp}(X)$, which gives the maps Q_n of total degree 1 (c.f. Section 16.3.3, Definition 16.9):

$$Q_n : (\mathcal{O}^\perp(n) \otimes X^{\otimes n})^{S_n} \to X. \qquad (A.19)$$

In particular, every $\mathcal{O}$-algebra X in the category $\underline{Ch_{\Bbbk}}$ is a $\mathcal{O}_\infty$-algebra. In fact, since $\mathcal{O}^\perp(1) = \Bbbk$ and $\mathcal{O}^\perp(2) = \mathcal{O}(2)[1]$ we can put $Q_1 : X \to X$ to be just the differential in X and $Q_2 : \mathcal{O}(2)[1] \otimes X^{\otimes 2} \to X$ to be the map determined by the $\mathcal{O}$-algebraic structure on X (the shift of $\mathcal{O}(2)$ provides the proper degree of Q_2). We also put $Q_n = 0$, $n \neq 1, 2$, and this completes the proof. We recommend the reader to check that the notions of **ASS**$_\infty$- and **LIE**$_\infty$-algebras, coincide with those of A_∞- and L_∞-algebras respectively (see Definitions 16.6 and 16.9 and discussion thereof).

On the other hand, the free operad $\mathbb{T}(\mathcal{O}^\perp[-1])$ can be endowed with a natural differential Q, induced from the cooperad structure on $\mathcal{O}^\perp$. If X is an $\mathcal{O}_\infty$-algebra, then taking the compositions of "Taylor coefficients" Q_n along the graph, representing an element in $\mathbb{T}(\mathcal{O}^\perp)(n)$ (see Fig. A.3, (ii) and also use the identification of invariants and coinvariants of S_n-action) determines the degree 0 maps

$$T(\mathcal{O}^\perp[-1])(n) \otimes_{S_n} (X^{\otimes n}) \to X,$$

which satisfy the associativity condition (hence X is an algebra over the operad $\mathbb{T}(\mathcal{O}^\perp[-1])$) and commute with the differentials in $\mathbb{T}(\mathcal{O}^\perp[-1])$ and X. The opposite is also true. Thus we see that the following is true:

Proposition A.13. *Let $\mathcal{O}$ be a quadratic operad. Then a graded vector space X is an $\mathcal{O}_\infty$-algebra in the sense of the Definition A.12 iff it is an algebra over the operad $\mathbb{T}(\mathcal{O}^\perp[-1])$.*

We will denote the operad $\mathbb{T}(\mathcal{O}^\perp[-1])$ by $\mathcal{O}_\infty$.

We suggest that the reader would again make a pause and use this proposition to check one more time that a space X is an **ASS**$_\infty$-algebra iff it is A_∞-algebra (see Definition 16.8) and that a space L is a **LIE**$_\infty$-algebra iff it is L_∞-algebra (see Definition 16.9).

Since every $\mathcal{O}$-algebra is automatically an $\mathcal{O}_\infty$-algebra, we conclude (see Remark A.4.3), that there exists a natural morphism of operads $e : \mathscr{O}_\infty \to \mathscr{O}$, which commutes with the differentials (the operad $\mathscr{O}$ is by default equipped with a zero differential, and the differential in $\mathscr{O}_\infty$ is Q). Now we can give the main definition of this section:

Definition A.14. *A quadratic operad $\mathscr{O}$ is said to satisfy* **Koszul property**, *or just called* **Koszul operad**, *if the natural morphism $e : \mathscr{O}_\infty \to \mathscr{O}$ induces an isomorphism in cohomology.*

As one sees, a quadratic operad $\mathscr{O}$ is Koszul, iff $\mathscr{O}_\infty$ is a resolution of $\mathscr{O}$. Indeed, $\mathscr{O}_\infty$ is free, so the only condition that remains for it to be resolution is precisely the condition of the Definition A.14. This construction gives a considerably smaller result than the bar-cobar resolution of Section A.2.2.

We will conclude this section by the following important theorem, due to Ginzburg and Kapranov; it constitutes a handy criterion for detecting the Koszul operads. To this end let $\mathscr{O}$ be a quadratic operad; consider the free $\mathscr{O}$-algebra $X = \mathbb{F}_\mathscr{O}(V)$, freely generated by a (graded) vector space V. Let $\mathbb{F}^*_{\mathscr{O}^\perp}(X)$ be the free $\mathscr{O}^\perp$-coalgebra, cogenerated by X. Since X is a $\mathscr{O}$-algebra, it has a natural structure of $\mathscr{O}_\infty$-algebra. So there is a differential Q on $\mathbb{F}^*_{\mathscr{O}^\perp}(X)$. On the other hand, the natural projection $X \to V$ ("projection to the tensor degree 1") can be extended to a map $q : \mathbb{F}^*_{\mathscr{O}^\perp}(X) \to V$, commuting with the differential Q. Then the following is true:

Theorem A.15 (Ginzburg, Kapranov). *The operad $\mathscr{O}$ is Koszul iff the map $q : \mathbb{F}^*_{\mathscr{O}^\perp}(X) \to V$ induces isomorphism in homology for all graded vector spaces V.*

Example A.4. The operad **ASS** is Koszul: for every (graded) vector space V, the free **ASS**-algebra, generated by V is just the tensor algebra $T(V)$. The differential Q on free coalgebra $\mathbb{F}^*_{\mathbf{ASS}^\perp}(T(V))$ is then identified with the usual (co)bar resolution of $T(V)$, and the result follows from Ginzburg and Kapranov's criterion, Theorem A.15.

In a similar manner one can show that the operads **COM** and **LIE** are also Koszul.

The general principle, related with this constructions, i.e. with the $\mathscr{O}_\infty$-algebras and Koszul property for operads, what one may call *meta-principle* behind this theory, is that $\mathcal{O}_\infty$-operad is related with the deformations of $\mathcal{O}$-algebras, see [Kon99] (or the more recent paper [CCN22], where this analogy has been moved to a whole new level) just like A_∞- and L_∞-algebraic structures are related with the deformations of associative algebras and Lie algebras. In particular, every quasi-isomorphism of differential graded $\mathcal{O}$-algebras can have homotopy inverse in the class of $\mathcal{O}_\infty$-homomorphisms. In other words, if an algebraic structure on X is governed by a quadratic operad $\mathcal{O}$, then $\mathcal{O}_\infty$-algebras play the role of strong homotopy algebras over $\mathcal{O}$. We are not going to expand on this subject here. Instead we will use the properties of Koszul operads to prove the formality theorem for polynomial algebra.

B Tamarkin's proof of formality theorem

In this section we give the proof of the Theorem 17.12, suggested by Tamarkin. In our exposition we closely follow the paper by Hinich [Hi03].

B.1 Intrinsic formality: Definition and criterion

Tamarkin's strategy for proving formality's theorem is to deduce this statement from the general properties of Koszul operads. Namely, fix a Koszul operad $\mathscr{O}$, and let X be an $\mathscr{O}_\infty$-algebra; in particular, X is a (co)chain complex with respect to the operator

$$Q_1 : X = \mathscr{O}^\perp(1) \otimes X \to X.$$

We shall denote by $H(X)$ the (co)homology of the complex (X, Q_1); it is clear that $H(X)$ is also an $\mathscr{O}$-algebra with respect to the map

$$H(Q_2) : (\mathscr{O}^\perp(2) \otimes H(X)^{\otimes 2})^{S_2} = \mathscr{O}(2) \otimes_{S_2} (H(X)^{\otimes 2}) \to H(X), \tag{B.1}$$

induced from the $\mathscr{O}_\infty$-structure on X and trivial differential (the compatibility of this map with higher operations in $\mathscr{O}$ follows from the definitions of $\mathscr{O}^\perp$); as we explained earlier every $\mathscr{O}$-algebra is automatically an $\mathscr{O}_\infty$-algebra. Then one can give the following definitions:

Definition B.1.

(i) An $\mathscr{O}_\infty$-algebra X is called **formal**, if there exists an $\mathscr{O}_\infty$-algebra F and $\mathscr{O}_\infty$-algebra morphisms $X \leftarrow F \to H(X)$, inducing isomorphisms in cohomology (in a word, $\mathscr{O}_\infty$-**quasisiomorphisms**);

(ii) A (graded) $\mathscr{O}$-algebra H is called **intrinsically-formal**, if any $\mathscr{O}_\infty$-algebra X with $H(X) = H$ (as $\mathscr{O}$-algebras) is formal.

We are now going to prove a criterion for intrinsic formality. To this end let H be a (graded) $\mathscr{O}$-algebra; consider the differential graded Lie algebra $\mathfrak{g}$ of coderivations of the (co)free $\mathscr{O}^\perp$-coalgebra $\mathbb{F}^*_{\mathscr{O}^\perp}(H)$: the differential in $\mathfrak{g}$ is induced from the natural

$\mathcal{O}_\infty$-codifferential Q on $\mathbb{F}^*_{\mathcal{O}^\perp}(H)$. Since $\mathbb{F}^*_{\mathcal{O}^\perp}(H)$ is cofree coalgebra, coderivations on it are uniquely determined by their "Taylor coefficients", i.e.

$$\mathfrak{g} = \mathrm{Hom}(\mathbb{F}^*_{\mathcal{O}^\perp}(H), H) = \mathrm{Hom}\left(\bigoplus_{n\geq 1}(\mathcal{O}^\perp(n) \otimes H^{\otimes n})^{S_n}, H\right) \quad (\mathrm{B}.2)$$

as linear space. Let $\mathfrak{g}_{\geq 1}$ be the differential graded subalgebra in $\mathfrak{g}$, defined as

$$\mathfrak{g}_{\geq 1} = \mathrm{Hom}\left(\bigoplus_{n\geq 2}(\mathcal{O}^\perp(n) \otimes H^{\otimes n})^{S_n}, H\right).$$

Then the following is true

Theorem B.2. *Suppose that the natural map in cohomology $H(\mathfrak{g}_{\geq 1}) \to H(\mathfrak{g})$ (induced by the inclusion $\mathfrak{g}_{\geq 1} \subseteq \mathfrak{g}$) is equal to zero. Then H is intrinsically formal.*

B.1.1 Proof of the criterion for intrinsic formality

In order to prove Theorem B.2, let us begin with the following Lemma:

Lemma B.3. *Let X be an $\mathcal{O}_\infty$-algebra; then there exists an $\mathcal{O}_\infty$-algebra structure on $H(X)$, such that $H(X)$ with this structure is equivalent to X; i.e. there exists an $\mathcal{O}_\infty$-algebra F and $\mathcal{O}_\infty$-quasi-isomorphisms $X \leftarrow F \to H(X)$ if we endow $H(X)$ with this structure.*

One should keep in mind that the $\mathcal{O}_\infty$-structure on $H(X)$, constructed in Lemma B.3 in general has little in common with the $\mathcal{O}_\infty$-structure induced from X, see Equation (B.1); for instance the higher "Taylor coefficients" of this structure need not vanish when X is an $\mathcal{O}$-algebra (the case, which we need here). Thus this lemma does not mean that every $\mathcal{O}_\infty$-algebra is formal.

Proof. Since we work over a characteristic zero field, there always exists an inclusion $i : H(X) \to X$, a projection $p : X \to H(X)$

and a cochain homotopy $h : X \to X[1]$ that satisfy the equations $pi = \mathbb{1}_{H(X)}$, $ip - \mathbb{1}_X = [d, h]$, $hh = ph = hi = 0$; since the operad $\mathcal{O}$ is Koszul and hence we can replace it by the free operad $\mathcal{O}_\infty$, this is sufficient to use the standard methods of homotopy perturbation theory, see [Sm01] (Hinich reformulates this construction in terms of *fibrant objects* and their properties, see [Hi03]). Remark, that the first two "Taylor coefficients" of such $\mathcal{O}_\infty$-structure (i.e. Q_1, Q_2) do in effect coincide with the $\mathcal{O}$-algebra structure on $H(X)$. In particular, evidently $Q_1 = 0$; to see the equality for Q_2 it is enough to observe that $\mathcal{O}_\infty(2) = \mathcal{O}^\perp(2) = \mathcal{O}(2)$ and the $\mathcal{O}$-structure on $H(X)$ is induced from X. □

Proof of Theorem B.2. Let H be an $\mathcal{O}$-algebra, and let X be an $\mathcal{O}_\infty$-algebra, such that $H = H(X)$ as $\mathcal{O}$-algebras. Let $Q = \{Q_n\}$, $Q_n : (\mathcal{O}^\perp(n) \otimes H^{\otimes n})^{S_n} \to H$ be the $\mathcal{O}_\infty$-algebraic structure on H, guaranteed by the Lemma B.3. Then the $\mathcal{O}$-algebra structure on H is given by $Q^0 = \{Q_n^0\}$, $Q_2^0 = Q_2$ and $Q_n^0 = 0$, $n \neq 2$.

Let λ be a real parameter (we assume that $\mathbb{R} \subseteq \Bbbk$); put $Q_n^\lambda = \lambda^{n-2} Q_n$. Then, since $Q_1 = Q_1^0 = 0$, we have $\{Q_n^\lambda\}|_{\lambda=0} = \{Q_n^0\}$ and $\{Q_n^\lambda\}|_{\lambda=1} = \{Q_n\}$. Moreover, it turns out that $\{Q_n^\lambda\}$ is a $\mathcal{O}_\infty$-algebra structure for all λ: to see this we recall that the condition that $\{Q_n^\lambda\}$ is $\mathcal{O}_\infty$-structure looks like a system of equations on $\{Q_n^\lambda\}$, $n = 1, 2, \ldots$:

$$[Q_1^\lambda, Q_n^\lambda] = P_n(Q_2^\lambda, \ldots, Q_{n-1}^\lambda), \qquad (B.3)$$

where P_n, $n \geq 3$ is a quadratic polynomial in Q_i^λ homogenous of degree $n + 1$ with respect to the indices of the coefficients (i.e. the sum of the indices of Q_i^λ that show up in monomials is always equal to $n+1$). Hence the expression on the right is homogenous of degree $n - 3$ as function of λ:

$$P_n(Q_2^\lambda, \ldots, Q_{n-1}^\lambda) = \lambda^{n-3} P_n(Q_2, \ldots, Q_{n-1}).$$

On the other hand, since $Q_1^\lambda \equiv 0$, the left-hand side of Equation (B.3) vanishes identically, so we need to show that

$$\lambda^{n-3} P_n(Q_2, \ldots, Q_{n-1}) \equiv 0,$$

which is true since $\{Q_n\}$ is an $\mathcal{O}_\infty$-algebra structure on H with trivial differential $Q_1 = 0$.

Thus we have two $\mathscr{O}_\infty$-structures Q and Q^0 on H, i.e. two coderivations of the free $\mathscr{O}^\perp$-coalgebra $\mathbb{F}^*_{\mathscr{O}^\perp}(H)$ and a 1-parameter family of such structures Q^λ, "connecting" them. Observe, that for all $\lambda \in \mathbb{R}$ the lower "Taylor coefficients" are the same: $Q_1^\lambda \equiv 0$, $Q_2^\lambda \equiv Q_2^0$. We are going to use the assumptions of the theorem to construct a family of $\mathscr{O}_\infty$-algebraic isomorphisms

$$\theta_\lambda : (\mathbb{F}^*_{\mathscr{O}^\perp}(H), Q^0) \to (\mathbb{F}^*_{\mathscr{O}^\perp}(H), Q^\lambda)$$

for all $\lambda \in [0,1]$, such that $\theta_0 = \mathrm{id}$ and the lower "Taylor coefficients" are not changed by θ_λ; this will prove the statement of the Theorem B.2. However, this point of view is not quite convenient, since constructing such θ_λ would involve working with power series in λ, which is fraught with convergence issues. Instead we are going to work with formal power series.

To this end let us take the algebraic standpoint: we regard Q^λ as a λ-linear codifferentiation of $\mathbb{F}^*_{\mathscr{O}^\perp}(H)[\lambda]$, connecting Q^0 and Q; then constructing θ_λ amounts to constructing an $\mathscr{O}^\perp$-coalgebra λ-linear isomorphism

$$\theta : (\mathbb{F}^*_{\mathscr{O}^\perp}(H)[\lambda], Q^0) \to (\mathbb{F}^*_{\mathscr{O}^\perp}(H)[\lambda], Q^\lambda),$$

equal to identity modulo λ. Then the isomorphism $\bar\theta : (\mathbb{F}^*_{\mathscr{O}^\perp}(H), Q^0) \to (\mathbb{F}^*_{\mathscr{O}^\perp}(H), Q)$ (the value of θ_λ at $\lambda = 1$) is obtained by tensoring θ with $\Bbbk[\lambda]/(\lambda - 1)$.

Again, since the coalgebras we deal with are cofree, all such isomorphisms (in fact all coalgebra homomorphisms between them) are uniquely determined by their "Taylor coefficients"; since we assume the λ-linearity, these coefficients are just maps

$$\theta_n : \mathscr{O}^\perp(n) \otimes_{S_n} H^{\otimes n} \to H[\lambda].$$

Clearly, we should take $\theta_1 = \mathrm{id}_H$, and we will look for θ_n in the form $\theta_n = \lambda^{n-1}\phi_n$ for some

$$\phi_n : \mathscr{O}^\perp(n) \otimes_{S_n} H^{\otimes n} \to H.$$

From a naïve point of view this means that we look for the θ_λ in the form of a (formal) power series

$$\theta_\lambda = \mathrm{id}_H + \lambda\phi_2 + \lambda^2\phi_3 + \ldots. \tag{B.4}$$

This choice of coefficients is also justified by the homogeneity principle: if we let the elements of the multiplicative group $\mathbb{k}^\times \ni \mu$ act on λ by the formula $\lambda^\mu = \mu\lambda$ and on the elements $x \in \mathcal{O}^\perp(n) \otimes_{S_n} H^{\otimes n}$ by the formula $x^\mu = \mu^n x$, so that the differentials Q, Q^λ (for all λ) are homogenous of degree -1 with respect to this action, i.e. satisfy the equation

$$\left(Q\left(x^{\mu^{-1}}\right)\right)^\mu = \mu^{-1} Q(x), \quad \left(Q^\lambda\left(x^{\mu^{-1}}\right)\right)^\mu = \mu^{-1} Q^\lambda(x),$$

then the assumptions we made are equivalent to the condition that θ_n are homogenous of degree 0.

Let us now construct the coefficients ϕ_n.[84] We are going to do it by induction on n, so that the base $n = 1$ is given by $\phi_1 = \mathrm{id}_H$. To this end let us denote the $\mathcal{O}^\perp$-coalgebra $\mathbb{F}^*_{\mathcal{O}^\perp}(H)$ by C; in this notation we can assume that the first n terms of the series (B.4) (with higher terms equal to 0) give the differential $\mathcal{O}^\perp$-coalgebra isomorphism

$$\theta_{(n)} : (C[\lambda]/(\lambda^n), Q^0) \to (C[\lambda]/(\lambda^n), Q^\lambda);$$

this is clearly true for $n = 1$. Then in order to extend this map to an isomorphism

$$\theta_{(n+1)} : (C[\lambda]/(\lambda^{n+1}), Q^0) \to (C[\lambda]/(\lambda^{n+1}), Q^\lambda)$$

we first extend $\theta_{(n)}$ tautologically to a linear isomorphism

$$\theta' : C[\lambda]/(\lambda^{n+1}) \to C[\lambda]/(\lambda^{n+1})$$

and use θ' to pull the differential Q^λ to a differential Q' the left-hand side, thus obtaining an isomorphism of differential $\mathcal{O}^\perp$-coalgebras

$$\theta' : (C[\lambda]/(\lambda^{n+1}), Q') \to (C[\lambda]/(\lambda^{n+1}), Q^\lambda).$$

Evidently, it follows from the definitions and the homogeneity of all maps that $Q' = Q^0 \mod \lambda^n$ so $Q'_k = Q^0_k$ for $k \leq n+1$ and that $Q'_{n+2} = \lambda^n \cdot z$ for some map

$$z : (O^\perp(n+2) \otimes H^{\otimes n+2})^{S_{n+2}} \to H.$$

---

[84] Compare this reasoning with the deformation/obstruction theory of associative algebras, see Section 9.

We can regard z as an element in $\mathfrak{g}_{\geq 1}$; then it is not difficult to see that z is a cocycle with respect to the differential Q^0 (it follows from the fact, that Q' is a differential). So by the assumptions made it follows that $z = du$ for an element in $\mathfrak{g}^0$. Then the map

$$\eta = \exp(\lambda^n \cdot u) : (C[\lambda]/(\lambda^{n+1}), Q^0) \to (C[\lambda]/(\lambda^{n+1}), Q')$$

is an isomorphism of differential $\mathscr{O}^\perp$-coalgebras. It remains to observe that as a formal power series in λ, the map η looks as

$$\eta = \mathrm{id}_H + \lambda^n \eta_{n+1} + \ldots,$$

where dots denote the terms of degrees higher than n in λ, so modulo λ^{n+1} we can replace η by a homogenous (in the previous sense) isomorphism $\kappa : (C[\lambda]/(\lambda^{n+1}), Q^0) \to (C[\lambda]/(\lambda^{n+1}), Q')$ with "Taylor coefficients" $\kappa_1 = \mathrm{id}_H$, $\kappa_{n+1} = \eta_{n+1}$, $\kappa_k = 0$, $k \neq 1, n+1$ and eventually we can take $\theta_{(n+1)}$ equal to the composition $\theta' \circ \kappa$. □

B.2 Gerstenhaber algebras and the corresponding operad

The idea behind the proof of formality theorem suggested by Tamarkin is to use the intrinsic formality of the algebra $\mathscr{T}^{\cdot}_{poly}(\mathbb{R}^d)[1][[\hbar]]$. However this cannot be done, if we regard it as just a differential graded Lie algebra: the criterion of Theorem B.2 does not hold in this case and, as much as the author knows, the question concerning the intrinsic formality of $\mathscr{T}^{\cdot}_{poly}(\mathbb{R}^d)$ is still open. So Tamarkin considered a richer algebraic structure on $\mathscr{T}^{\cdot}_{poly}(\mathbb{R}^d)$, which we are now going to explain.

To this end we recall that there are two natural algebraic operations on polyvector fields: in addition to the Schouten bracket, which gives the structure of graded Lie algebra, there is also wedge product of polyvectors. The structure that we need takes advantage of both these structures and is called *Gerstenhaber algebra structure*; as we explained in Appendix A, see Section A.2, this structure is governed by a certain quadratic operad. Here are formal definitions of these objects.

Definition B.4.

(i) *One says that a graded vector space $\mathcal{A}$ is a **Gerstenhaber algebra** if it is equipped with a graded commutative and associative product $m : \mathcal{A} \otimes \mathcal{A} \to \mathcal{A}$ of degree 0 and a graded anti-symmetric bracket $\ell : \mathcal{A} \otimes \mathcal{A} \to \mathcal{A}$ of degree -1, satisfying the graded Jacobi identity and graded Leibniz rule with respect to m;*

(ii) *The **Gerstenhaber algebra operad** $\mathcal{G}$ is the quadratic operad with generators $m \in \mathcal{G}^0(2)$, $\ell \in \mathcal{G}^{-1}(2)$, whose space of relations $\mathcal{R} \in \mathcal{G}(3)$ is generated by the following conditions:*

  - *m is graded associative and commutative;*
  - *ℓ is graded Lie bracket;*
  - *ℓ satisfies graded Leibniz rule with respect to m.*

Clearly, a vector space V is a Gerstenhaber algebra, iff it is a $\mathcal{G}$-algebra, hence the name. Polyvector fields on a manifold is an example of Gerstenhaber algebra. Another typical example of $\mathcal{G}$-algebra is given by the free commutative algebra generated by a shifted differential graded Lie algebra: if $\mathfrak{g}$ is a differential Lie algebra, then

$$\mathbb{F}_{\underline{\mathrm{COM}}}(\mathfrak{g}[-1]) = S(\mathfrak{g}[-1]) \tag{B.5}$$

has a natural structure of Gerstenhaber algebra, extending the Lie bracket in $\mathfrak{g}$. More generally, the same construction can be applied to a Lie algebra $\mathfrak{g}$ over a commutative ring $\mathcal{A}$ on which $\mathfrak{g}$ acts by derivations: the free commutative algebra of $\mathfrak{g}$ over $\mathcal{A}$, $S_\mathcal{A}(\mathfrak{g}[-1])$ has a natural structure of Gerstenhaber algebra. If we take $\mathfrak{g} = \mathcal{T}^1_{poly}(M)$, $\mathcal{A} = C^\infty(M)$, we obtain precisely the Gerstenhaber algebra $\mathcal{T}^{\cdot}_{poly}(M)$.

Since the operad $\mathcal{G}$ is quadratic, we can define its quadratic dual $\mathcal{G}^\perp$ and get the operad $\mathcal{G}_\infty = \mathbb{T}(\mathcal{G}^\perp[-1])$, responsible for the deformations of Gerstenhaber algebras.

Remark B.5. In order to take advantage of the theory we developed and prove the formality theorem, we need to prove three statements:

- That the operad $\mathcal{G}$ is Koszul;

- That $\mathcal{T}_{poly}^{\cdot}(\mathbb{R}^d)$ satisfies the intrinsic formality criterion of Theorem B.2;

- That the algebra of (differentiable) Hochschild cochains on $\mathbb{R}^d$, i.e. $\mathcal{D}_{poly}^{\cdot}(\mathbb{R}^d)[[\hbar]]$ bears a structure of $\mathcal{G}_\infty$-algebra such that its cohomology is isomorphic to $\mathcal{T}_{poly}^{\cdot}(\mathbb{R}^d)[[\hbar]]$ as $\mathcal{G}$-algebras.

In order to avoid the analytic difficulties we will prove the following algebraic analog of Kontsevich's Theorem 17.12 in which smooth functions are replaced by polynomials and polydifferential operators by Hochschild cochains:

Theorem B.6 (Algebraic formality theorem). *Let $\Bbbk$ be a characteristic zero field and $A = \Bbbk[x^1, \ldots, x^d]$ be the polynomial algebra over $\Bbbk$, let $C^{\cdot} = C^{\cdot}(A, A)$ be the Hochschild cochains complex of A. Then the differential Lie algebra $C^{\cdot}[1]$ (with respect to the Gerstenhaber brackets) is formal; in effect $C^{\cdot}$ is formal as a $\mathcal{G}_\infty$-algebra.*

To prove this theorem we will follow the same steps as listed above (see Remark B.5): one just replaces the algebras $\mathcal{T}_{poly}^{\cdot}(\mathbb{R}^d)$ and $\mathcal{D}_{poly}^{\cdot}(\mathbb{R}^d)$ with polynomial polyvector fields on $\Bbbk^d$ and Hochschild cochain complex $C^{\cdot}(A, A)$ of the polynomial algebra $A = \Bbbk[x^1, \ldots, x^d]$ respectively.

The "analytic version" of formality theorem can be proved similarly: the only place, where there will be a slight difference in the proof is when we check the intrinsic formality criterion; one should replace the algebraic Hochschild complex and polynomial polyvectors by smooth polydifferential operators on $\mathbb{R}^d$ and check that the conclusions remain unchanged. We leave this modification as an exercise for the reader.

B.3 Proof of the formality theorem

The remaining part of this Appendix deals with the proof of the Theorem B.6. We do it by consequently examining and proving the statements, listed in Remark B.5 (with polyvectors and polydifferential operators replaced with their algebraic counterparts, as explained above).

B.3.1 Koszulity of the operad $\mathcal{G}$

We are going to use the test of Ginzburg and Kapranov (see Theorem A.15): let V be any graded vector space, we need to show that the natural map $q : \mathbb{F}^*_{\mathcal{G}^\perp}(\mathbb{F}_{\mathcal{G}}(V)) \to V$ induces an isomorphism in cohomology.

To do so we need to identify the free Gerstenhaber algebras and coalgebras (co)generated by V with something familiar; so recalling the construction of Gerstenhaber algebra of a Lie algebra, see (B.5), on one hand, and on the other hand observing that free Gerstenhaber algebra of V is *par excellence* free (shifted) Lie algebra and free commutative algebra, we conclude that

$$\mathbb{F}_{\mathcal{G}}(V) = \mathbb{F}_{\underline{COM}}(\mathbb{F}_{\underline{LIE}[1]}(V)) = \mathbb{F}_{\underline{COM}}(\mathbb{F}_{\underline{LIE}}(V[-1])[-1])[2]. \tag{B.6}$$

Further, since $\mathcal{G}$ is finite-dimensional as $\mathbb{S}$-space, its dual $\mathbb{S}$-space $\mathcal{G}^*$ has a natural cooperad structure and by a straightforward computation we have

$$\mathcal{G}^\perp = \mathcal{G}^\dagger[2]. \tag{B.7}$$

Here $\dagger$ denotes the dual space cooperad. So we can obtain the description of free $\mathcal{G}^\perp$-coalgebras by dualising (B.6) (see Equations (A.17), (A.17))

$$\mathbb{F}^*_{\mathcal{G}^\perp}(V) = \mathbb{F}^*_{\mathcal{G}^\dagger[2]}(V) = \mathbb{F}^*_{\underline{COM}^\dagger}(\mathbb{F}^*_{\underline{LIE}^\dagger}(V[1])[1])[-2] = \mathbb{F}^*_{\underline{LIE}^\perp}(\mathbb{F}^*_{\underline{COM}^\perp}(V)). \tag{B.8}$$

The representation (B.8) now can be used to describe the differential Q in the complex $\mathbb{F}^*_{\mathcal{G}^\perp}(\mathbb{F}_{\mathcal{G}}(V))$: let $\mathbb{F}_{\mathcal{G}}(V) = X$ so

$$\mathbb{F}^*_{\mathcal{G}^\perp}(\mathbb{F}_{\mathcal{G}}(V)) = \mathbb{F}^*_{\underline{LIE}^\perp}(\mathbb{F}^*_{\underline{COM}^\perp}(X));$$

considering separately the free $\underline{LIE}^\perp$- and free $\underline{COM}^\perp$-coalgebra constructions we get a bigrading on the complex $\mathbb{F}^*_{\mathcal{G}^\perp}(\mathbb{F}_{\mathcal{G}}(V))$:

$$\mathbb{F}^*_{\mathcal{G}^\perp}(\mathbb{F}_{\mathcal{G}}(V)) = \bigoplus_{p,q=0}^{\infty} \mathbb{F}^{*1+p}_{\underline{LIE}^\perp}(\mathbb{F}^{*q}_{\underline{COM}^\perp}(X)) = \bigoplus_{p,q=0}^{\infty} \mathbb{F}^{*1+p}_{\underline{COM}^\dagger}(\mathbb{F}^{*q}_{\underline{LIE}^\dagger}(V[1])[1])[-2]. \tag{B.9}$$

The differential Q in $\mathbb{F}^*_{\mathcal{G}^\perp}(X)$ is now decomposed into the sum of operators Q_m and Q_ℓ compatible with the bigrading: the former is the differential induced by commutative multiplication, i.e. Q_m is the differential in so-called *Harrison complex of the (free) commutative algebra* (see a description of this complex below in Section B.3.2) and the latter is induced by the Lie bracket. But since both operads COM and LIE are Koszul, the cohomology of Q_m and Q_ℓ can be easily computed: for instance

$$H(\mathbb{F}^*_{\mathcal{G}^\perp}(X), Q_m) = \mathbb{F}^*_{\text{LIE}^\perp}(\mathbb{F}_{\text{LIE}}(V))$$

and the operator induced by Q on this cohomology is Q_ℓ. We can now use the spectral sequence for bicomplexes to complete the proof.

B.3.2 Intrinsic formality of polynomial polyvectors in $\mathbb{R}^d$

The next step is to check the intrinsic formality test of Theorem B.2 for the Hochschild cohomology of the $H = H(A, A)$ of the polynomial algebra $A = S(\mathbb{R}^d) = \Bbbk[x^1, \ldots, x^k]$, viewed as the Gerstenhaber algebra; by Hochschild-Kostant-Rosenberg Theorem 8.9,[85] $H = A \otimes \Lambda^*(\mathbb{R}^d)$ as Gerstenhaber algebra.

In order to prove the intrinsic formality of H we shall use the expression (B.8) for the cofree coalgebra cogenerated by H:

$$\mathbb{F}^*_{\mathcal{G}^\perp}(H) = \mathbb{F}^*_{\text{COM}^\dagger}(\mathbb{F}^*_{\text{LIE}^\dagger}(H[1])[1])[-2].$$

Hence the Lie algebra $\mathfrak{g}$ of coderivations on $\mathbb{F}^*_{\mathcal{G}^\perp}(H)$ (see Equation (B.2)) is equal to

$$\mathfrak{g} = \text{Hom}\left(\mathbb{F}^*_{\text{COM}^\dagger}(\mathbb{F}^*_{\text{LIE}^\dagger}(H[1])[1]), H[2]\right).$$

The bigrading (B.9) in $\mathbb{F}^*_{\text{COM}^\dagger}(\mathbb{F}^*_{\text{LIE}^\dagger}(H[1])[1])$ induces a bigrading in $\mathfrak{g}$ so that[86]

$$\mathfrak{g}^{pq} = \text{Hom}\left(\mathbb{F}^{*1+p}_{\text{COM}^\dagger}(\mathbb{F}^{*q}_{\text{LIE}^\dagger}(H[1])[1]), H[2]\right).$$

Then $\mathfrak{g}_{\geq 1} = \bigoplus_{(p,q) \neq (0,0)} \mathfrak{g}^{pq}$. Now we are going to use the spectral sequence for this bicomplex to check the test of Theorem B.2. We

---

[85] Here we rather need algebraic version of the HKR theorem, where smooth functions are replaced by polynomials: see Section 7.4.3. where the homological version of this theorem is discussed.

[86] Here and below we use the following notation $\mathbb{F}^{*p}_{\mathscr{C}}(X) = (\mathscr{C}(p) \otimes X^{\otimes p})^{S_p}$.

observe that
$$Har(H) = \mathbb{F}^*_{\text{LIE}^\dagger}(H[1]) \otimes H$$
is the homological Harrison complex of H, so we can write
$$\mathfrak{g}^{0q} = \text{Hom}\left(\mathbb{F}^{*1+q}_{\text{LIE}^\dagger}(H[1])[1], H[2]\right) = \text{Hom}_H\left(\mathbb{F}^{*1+q}_{\text{LIE}^\dagger}(H[1]) \otimes H[1], H[2]\right)$$
$$= \text{Hom}_H(Har(H)[1], H[2]).$$

In fact, this is the isomorphism of chain complexes:
$$(\mathfrak{g}^{0q}, Q_m) = \text{Hom}_H(Har(H)[1], H[2]).$$

It follows that for generic indices we have
$$(\mathfrak{g}^{pq}, Q_m) = \text{Hom}_H(S_H^{1+p}(Har(H)[1]), H[2]).$$

Now it is known[87] (see [Lo98] and Section 7.4.3 for definitions and some details) that in case $H = S(\mathbb{R}^d) \otimes \Lambda^*(\mathbb{R}^d)$ the Harrison complex $Har(H)$ is (up to a shift) quasi-isomorphic to the module of Kähler differentials on H, standing in degree $q = 0$, i.e. $Har(H) \cong \Omega(H)[1]$; so we have the following description of the first sheet of the spectral sequence
$$E_1^{pq} = \begin{cases} \text{Hom}_H(S_H^{1+p}(\Omega(H)[2]), H[2]), & q = 0, \\ 0, & q > 0. \end{cases}$$

As we see E_1^{pq} can be regarded as a subcomplex in $\mathfrak{g}^{pq}$. On the other hand since $H = S(\mathbb{R}^d) \otimes \Lambda^*(\mathbb{R}^d)$, the module $\Omega(H)$ can be described as follows
$$\Omega(H) \cong H \otimes \left(\mathbb{R}^d[-1] \oplus (\mathbb{R}^d)^*\right) = H \otimes_{S(\mathbb{R}^d)} \left(T_{S(\mathbb{R}^d)}[-1] \oplus T^*_{S(\mathbb{R}^d)}\right),$$
where $T_{S(\mathbb{R}^d)} \cong S(\mathbb{R}^d) \otimes \mathbb{R}^d$ is the module of derivations of $S(\mathbb{R}^d)$. Hence, applying the dualisation we obtain
$$\left((T_{S(\mathbb{R}^d)}[-1] \oplus T^*_{S(\mathbb{R}^d)})[2]\right)^* = \left(T_{S(\mathbb{R}^d)}[1] \oplus T^*_{S(\mathbb{R}^d)}[2]\right)^*$$
$$= T^*_{S(\mathbb{R}^d)}[-1] \oplus T_{S(\mathbb{R}^d)}[-2] = \left(T_{S(\mathbb{R}^d)}[-1] \oplus T^*_{S(\mathbb{R}^d)}\right)[-1].$$

---

[87]In effect, one can regard this as a version of the Hochschild-Kostant-Rosenberg theorem, see Section 7.4.3.

So eventually

$$E_1^{p0} = S_{S(\mathbb{R}^d)}^{1+p}\left((T_{S(\mathbb{R}^d)}[-1] \oplus T^*_{S(\mathbb{R}^d)})[-1]\right) \otimes_{S(\mathbb{R}^d)} H[2]$$
$$\cong S_H^{1+p}(\Omega(H)[-1])[2] \cong \Omega^{1+p}(H)[2],$$

i.e. $E_1^{\cdot 0}$ can be identified with the algebra of Kähler differential forms on H. Similarly the differential in $E_1^{\cdot 0}$ is now induced from Q_ℓ; it coincides with (shifted) de Rham differential in $\Omega^{\cdot}(H)$. Now the intrinsic formality follows from the fact that the algebra H is acyclic, i.e. its cohomology is concentrated in degree $p = 0$.

B.3.3 The $\mathcal{B}_\infty$-operad and its action on Hochschild complex

The remaining part of this appendix is reserved for the proof of the third statement, listed in Remark B.5: that the Hochschild complex of the polynomial algebra can be endowed with a structure of a $\mathcal{G}_\infty$-algebra.

This turns out to be the hardest part of the proof. In effect it is a version of the so-called *Deligne's conjecture*. Recall, that Hochschild cohomology $HH^{\cdot}(A)$ of every algebra A bears the structure of Gerstenhaber algebra, induced from the $\cup$-product and the Gerstenhaber bracket of the cochains, see Corollary 8.8, part (ii). On the other hand it has long been known, e.g. from the papers [Ar69], and especially [Co76], that the cohomology with coefficients in $\mathbb{k}$ of the little squares operad $\mathcal{E}_2$ (see the end of Section A.1 for the definitions) is equal to the operad $\mathcal{G}$. On the other hand, applying a suitable cochain functor to $\mathcal{E}_2$, we can turn it into an operad in the category $Ch_{\mathbb{k}}$. Now, Deligne's conjecture (formulated in his letter to Jim Stasheff, Murray Gerstenhaber, John Peter May, Pavel Schechtman and Vladinir Drinfeld in 1993) states that *the Hochschild cohomology complex of an algebra A can be endowed with a structure of the $C.(\mathcal{E}_2)$-algebra that induces the Gerstenhaber algebra structure on $HH^{\cdot}(A)$*; more generally, we can replace here $C.(\mathcal{E}_2)$ by (in Deligne's own words) any "suitable version of it", i.e. by a homotopy equivalent operad.

The operad $\mathcal{G}_\infty$ being free, it can be mapped into $C.(\mathcal{E}_2)$ in a way that induces an isomorphism in cohomology. Thus, the proof of Deligne's conjecture would entail the proof of formality result. In effect, the opposite is also true since as one can show the operad $\mathcal{G}_\infty$ is homotopy-equivalent to $C.(\mathscr{E}_2)$. So Tamarkin's proof of formality theorem does in effect contain a proof of Deligne's conjecture, moreover the first known such proof, "modulo" the identification of $\mathcal{G}_\infty$ and the chain complex of $\mathscr{E}_2$. Other approaches to this conjecture and its various natural generalisations were later independently found by McClure and Smith [McCS02], Kontsevich and Soibelman [KS00] and many other authors.

In order to prove that Hochschild complex bears the structure of $\mathcal{G}_\infty$-algebra, we begin with constructing an algebraic operad $\mathcal{B}_\infty$, which acts naturally on the Hochschild complex of any algebra A. We begin with setting the following notation: let us denote any cochain $f \in C^p(A, A)$ by a box with p inputs (drawn above the box) and one output (below the box):

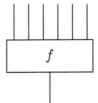

In this notation we can define the *brace* operations $f\{g_1, g_2, \ldots, g_n\}$ for all $n \leq p$ and all $g_i \in C^{p_i}(A, A)$:

$$f\{g_1, g_2, \ldots, g_n\} = \sum_{\text{all insertions}} (-1)^\epsilon \boxed{},$$

where the summation is over all possible ways to insert g_i, $i = 1, \ldots, n$ into f and the sign $(-1)^\epsilon$ is determined by the Koszul sign rule so that $f\{g\} = f \circ g$, c.f. Equation (8.12). We put $f\{g_1, \ldots, g_n\} = 0$ if $p < n$.

Also recall that an associative algebra B is said to have *bialgebra* structure, if it is equipped with a coassociative coproduct $\Delta : B \to B \otimes B$, which is a homomorphism of algebras at the same time; in many books one also asks for the existence of counit $\epsilon : B \to \Bbbk$ of B (viewed as a coalgebra), which should be an algebra homomorphism now. Dually, a coalgebra C has bialgebra structure if it is equipped with an associative product $m : C \otimes C \to C$, which is a homomorphism of coalgebras, and with a unit $1 \in C$ for which $\Delta(1) = 1 \otimes 1$. Now we can give the following definition:

Definition B.7. *One says that a graded vector space X is a $\mathcal{B}_\infty$-algebra if the free $\underline{\mathrm{ASS}}^\perp$-coalgebra $\mathbb{F}^*_{\underline{\mathrm{ASS}}^\perp}(X)$ can be equipped with a differential graded bialgebra structure such that the coproduct in it coincides with the standard cofree one.*

In order to describe the operad that encodes these structures, let us recall that the $\underline{\mathrm{ASS}}^\perp$-coalgerbra $\mathbb{F}^*_{\underline{\mathrm{ASS}}^\perp}(X) \cong \bigoplus_{n \geq 1}(X[1])^{\otimes n}$ is cofree and hence all coalgebra homomorphisms and coderivations with values in it, are uniquely determined by their "ground floor", i.e. by their composition with projection onto $(X[1])^{\otimes 1} = X[1]$.

Hence the structure described in the Definition B.7 is determined by the following maps:

(i) the maps $m_n : (X[1])^{\otimes n} \to X[1]$, $n = 1, 2, \ldots$ of degree 1, that make up the differential Q in the cofree coalgebra $\mathbb{F}^*_{\underline{\mathrm{ASS}}^\perp}(X[1])$, or equivalently, $m_n : X^{\otimes n} \to X$ of degree $2 - n$;

(ii) the maps $m_{pq} : (X[1])^{\otimes p} \otimes (X[1])^{\otimes q} \to X[1]$, $p, q = 1, 2 \ldots$ of degree 0, that make up the associative product M in $\mathbb{F}^*_{\underline{\mathrm{ASS}}^\perp}(X[1])$, or equivalently, the maps $m_{pq} : X^{\otimes p} \otimes X^{\otimes q} \to X$ of degree $1 - p - q$.

The associativity condition for M, the equation $Q^2 = 0$ and Leibniz rule for M and Q give certain equations that should be satisfied by $\{m_n\}$ and $\{m_{pq}\}$. Observe that the maps $\{m_n\}$ alone give an A_∞-algebra structure on X.

Definition B.8. *Operad $\mathcal{B}_\infty$ is the graded operad, generated by the elements $m_n \in \mathcal{B}^{2-n}(n)$, $n \geq 1$ and $m_{pq} \in \mathcal{B}^{1-p-q}(p+q)$, $p, q \geq 1$, that satisfy the relations, mentioned above. Graded vector space X is $\mathcal{B}_\infty$-algebra (in the sense of Definition B.7) if and only if it is algebra over the operad $\mathcal{B}_\infty$.*

Warning! The operad $\mathcal{B}_\infty$ is not equal to the free operad, constructed from the quadratic dual cooperad of any quadratic operad; the notation $\mathcal{B}_\infty$ was suggested by Ezra Getzler and James

Jones in their influential, but unpublished paper "Operads, homotopy algebra and iterated integrals for double loop spaces" (see https://arxiv.org/abs/hep-th/9403055) independently of the theory of Koszul operads.

It remains to remark that $\mathcal{B}_\infty$ acts on the Hochschild complex $\mathcal{C} = C^{\cdot}(A, A)$ of any algebra A if we put

- $m_1 : \mathcal{C} \to \mathcal{C}$ equal to the differential in $\mathcal{C}$;
- $m_2 : \mathcal{C} \otimes \mathcal{C} \to \mathcal{C}$ equal to the $\cup$-product in $\mathcal{C}$;
- $m_n = 0$, $n > 2$;
- $m_{1n} : \mathcal{C} \otimes \mathcal{C}^{\otimes n} \to \mathcal{C}$ equal to the "brace", defined above: $m_{1n}(f; g_1, \ldots, g_n) = f\{g_1, \ldots, g_n\}$;
- $m_{kn} = 0$, $k > 1$.

The fact that relations on these operations hold is pretty evident; for example the relations on m_n follow from the associativity of the $\cup$-product. The rest follows by a straightforward combinatoric computations similar to the proof of the properties of Gerstenhaber brackets, see Proposition 8.7 and we omit it.

The operad $\mathcal{B}_\infty$ is well-defined and its action on $C^{\cdot}(A, A)$ is natural, but the problem is that the relation of $\mathcal{B}_\infty$ and $\mathcal{G}_\infty$ is not very evident. In order to construct an operad morphism $\mathcal{G}_\infty \to \mathcal{B}_\infty$ that will induce the $\mathcal{G}$-algebra structure on Hochschild cohomology (which is enough to prove the Deligne conjecture) Tamarkin constructs a third operad $\widetilde{\mathcal{B}}$ (see Section B.3.4), such that the natural map $\mathcal{G}_\infty \to \mathcal{G}$ is equal to the composition of two operad morphisms $\mathcal{G}_\infty \to \widetilde{\mathcal{B}}$ and $\widetilde{\mathcal{B}} \to \mathcal{G}$. After this he shows that $\widetilde{\mathcal{B}}$ is in effect isomorphic to the operad $\mathcal{B}_\infty$. This isomorphism will be proved in Section B.3.6, with the help of Kazhdan and Etingof theorem about the quantisation of Lie bialgebras, which in its turn is based on the existence theorem for Drinfeld associators.

B.3.4 The operad $\widetilde{\mathcal{B}}$

Just as the definition of $\mathcal{B}_\infty$ operad was based on the description of $\mathcal{B}_\infty$-algebras, so the definition of the operad $\widetilde{\mathcal{B}}$ relies on the definition $\widetilde{\mathcal{B}}$-algebras:

Definition B.9. *A graded vector space X is said to be $\widetilde{\mathcal{B}}$-algebra, if the cofree $\underline{\text{COM}}^\perp$-coalgebra $\mathbb{F}^*_{\underline{\text{COM}}^\perp}(X)$ is equipped with a differential graded Lie bialgebra structure, extending the standard (cofree) coproduct in it.*

Remark B.10. The difference between a *bialgebra* and a **Lie** *bialgebra* structure on a (graded) vector space V is that in the latter case product and coproduct connect V with the graded wedge product $\Lambda^2 V = V \wedge V$. One can rephrase it by saying that the coproduct in V is a (graded) skew-cocommutative map $V \to V \otimes V$ (c.f. the discussion at page 251, before the Proposition 16.7) and the product is a (graded) skew-symmetric map $V \otimes V \to V$. In order to account for these conditions one calls these maps "the graded Lie bracket" and "graded Lie cobracket" respectively. These maps should satisfy the Jacobi identity and coassociativity conditions, as well as an analog of the compatibility, similar to the compatibility in usual bialgebras; compare the Definition B.11.

As before, the cofreeness of $\mathbb{F}^*_{\underline{\text{COM}}^\perp}(X[1])$ guarantees that the corresponding maps are determined by their "Taylor coefficients": the differential in it is determined by the maps

$$\ell_n : \mathbb{F}^{*n}_{\underline{\text{COM}}^\perp}(X) \to \mathbb{F}^{*1}_{\underline{\text{COM}}^\perp}(X) \cong X[1]$$

of degree 1, and the Lie bracket by skew-symmetric maps

$$\ell_{pq} : \mathbb{F}^{*p}_{\underline{\text{COM}}^\perp}(X) \otimes \mathbb{F}^{*q}_{\underline{\text{COM}}^\perp}(X) \to X[1]$$

of total degree 0. These maps are subject to a series of quadratic relations. So just like in case of the operad $\mathcal{B}_\infty$ we can define the operad $\widetilde{\mathcal{B}}$ as the free operad generated by elements ℓ_n, ℓ_{pq} factorised over their natural relations. We are now going to construct the operad morphisms $\widetilde{\mathcal{B}} \to \mathcal{G}$ and $\mathcal{G}_\infty \to \widetilde{\mathcal{B}}$.

First we recall (see Remark A.4.3) that if every $\mathcal{G}$-algebra can be endowed with a natural $\widetilde{\mathcal{B}}$-algebra structure, then there exists an operad morphism $\widetilde{\mathcal{B}} \to \mathcal{G}$, inducing these structures. But since any Gerstenhaber algebra X is in particular a commutative algebra, we get a differential on $\mathbb{F}^*_{\underline{\text{COM}}^\perp}(X)$ induced from the commutative product, encoded by the operations ℓ_n (recall that every commutative algebra is in effect $\underline{\text{COM}}_\infty$-algebra). On the other hand, the Lie bracket on $X[1]$ (which exists by the definition of Gerstenhaber algebras) extends to a unique bracket in $\mathbb{F}^*_{\underline{\text{COM}}^\perp}(X)$ of degree 0, which respects the coalgebra structure: one can do it similarly to the way the Lie bracket can be extended from a Lie algebra $\mathfrak{g}$ to the exterior algebra $\Lambda\mathfrak{g}$. The compatibility of these two structures is evident from the Leibniz rule for Gerstenhaber algebras.

Similarly, let X be a $\widetilde{\mathcal{B}}$-algebra. Then the free $\underline{\text{COM}}^\perp$-coalgebra $L = \mathbb{F}^*_{\underline{\text{COM}}^\perp}(X)$ has the structure of differential graded Lie algebra (the bracket being given by the maps ℓ_{pq}). This structure induces a differential on the free $\underline{\text{LIE}}^\perp$-coalgebra $\mathbb{F}^*_{\underline{\text{LIE}}^\perp}(L)$, and hence on $\mathbb{F}^*_{\mathcal{G}^\perp}(X)$, since

$$\mathbb{F}^*_{\underline{\text{LIE}}^\perp}(L) = \mathbb{F}^*_{\underline{\text{LIE}}^\perp}(\mathbb{F}^*_{\underline{\text{COM}}^\perp}(X)) = \mathbb{F}^*_{\mathcal{G}^\perp}(X),$$

by Equation (B.8). It means that X is a $\mathcal{G}_\infty$-algebra, and we obtain an operad morphism $\mathcal{G}_\infty \to \widetilde{\mathcal{B}}$. If we "compose" the constructions of these two maps, we see that the $\mathcal{G}_\infty$-structure that proceeds from the Gerstenhaber algebra structure "via" the $\widetilde{\mathcal{B}}$-structure, is just the canonical $\mathcal{G}_\infty$-structure of Gerstenhaber algebra, we conclude that the composition $\mathcal{G}_\infty \to \widetilde{\mathcal{B}} \to \mathcal{G}$ is equal to the canonical projection $\mathcal{G}_\infty \to \mathcal{G}$.

B.3.5 The theorem of Kazhdan and Etingof

The isomorphism between $\mathcal{B}_\infty$ and $\widetilde{\mathcal{B}}$ is in effect the most difficult part of Tamarkin's proof. Its construction is based on Kazhdan and Etingof's theorem, see [EK98], which describes the quantisation of Lie bialgebras. So we begin with formulating this theorem; in order to make this subject more accessible to a reader not very familiar with category theory, we give here just a short compilation of the statements, that are important for our purposes.

Let $R = \Bbbk[[\hbar]]$ be the algebra (commutative ring) of formal power series over $\Bbbk$. Then

Definition B.11. *We will say that a (graded) R-module $\mathfrak{g}$ is a (graded) topologically-free R-Lie bialgebra if*

(i) $\mathfrak{g} = V \otimes \Bbbk[[\hbar]]$ *as R-module for some vector space V;*

(ii) $\mathfrak{g}$ is equipped with the R-linear coassociative and (graded) anti-cocommutative coproduct $\delta : \mathfrak{g} \to \mathfrak{g} \otimes_R \mathfrak{g}$, called **cobracket**;

(iii) is equipped with the Lie bracket $[,] : \mathfrak{g} \otimes_R \mathfrak{g} \to \mathfrak{g}$;

(iv) these maps are compatible and satisfy the coassociativity condition and Jacobi identity.

If the condition of anti-cocommutativity is omitted and the Lie bracket is replaced by an associative product, we get the definition of a topologically-free R-bialgebra.

The condition (i) here will be referred to as the *topological freeness*. Observe that the bracket $[,]$ and cobracket δ in this situation are given by a formal power series:

$$[a,b] = [a,b]_0 + \hbar[a,b]_1 + \hbar^2[a,b] + \ldots,$$
$$\delta(a) = \delta_0(a) + \hbar\delta_1(a) + \hbar^2\delta_2(a) + \ldots,$$

where $[,]_i : \mathfrak{g} \otimes \mathfrak{g} \to \mathfrak{g}$, $\delta_j : \mathfrak{g} \to \mathfrak{g} \otimes \mathfrak{g}$ are some R-linear maps. Let $\mathfrak{g}$ be an R-bialgebra such that $\delta_0 = 0$. Let $\mathtt{LBA}_0(R)$ denote the category of such R-bialgebras and R-linear morphisms between them; for any $\mathfrak{g}$ in this category we let $\bar{\mathfrak{g}}$ denote $\mathfrak{g}$, regarded as a Lie algebra with respect to $[,]_0$.

Recall that a R-Hopf algebra is a R-bialgebra H, equipped with an anti-automorphism $S : H \to H$, called *the antipode*, such that

$$S(a_{(1)})a_{(2)} = \epsilon(a), \quad \forall a \in H,$$

where $\epsilon : H \to \Bbbk$ is the counit in bialgebra and

$$\Delta : H \to H \otimes H, \quad \Delta(a) = a_{(1)} \otimes a_{(2)} \qquad (B.10)$$

denotes the coproduct in H. Let $\mathtt{HA}_0(R)$ be the category of R-Hopf algebras which are cocommutative modulo $\hbar$; for an object H of $\mathtt{HA}_0(R)$ we will denote by $\overline{H}$ the same R-Hopf algebra with the product and coproduct given by their reduction modulo $\hbar$ (it is an easy exercise to check that this reduction preserves the Hopf algebra relations).

Then the following is due to Kazhdan and Etingof, see [EK98] and [Hi03]:

Theorem B.12 (Kazhdan, Etingof). *There exists a functor*

$$\mathcal{Q} : \mathrm{LBA}_0(R) \to \mathrm{HA}_0(R)$$

such that

(i) $\mathcal{Q}$ *is an equivalence of categories;*

(ii) $\overline{\mathcal{Q}(\mathfrak{g})} = U_\hbar \bar{\mathfrak{g}}$, *in particular by the PBW theorem* $\mathcal{Q}(\mathfrak{g}) \cong U_\hbar \mathfrak{g} \cong S_\hbar(\mathfrak{g})$ *as R-modules (here $S_\hbar(\mathfrak{g})$ denotes the symmetric algebra of the R-module);*

(iii) $\Delta(a) - \Delta'(a) = \delta(a) + O(\hbar^2)$ *for any $a \in \bar{\mathfrak{g}} \cong \mathfrak{g}$, where $\Delta'(a) = a_{(2)} \otimes a_{(1)}$ (c.f. Equation (B.10)) and we identify $\bar{\mathfrak{g}}$ with its image in $U_\hbar \bar{\mathfrak{g}}$;*

(iv) *the functor $\mathcal{Q}$ is given by universal formulas.*

It might be useful to explain that when we speak about "universal formulas" in part (iv) of this theorem, we mean that if one identifies $\mathcal{Q}(\mathfrak{g})$ with the symmetric algebra $S(\mathfrak{g})$ (see part (i)), then the product m and coproduct Δ in it can be expressed as certain universal formal power series of bracket and cobracket in $\mathfrak{g}$, independent on the particular choice of $\mathfrak{g}$. However, these formulas are not unique.

We are not going to discuss the proof of this theorem here: it is one of the "black boxes" that we leave in our lecture notes. We just remark that the eventual construction of $\mathcal{Q}$ relies on the choice of a *Drinfeld associator*, which allows one introduce the product and coproduct in $\mathcal{Q}(\mathfrak{g})$: it is this choice that determines the universal formulas for these maps.

For our purposes we will use the following modification of Kazhdan and Etingof's constructions: consider the action of the group $\Bbbk^\times$ of invertible elements in $\Bbbk$ on R by the formula $\mu(\hbar) = \mu^{-1}\hbar$. For a topologically free $\mathfrak{g} = V \otimes R$ we extend this action to $\mathfrak{g}$ by setting $\mu(v) = \mu \cdot v$ for any $v \in V$. This action can now be extended to the free Lie and free associative algebras, generated by V and $\mathfrak{g}$, i.e. on $\mathbb{F}_{\underline{\mathrm{LIE}}}(V)$, $\mathbb{F}_{\underline{\mathrm{ASS}}}(V)$, $\mathbb{F}_{\underline{\mathrm{LIE}}}(V)[[\hbar]]$ and $\mathbb{F}_{\underline{\mathrm{ASS}}}(V)[[\hbar]]$. Then we have the following corollary of the Theorem B.12 (one may call it "equivariant version of the theorem"):

Corollary B.13. *The functor Q induces an equivalence between the category $\texttt{LBA}_0(V,R)$ of Lie bialgebras $\mathfrak{g}$ from $\texttt{LBA}_0(R)$ with $\bar{\mathfrak{g}} \cong \mathbb{F}_{\underline{\texttt{LIE}}}(V)[[\hbar]]$ and the category $\texttt{HA}_0(V,R)$ of associative bialgebras H from $\texttt{HA}_0(R)$, such that $\overline{H} \cong \mathbb{F}_{\underline{\texttt{ASS}}}(V)[[\hbar]]$; this equivalence sends equivariant (with respect to the $\Bbbk^\times$-action) structures to equivariant.*

As a good exercise, that can help understanding this result, we recommend the reader to show that the universal enveloping algebra of a free Lie algebra of a vector space V is equal to the free associative algebra of V.

B.3.6 The isomorphism $\widetilde{\mathcal{B}} \cong \mathcal{B}_\infty$

We are eventually able to sketch the last step in the proof of formality theorem. To this end it is enough to show that every $\mathcal{B}_\infty$-algebra has a structure of $\widetilde{\mathcal{B}}$-algebra and vice-versa, so that these constructions are mutually-inverse.

We begin with the following observation: instead of the $\mathcal{B}_\infty$- and $\widetilde{\mathcal{B}}$-algebras we can prove the same result for dual $\mathcal{B}_\infty^\dagger$- and $\widetilde{\mathcal{B}}^\dagger$-coalgebras: dualisation works in our situation, since both operads $\mathcal{B}_\infty$ and $\widetilde{\mathcal{B}}$ are generated by finite-dimensional vector spaces, so $\mathcal{B}_\infty$ and $\widetilde{\mathcal{B}}$ are graded operads, for which every graded component is finite-dimensional. Then the definitions of the corresponding coalgebra structures are easy to "read off" off the dual Definitions B.7 and B.9 respectively; we formulate them in the form of a proposition:

Proposition B.14.

(i) $\mathcal{B}_\infty^\dagger$-*coalgebra structure on a graded vector space X is given by a differential graded associative bialgebra structure on the completion of the free $\underline{\texttt{ASS}}$-algebra, generated by $X[1]$, compatible with the $\underline{\texttt{ASS}}$-algebra structure on it:*

$$\widehat{\mathbb{F}}_{\underline{\texttt{ASS}}}(X[1]) = \prod_{n=0}^{\infty} \mathbb{F}^n_{\underline{\texttt{ASS}}}(X[1]) \cong \prod_{n=0}^{\infty} (X[1])^{\otimes n},$$

(ii) $\widetilde{\mathcal{B}}^\dagger$-*coalgebra structure on a graded vector space X is given by a differential graded Lie bialgebra structure on the completion* of the free LIE-algebra, generated by $X[1]$, compatible with the LIE-algebra structure on it:

$$\widehat{\mathbb{F}}_{\underline{\text{LIE}}}(X[1]) = \prod_{n=0}^{\infty} \mathbb{F}_{\underline{\text{LIE}}}^n(X[1]) \cong \prod_{n=0}^{\infty} (\underline{\text{LIE}}(n) \otimes X[1]^{\otimes n})^{S_n}.$$

Let now δ be a $\Bbbk^\times$-equivariant cobracket on $\mathbb{F}_{\underline{\text{LIE}}}(V)[[\hbar]]$, then $\delta = \sum_{p,q} \sum_r \hbar^r \delta^r_{pq}$, where

$$\delta^r_{pq} : V \to \mathbb{F}^p_{\underline{\text{LIE}}}(V) \otimes \mathbb{F}^q_{\underline{\text{LIE}}}(V).$$

The map δ then is equivariant with respect to $\Bbbk^\times$-action, iff $\delta^r_{pq} = 0$ for all p, q, r, $r \neq p+q-1$. Then the coefficients δ^{p+q-1}_{pq} of this map determine the cobracket in $\widehat{\mathbb{F}}_{\underline{\text{LIE}}}(V)$, which is compatible with the LIE-algebra structure on it, and vice-versa. Similarly, coproduct Δ in $\widehat{\mathbb{F}}_{\underline{\text{LIE}}}(V)$, compatible with ASS-algebra structure is uniquely determined by an associative coproduct on $\mathbb{F}_{\underline{\text{ASS}}}(V)[[\hbar]]$ and vice-versa. All these results can be extended straightforwardly to differential graded case.

Finally, taking $V = X[1]$ for a (co)chain complex X and using the Corollary B.13, we see that any $\widetilde{\mathcal{B}}^\dagger$-coalgebra structure on X gives rise to an equivariant dg Lie-bialgebra structure on $\mathfrak{g} = \mathbb{F}_{\underline{\text{LIE}}}(X[1])[[\hbar]]$, which induces a unique associative equivariant dg bialgebra (H, m, Δ) in $\text{HA}_0(R, V)$. It remains to show that $H \cong \mathbb{F}_{\underline{\text{ASS}}}(X[1])[[\hbar]]$ as algebras; but it follows from the fact that the homomorphism $\mathbb{F}_{\underline{\text{ASS}}}(X[1])[[\hbar]] \to H$, induced by the inclusion of $X[1]$ into $\mathfrak{g}$, is identity modulo $\hbar$. Hence, we get a dg associative coalgebra structure on $\mathbb{F}_{\underline{\text{ASS}}}(X[1])[[\hbar]]$, compatible with the product. As we observed earlier, this is equivalent to a $\mathcal{B}^\dagger_\infty$-coalgebra structure on X.

List of references

Textbooks, surveys and monographs

[Ar89] Arnold, Vladimir I., "Mathematical methods of classical mechanics", (2nd ed.) Springer-Verlag, Berlin-New York, 1989.

[AG85] Arnold, Vladimir; Givental, Alexander, "Symplectic geometry", in *Dynamical systems–4, Symplectic geometry and its applications, Itogi Nauki i Tekhniki. Ser. Sovrem. Probl. Mat. Fund. Napr.* **4** VINITI, Moscow, 1985, 5–135, in Russian. For the English translation see *Dynamical Systems IV. Symplectic Geometry and its Applications* (Arnold, V., Novikov, S., eds.), *Encyclopaedia of Math. Sciences* **4**, Springer, Berlin-New York, 1990, 1–138.

[At64] Atiyah, Michael F., "K-theory" W. A. Benjamin, New York, 1964.

[BV73] Boardman, Michael, Vogt, Rainer, "Homotopy invariant algebraic structures on topological spaces", *Lecture Notes in Mathematics* **347** Springer-Verlag, Berlin-Heidelberg-New York, 1973.

[BS83] Bogoliubov, Nikolai N.; Shirkov, Dmitriy V., "Quantum Fields" Benjamin, London, 1983.

[BT82] Bott, Raoul; Tu, Loring W., "Differential forms in algebraic topology", *Graduate Texts in Mathematics* **82** Springer, New York, 1982.

[BGM03] Bruzzo, Ugo; Gorini, Vittorio; Moschella, Ugo eds., "Geometry and Physics of Branes" in *Series in High Energy Physics, Cosmology and Gravitation*, Institute of Physics, Bristol-Philadelphia, 2003.

[CdS08] Cannas da Silva, Ana, "Lectures on Symplectic Geometry", *Lecture notes in Mathematics* **1764**, Springer, New York, 2008.

[CMOPRl17] Cardona, Alexander; Morales, Pedro; Ocampo, Hernán; Paycha, Sylvie; Reyes Lega, Andrés F. (Eds.), "Quantization, Geometry and Noncommutative Structures in Mathematics and Physics", in *Mathematical Physics Studies*, Springer International Publishing AG, Berlin-Heidelberg-New York, 2017.

[CGP11] Cattaneo, Alberto S.; Giaquinto, Anthony; Ping Xu (Eds.), "Higher Structures in Geometry and Physics. In Honor of Murray Gerstenhaber and Jim Stasheff", Birkhäuser, Springer, New York-Dordrecht-Heidelberg-London, 2011.

[CKTB06] Cattaneo, Alberto S.; Keller, Bernhard; Torossian, Charles; Bruguières, Alain, "Déformation, Quantification, Théorie de Lie", *Panoramas et Synthèses* **20**, Société Mathémathique de France, Paris, 2006.

[Ci06] Ciccoli, Niccola, "From Poisson to Quantum Geometry", notes taken by Paweł Witkowski; available online, 2006.

[CLM76] Cohen, Frederic R.; Lada, Thomas J.; May, John P., "The homology of iterated loop spaces", *Lecture Notes in Mathematics* **533** Springer-Verlag, Berlin-Heidelberg-New York, 1976.

[CFM21] Crainic, Marius; Fernandes, Rui Loja; Mărcuţ, Ioan, "Lectures on Poisson Geometry", *Grad. Studies in Math.* **217**, AMS, Providence, 2021.

[Del99] P. Deligne *et al.*, "Quantum Fields And Strings: A Course For Mathematicians. Vol. 1, 2," AMS, Providence, 1999.

[Di30] Dirac, Paul, "The Principles of Quantum Mechanics", Oxford University Press, Oxford 1930. For the recent edition, see: Dirac, Paul, ' 'The Principles of Quantum Mechanics" (4th ed.) Oxford Science Publications, Oxford, 1981.

[Dix77] Dixmier, Jacques, "Enveloping algebras", *North Holland Mathematical library* **14**, North Holland Publishing Co., Amsterdam-New York-Oxford, 1977.

[EMS04] Esposito, Giampiero; Marmo, Giuseppe; Sudarshan, George, "From Classical to Quantum Mechanics. An Introduction to the Formalism, Foundations and Applications", Cambridge University Press, Cambridge, 2004.

[Fed91] Fedosov, Boris V., "Index theorems", in *Partial differential equations–8, Itogi Nauki i Tekhniki. Ser. Sovrem. Probl. Mat. Fund. Napr.* **65** VINITI, Moscow, 1991, 165–268, in Russian. For the English translation see: *Partial Differential Equations VIII. Overdetermined Systems, Dissipative Singular Schrödinger Operator, Index Theory* (Shubin M. ed.), *Encyclopaedia of Math. Sciences* **65**, Springer, Berlin-New York, 1996, 155–251.

[Fed96] Fedosov, Boris V., "Deformation Quantization and Index Theory", *Mathematical Topics* **9**, Akademie Verlag, Berlin, 1996.

[Fey05] Feynmann, Richard P.; Leighton, Robert B.; Sands, Matthew, "The Feynman Lectures on Physics: The Definitive and Extended Edition" vol. 1-3 (2nd ed.) Addison Wesley, Reading, 2005.

[Fu86] Fuks, Dmitriy B., "Cohomology of infinite-dimensional Lie algebras", *Monographs in Contemporary Mathematics*, Consultants Bureau, New York-London, 1986 (see also later reprints by Springer).

Murray; Shack, Samuel D., "Algebraic
d deformation theory" in *Deformation
ebras and Structures and Applications*
M., Gerstenhaber M. eds.) *NATO ASI
ries C, Mathematical and physical sciences*
luwer Academic Publishers, Dordrecht-Boston-
don, 1988.

[GM03] Gelfand, Sergei I.; Manin, Yuri I., "Methods of Homological Algebra", (2nd ed.) *Springer Monographs in Mathematics*, Springer, Berlin-Heidelberg, 2003.

[Go58] Godement, Roger, "Topologie algébrique et Théorie des faisceaux", *Act. Sci. Ind.* **1252** Hermann, Paris, 1958.

[G80] Goldstein, H., "Classical mechanics", 2nd ed., Addison-Wesley, Reading, 1980.

[GVF01] Gracia-Bondía, José M.; Várilly, Joseph C.; Figuera, Héctor, "Elements of Noncommutative Geometry", *Birkhäuser Advanced Texts*, Springer, 2001.

[GrHa78] Griffiths, Phillip; Harris, Joseph, "Principles of algebraic geometry", Wiley, New York, 1978.

[He78] Helgason, Sigurdur, "Differential Geometry, Lie groups and Symmetric spaces", *Grad. Studies in Math.* **34**, AMS, Providence, 1978 (repr. 2012).

[He84] Helgason, Sigurdur, "Groups and Geometric Analysis", Academic Press, Inc., Orlando, 1984.

[Hu78] Humphreys, James E., "Introduction to Lie Algebras and Representation Theory by", Springer, Berlin-Heidelberg-New York, 1978.

[K93] Kaku, Michio, "Quantum field theory: A modern introduction", Oxford University Press, New York-Oxford, 1993.

[Ka78]　　　Karoubi, Max, "K-theory, An ... Grundlehren der Mathematischen Wiss. **226**, Springer-Verlag, Berlin-Heidelberg-New 1978; reprinted in *Classics in Mathematics*, Springer, Berlin-Heidelberg, 2008.

[Ka87]　　　Karoubi, Max, "Homologie Cyclique et K-théorie", *Astérisque* **149** Société mathématique de France, Paris, 1987.

[Ki85]　　　Kirillov, Alexandre, "Geometric quantization" in *Dynamical systems–4, Symplectic geometry and its applications, Itogi Nauki i Tekhniki. Ser. Sovrem. Probl. Mat. Fund. Napr.* 4 VINITI, Moscow, 1985, 141–176, in Russian. For the English translation see: *Dynamical Systems IV. Symplectic Geometry and its Applications* (Arnold, V., Novikov, S., eds.), *Encyclopaedia of Math. Sciences* **4**, Springer, Berlin-New York, 1990, 139–176.

[KN63]　　　Kobayashi, Shoshichi; Nomizu, Katsumi, "Foundations of Differential Geometry", vol. 1-2, Interscience Publishers, New York-London, 1963.

[KM95]　　　Kříž, Igor; May, John P., "Operads, algebras, modules and motives", *Astérisque* **233** Société mathématique de France, Paris, 1995.

[Ku18]　　　Kumar, Ajit, "Fundamentals of Quantum Mechanics", Cambridge University Press, Cambridge, 2018.

[LaVo14]　　Lambrechts, Pascal; Volić, Ismar, "Formality of the little N-disks operad" *Mem. Am. Math. Soc.* **230** (1079) AMS, Providence, 2014.

[LL77]　　　Landau, Lev D.; Lifshitz, Evgenii M., "Course of Theoretical Physics. Volume 3 - Quantum Mechanics: Non-Relativistic Theory", edited by Pitaevskii L. P. Translated by J. B. Sykes and J. S. Bell, 3rd edition, revised and enlarged ed., Pergamon Press, Oxford, 1977.

[Lo98] Loday, Jean-Louis, "Cyclic homology", (2nd ed.) *Grundlehren der Mathematischen Wissenschaften* **301**, Springer-Verlag, Berlin-Heidelberg-New York, 1998.

[LoVa12] Loday, Jean-Louis; Vallette, Bruno, "Algebraic Operads", *Grundlehren der Mathematischen Wissenschaften* **346** Springer, Heidelberg, 2012.

[Lu09] Lurie, Jacob, "Higher Topos Theory", *Annals of Mathematics Studies* **170**, Princeton University Press, Princeton-Oxford, 2009.

[Man16] Manoukian, Edouard B., "Quantum Field Theory, vol. I and II", *Graduate texts in Physics*, Springer, Basel, 2016.

[MSS02] Markl, Martin; Shnider, Steve; Stasheff, James D., "Operads in Algebra, Topology and Physics", *Mathematical Surveys and Monographs Volume* **96**, AMS, Providence, 2002.

[May72] May, John P., "The geometry of iterated loop spaces", *Lecture Notes in Mathematics* **271**, Springer-Verlag, Berlin-Heidelberg-New York, 1972.

[MCl01] McCleary, John, "A User's Guide to Spectral Sequences", (2nd ed.) *Cambridge Studies in Advanced Mathematics* **58**, Cambridge University Press, Cambridge, 2001.

[MS74] Milnor, John W.; Stasheff, James D., "Characteristic classes", *Annals of Mathematics Studies* **76** Princeton University Press and University of Tokyo Press, Princeton, 1974.

[Mor07] Morin, David, "Introduction to Classical Mechanics. With problems and solutions", Cambridge University Press, Cambridge, 2007.

[vN32] von Neumann, Johann, "Mathematische Grundlagen der Quantenmechanik", in *Die Grundlehren der Mathematischen Wissenschaften, Band XXXVIII* Springer, Berlin 1932. For the modern English translation, see: von Neumann, John; Nicholas A. Wheeler (ed.) "Mathematical Foundations of Quantum Mechanics", translated by Robert T. Beyer, Princeton University Press, Princeton, 2018.

[OV90] Onishchik, Arkadij L.; Vinberg, Ernest B., "Lie Groups and Algebraic Groups", *Springer Series in Soviet Mathematics*, Springer, Berlin-Heidelberg-New York, 1990.

[Pa65] Palais, Richard S. (ed.) "Seminar on Atiayh-Singer index theorem", *Annals of Mathematics Studies* **57** Princeton University Press, Princeton, 1965.

[Ra01] Ramond, Pierre, "Field theory: A modern primer" *Frontiers in Physics* **74** (2nd ed.) Westview Press, Boulder, 2001.

[Sa89] Saunders, D.J., "The geometry of jet bundles", *London Mathematical Society Lecture Note Series* **142** Cambridge Univ. Press, Cambridge, 1989.

[Se65] Serre, Jean-Pierre, "Lie Algebras and Lie Groups", W. A. Benjamin, New York, 1965.

[Sh11] Shankar, Ramamurti, "Principles of Quantum Mechanics", (2nd ed.) Plenum Press, New York-London, 2011.

[Sh01] Shubin, Mikhail A., "Pseudodifferential Operators and Spectral Theory", (2nd ed.) Springer, New York, 2001.

[Sm01] Smirnov, Vladimir A., "Simplicial and Operad Methods in Algebraic Topology", *Translations of Mathematical Monographs* **198**, AMS, Providence, 2001.

[Sou70] Souriau, Jean-Marie, "Structure des Systèmes Dynamiques. Maîtrises de mathématiques" Dunod, Paris 1970; see also the English translation: Souriau, Jean-Marie, "Structure of Dynamical Systems. A Symplectic View of Physics", *Progress in Mathematics* **149**, Birkhäuser, Boston, 1997 (reprinted by Springer in 1999).

[Sp66] Spanier, Edwin H., "Algebraic Toplogy", McGraw-Hill, New York, 1966 (see also later reprints by Springer).

[Sta70] Stasheff, James D., "H-Spaces from a Homotopy Point of View", *Lecture notes in Mathematics* **161** Springer-Verlag, Berlin-Heidelberg-New York, 1970.

[Sw69] Sweedler, Moss E., "Hopf algebras", W. A. Benjamin, New York, 1969.

[Sza00] Szabo, Richard J., "Equivariant cohomology and localization of path integrals", *Lecture Notes in Physics Monographs* **63** Springer, Berlin-Heidelberg, 2000.

[Tu20] Tu, Loring, "Introductory Lectures on Equivariant Cohomology", *Annals of Mathematics Studies* **204** Princeton University Press, Princeton, 2020.

[Wa71] Warner, Frank, "Foundations of Differentiable Manifolds and Lie Groups", Scott, Foresman and Co., Glenview 1971; see also later reprint as *Graduate texts in Mathematics* **94** Springer, Berlin-Heidelberg-New York, 1983.

[Wei94] Weibel, Charles A., "An introduction to Homological Algebra", *Cambridge Studies in Advanced Mathematics* **38**, Cambridge University Press, Cambridge, 1994.

[Wey31] Weyl, Hermann, "Quantenmechanik und Gruppentheorie", Leipzig, Hirzel, 1931. English version: "Quantum Mechanics and Group Theory", translation by R. P. Robertson, New York, Dutten, 1931.

[Wi19] Witherspoon, Sarah, "Hochschild cohomology for algebras", *Grad. Studies in Math* **204**, AMS, Providence, 2019.

Original papers

[Ar69] Arnold, Vladimir I., "The Cohomology Ring of the Colored Braid Groups", *Mat. Zametki* **5** 2 (1969) 227–231 (in Russian); for the English translation see *Math. Notes* **5** 2 (1969) 138–140.

[At57] Atiyah, Michael F., "Complex analytic connections in fibre bundles", *Trans. Amer. Math. Soc.* **85** (1957) 181–207.

[APS75$_1$] Atiyah, Michael F.; Patodi, Vijay K.; Singer, Isadore M., "Spectral asymmetry and Riemannian Geometry I", *Math. Proc. Cambridge Phil. Soc.* **77** (1975) 43–69.

[APS75$_1$] Atiyah, Michael F.; Patodi, Vijay K.; Singer, Isadore M., "Spectral asymmetry and Riemannian Geometry II", *Math. Proc. Cambridge Phil. Soc.* **78** (1975) 405–432.

[APS76] Atiyah, Michael F.; Patodi, Vijay K.; Singer, Isadore M., "Spectral asymmetry and Riemannian Geometry III", *Math. Proc. Cambridge Phil. Soc.* **79** (1976) 71–99.

[AS63] Atiyah, Michael F.; Singer, Isadore M., "The index of elliptic operators on compact manifolds", *Bull. A.M.S.* **69** (1963) 422–433.

[AS68$_1$] Atiyah, Michael F.; Singer, Isadore M., "The index of elliptic operators. I" *Annals of Math.* **87** 3 (1968) 484–530 and 546–604.

[AS68$_2$] Atiyah, Michael F.; Singer, Isadore M., "The index of elliptic operators. III" *Annals of Math.* **87** 3 (1968) 546–604.

[AS71] Atiyah, Michael F.; Singer, Isadore M., "The index of elliptic operators. IV" *Annals of Math.* **93** 1 (1971) 119–138.

[BFFLS78$_1$] Bayen, F.; Flato, M.; Fronsdal, C.; Lichnerowicz, A.; Sternheimer, D., "Deformation theory and quantization I. Deformations of symplectic structures" *Ann. Physics* **111** (1978) 61–110.

[BFFLS78$_2$] Bayen, F.; Flato, M.; Fronsdal, C.; Lichnerowicz, A.; Sternheimer, D., "Deformation theory and quantization II. Physical applications" *Ann. Physics* **111** (1978) 111–151.

[Be67] Berezin, Felix A., "Some remarks on the associative envelope of a Lie algebra", *Funkcion. Anal. Priloz.* **1** (1967) 1–14 (in Russian).

[BCGRS06] Bieliavsky, P.; Cahen, M.; Gutt, S.; Rawnsley, J.; Schwachhöfer, L., "Symplectic connections", *Int. J. Geom. Methods Mod. Phys.* **3** (2006) 375–420.

[Bo08] Bordemann, Martin, "Deformation Quantization: a survey", *Journal of Physics: Conference Series* **103** INTERNATIONAL CONFERENCE ON NON-COMMUTATIVE GEOMETRY AND PHYSICS 23–27 April 2007, Université Paris XI, Orsay, France, 1–31.

[BG96] Braverman, Alexander; Gaitsgory, Dennis, "Poincaré-Birkhoff-Witt theorem for quadratic algebras of Koszul type", *J. Algebra* **181** 2 (1996) 315–328.

[BK04] Bezrukavnikov, Roman V.; Kaledin, Dmitry, "Fedosov quantization in algebraic context", *Mosc. Math. J.* **4** (3) (2004) 559–592.

[BNT02$_1$] Bressler, Paul; Nest, Ryszard; Tsygan, Boris, "Riemann-Roch theorems via deformation quantization. I", *Adv. Math.* **167** 1 (2002) 1–25.

[BNT02$_2$] Bressler, Paul; Nest, Ryszard; Tsygan, Boris, "Riemann-Roch theorems via deformation quantization. II", *Adv. Math.* **167** 1 (2002) 26–73.

[BDW12] Burzstyn, Henrique; Dolgushev, Vassiliy A.; Waldmann, Stefan, "Morita equivalence and characteristic classes of star products", *Crelle's J. reine angew. Math.* **662** (2012) 95–163.

[CG82] Cahen, Michel; Gutt, Simone, "Regular *-representations of Lie algebras", *Lett. Math. Phys.* **6** (1982) 395–404.

[CDG80] Cahen, Michel; De Wilde, Marc; Gutt, Simone, "Local cohomology of the algebra of C^∞ functions on a connected manifold", *Lett. Math. Phys.* **4** (1980) 157–167.

[Ca05] Calaque, Damien, "Formality for Lie algebroids", *Comm. Math. Phys.* **257** 3 (2005), 563–578.

[CCN22] Calaque, Damien; Campos, Ricardo; Nuiten, Joost, "Moduli problems for operadic algebras", *J. London Math. Soc.* **2** (2022) 1–95.

[CDH07] Calaque, Damien; Dolgushev, Vassiliy; Halbout, Gilles, "Formality theorems for Hochschild chains in the Lie algebroid setting", *J. Reine Angew. Math.* **612** (2007) 81–127.

[CR11] Calaque, Damien; Rossi, Carlo A., "Compatibility with Cap-Products in Tsygan's Formality and Homological Duflo Isomorphism", *Lett. Math. Phys.* **95** (2011) 135–209.

[CvdB10] Calaque, Damien; van den Bergh, Michel, "Hochschild cohomology and Atiyah classes", *Advances in Math* **224** (2010) 1839–1889.

[CF00] Cattaneo, Alberto S.; Felder, Giovanni, "A path integral approach to the Kontsevich quantization formula" *Comm. Math. Phys.* **212** (2000) 591–611.

[CF01]   Cattaneo, Alberto S.; Felder, Giovanni, "On the globalization of Kontsevich's star product and the perturbative Poisson sigma model", *Prog. Theor. Phys. Suppl.* **144** (2001) 38–53.

[CFT02]   Cattaneo, Alberto S.; Felder, Giovanni; Tomassini, Lorenzo, "From local to global deformation quantization of Poisson manifolds", *Duke Math. Journ.* **115** 2 (2002) 329–352.

[Co76]   Cohen, Frederic R., "The homology of $\mathcal{C}_{n+1}$-spaces, $n \geq 0$", from the book [CLM76], 207–351.

[CM98]   Connes, Alain; Moscovici, Henri, "Hopf Algebras, Cyclic Cohomology and the Transverse Index Theorem", *Communications in Mathematical Physics* **198** (1998) 199–246.

[DGL84]   De Wilde, Marc; Gutt, Simone; Lecomte, Pierre B.A., "À propos des deuxième et troisième espaces de cohomologie de l'algèbre de Lie de Poisson d'une variété symplectique", *Annales de l'I. H. P., section A* **40** 1 (1984) 77–83.

[DL83$_1$] De Wilde, Marc; Lecomte, Pierre B. A., "Star-products on cotangent bundles", *Lett. Math. Phys.* **7** (1983) 235–241.

[DL83$_2$] De Wilde, Marc; Lecomte, Pierre B. A., "Existence of Star-Products and of Formal Deformations of the Poisson Lie Algebra of Arbitrary Symplectic Manifolds", *Lett. Math. Phys.* **7** (1983) 487–496.

[DL85]   De Wilde, Marc; Lecomte, Pierre B. A., "Existence of star-products on exact symplectic manifolds", *Annales de l'institut Fourier*, **35** 2 (1985) 117–143.

[Di15]   Dito, Giuseppe, "The Necessity of Wheels in Universal Quantization Formulas", *Commun. Math. Phys.* **338** (2015) 523–532.

[Do05] Dolgushev, Vassiliy A., "Covariant and equivariant formality theorems", *Adv. Math.* **191** 1 (2005) 147–177.

[DW15] Dolgushev, Vasily; Willwacher, Thomas, "Operadic twisting — with an application to Deligne's conjecture", *J. Pure Appl. Algebra* **219** (5) (2015) 1349–1428.

[Du70] Duflo, Michel, "Caractères des groupes et des algèbres de Lie résolubles" *Annales scient. de l'É.N.S. 4e série* **3** 1 (1970) 23–74.

[Du77] Duflo, Michel, "Opérateurs différentiels bi-invariants sur un groupe de Lie", *Annales scient. de l'É.N.S. 4e série* **10** 2 (1977) 265–288.

[EK98] Etingof, Pavel; Kazhdan, David, "Quantization of Lie bialgebras, II", *Selecta Math., New Series* **4** (1998) 213–231.

[Fed86] Fedosov, Boris V., "Quantisation and index", *Dokl. AN SSSR* **291** 1 (1986) 82–86 (in Russian).

[Fed94] Fedosov, Boris V., "A simple geometrical construction of deformation quantization", *J. Differential Geom.* **40** 2 (1994) 213–238.

[Fl82] Flato, Moshe, "Deformation view of physical theories", *Czechoslovak J. Phys.* **B32** 4 (1982) 472–475.

[FM94] Fulton, William; MacPherson, Robert, "Compactification of configuration spaces" *Ann. of Math.* **139** (1994) 183–225.

[GGL75] Gabrièlov, A. M.; Gel'fand, I. M.; Losik, M. V., "Combinatorial calculus of characteristic classes", *Funktsional. Anal. i Prilozhen.* **9** (3) (1975) 5–26 (in Russian); for the English translation see: *Funct. Anal. Appl.* **9** (3) (1975) 186–202.

[GvSt10] Garay, Mauricio D.; van Straten, Duco, "Classical and quantum integrability", *Mosc. Math. J.* **10** (3) (2010) 519–545.

[GF69]　　Gel'fand, Israël M.; Fuks, Dmitriy B., "Cohomologies of Lie algebra of tangential vector fields of a smooth manifold", *Funkcion. Anal. Prilozh.* **3** 3 (1969) 32–52 (in Russian).

[GF70$_1$]　　Gel'fand, Israël M.; Fuks, Dmitriy B., "Cohomologies of Lie algebra of tangential vector fields of a smooth manifold. II", *Funkcion. Anal. Prilozh.* **4** 2 (1970) 23–31 (in Russian).

[GF70$_2$]　　Gel'fand, Israël M.; Fuks, Dmitriy B., "Cohomologies of Lie algebra of formal vector fields", *Izv. AN SSSR, seriya matem.* **34** 2 (1970) 322–337 (in Russian).

[Ge63]　　Gerstenhaber, Murray, "The cohomology structure of associative ring", *Ann. of Math., Second Series* **78**, No. 2 (1963) 267–288.

[Ge64]　　Gerstenhaber, Murray, "On the Deformation of Rings and Algebras", *Ann. of Math., Second Series* **79** (1964) 59–103.

[Ge66]　　Gerstenhaber, Murray, "On the Deformation of Rings and Algebras II", *Ann. of Math., Second Series* **84** (1966) 1–19.

[Ge68]　　Gerstenhaber, Murray, "On the Deformation of Rings and Algebras III", *Ann. of Math., Second Series* **88** (1968) 1–34.

[Ge74]　　Gerstenhaber, Murray, "On the Deformation of Rings and Algebras IV", *Ann. of Math., Second Series* **99** (1974) 257–276.

[GK95]　　Ginzburg, Victor; Kapranov, Mikhail, "Koszul duality for operads", *Duke Math. J.* **76** 1 (1994) 203–272.

[Gr46]　　Groenewold, Hilbrand J., "On the principles of elementary Quantum Mechanics", thesis, Univ. of Utrecht, Springer (Netherlands) 1946.

[Gro57] Grothendieck, Alexander, "Sur quelques points d'algèbre homologique, I", *Tohoku Math. J. (2)* **9** (2) (1957) 119–221.

[Gu80] Gutt, Simone, "Second et troisieme espaces de cohomologie différentiable de l'algèbre de Lie de Poisson d'une variété symplectique", *Ann. Inst. H. Poincaré Sect. A (NS)* **33** (1980) 1–31.

[Gu83] Gutt, Simone, "An Explicit ∗-product on the cotangent bundle of a Lie group", *Lett. Math. Phys.* **7** (1983), 249–258.

[Gu11] Gutt, Simone, "Deformation quantisation of Poisson manifolds", *Geometry & Topology Monographs* **17** (2011) 171–220.

[GR99] Gutt, Simone; Rawnsley, John, "Equivalence of star products on a symplectic manifold: an introduction to Deligne's Cech cohomology classes", *Journal of Geometry and Physics* **29** (1999) 347–392.

[Hi03] Hinich, Vladimir, "Tamarkins proof of Kontsevich formality theorem", *Forum Math.* **15** 4 (2003) 591–614.

[Kon93] Kontsevich, Maxim, "Formal (non)-commutative symplectic geometry", in *The Gelfand Mathematical Seminars*, 1990-1992, Eds. L. Corwin, I. Gelfand, J. Lepowsky, Birkhauser 1993, 173–187.

[Kon99] Kontsevich, Maxim, "Operads and Motives in Deformation Quantization", *Lett. Math. Phys.* **48** (1999) 35–72.

[Kon01] Kontsevich, Maxim, "Deformation quantization of algebraic varieties", *Lett. Math. Phys.* **56** (3) (2001) 271–294.

[Kon03] Kontsevich, Maxim, "Deformation Quantization of Poisson Manifolds" *Lett. Math. Phys.* **66** (2003) 157–216.

[KS00]　　Kontsevich, Maxim; Soibelman, Yan, "Deformations of algebras over operads and Deligne conjecture", *Lett. Math. Phys.* **21** 1 (2000) 255–307.

[KS18]　　Konyaev Andrey Yu.; Sharygin Georgy I., "Survey of the Deformation Quantization of Commutative Families", in: Buchstaber, Victor M.; Konstantinou-Rizos, Sotiris; Mikhailov, Alexander V. (Eds.) *Proceedings of Kezenoy-Am conference on Mathematical Physics, 2016* in *Springer Proceedings in Mathematics and Statistics*, Springer, 2018.

[Kos70]　　Kostant, Bertram, "Quantization and Unitary Representations", in the book *Lecture Notes in Mathematics* **170** Springer, Berlin, 1970, 87–207.

[McCS02]　　McClure, Jame E.; Smith, Jeffrey H., "A solution of Deligne's Hochschild cohomology conjecture", in: "Proceedings of the JAMI conference on Homotopy Theory" *Contemp. Math.* **293** (2002) 153–193.

[Mo49]　　Moyal, Jose E., "Quantum mechanics as a statistical theory", *Math. Proc. of the Cambridge Philosophical Society*, **45** 1 (1949) 99–124.

[Na97]　　Nakanishi, Nobutada, "Poisson Cohomology of Plane Quadratic Poisson Structures", *Publ. RIMS, Kyoto Univ.*, **33** (1997) 73–89.

[NV81]　　Neroslavski, O.; Vlassov, A.T., "Sur les déformations de l'algèbre des fonctions d'une variété symplectique", *C. R. Acad. Sc. Paris, Sér. I* **292** (1981) 71–73.

[NT01]　　Nest, Ryszard; Tsygan, Boris, "Deformations of symplectic Lie algebroids, deformations of holomorphic symplectic structures, and index theorems", *Asian Journal of Math.* **5** 4 (2001) 599–636.

[Pe59]　　Peetre, Jaak, "Une caractérisation abstraite des opérateurs différentiels", *Math. Scand.* **7** (1959), 211–218.

[Pe60] Peetre, Jaak, "Rectifications à l'article 'Une caractérisation abstraite des opérateurs différentiels'", *Math. Scand.* **8** (1960) 116–120.

[RW16] Reichert, Thorsten; Waldmann, Stefan, "Classification of Equivariant Star Products on Symplectic Manifolds", *Lett. Math. Phys.* **106** (2016) 675–692.

[ŠW] Ševera, Pavol; Willwacher, Thomas, "Equivalence of formalities of the little discs operad", *Duke Math. J.* **160** (1) (2011) 175–206.

[Sha17] Sharygin, Georgy I., "Deformation quantization and the action of Poisson vector fields", *Lobachevskii J. Math.* **38** (6) (2017) 1093–1107.

[ST17] Sharygin, Georgy I.; Talalaev Dmitrii V., "Deformation quantization of integrable systems", *J. of Noncommut. Geom.* **11** (2) (2017) 741–756.

[Sh03] Shoikhet, Boris, "A proof of the Tsygan formality conjecture for chains", *Adv. Math.* **179** 1 (2003) 7–37.

[Sta63$_1$] Stasheff, James D., "Homotopy associativity of H-spaces, I", *Trans. Amer. Math. Soc.* **108** (1963) 275–292.

[Sta63$_2$] Stasheff, James D., "Homotopy associativity of H-spaces, II", *Trans. Amer. Math. Soc.* **108** (1963) 293–312.

[vdB06] Van den Bergh, Michel, "On global deformation quantization in the algebraic case", *J. Algebra* **315** (2006) 326–395.

[Vey75] Vey, Jacques, "Déformation du crochet de Poisson sur une variété symplectique", *Comment. Math. Helv.* **50** (1975) 421–454.

[Wei83] Weinstein, Alan, "The local structure of Poisson manifolds", *J. Differential Geom.* **18**, 3 (1983) 523–557.

[Wey27] Weyl, Hermann, "Quantenmechanik und Gruppentheorie", *Zeitschrift für Physik*, **46** (1927) 1–46.

[Wi15] Willwacher, Thomas, "M. Kontsevich's graph complex and the Grothendieck–Teichmüller Lie algebra" *Invent. Math.* **200** (3) (2015) 671–760.

[Ye08] Yekutieli, Amnon, "Twisted Deformation Quantization of Algebraic Varieties", *Advances in MAth.* **268** (2015) 241–305.

Index

(p,q)-shuffles, 102
(p,q)-unshuffles, 252
A_∞-algebra, 248, 253
    A_∞-morphism, 258
A_∞-space, 246
    associahedron, 245
    Stasheff's pentagon, 244
    Stasheff's polytopes, 245
L_∞-algebra, 253
    L_∞-morphism, 257, 259
        tangent L_∞-morphism, 312, 314
    homotopy equivalence of L_∞-algebras, 260
    homotopy of L_∞-morphisms, 260
$\mathbb{R}^d_{formal}$, 300
$\mathcal{B}_\infty$-algebra, 370
$\widetilde{\mathcal{B}}$-algebra, 372

admissible graphs, 279
    binary admissible graph, 280
    set $G_{n,m}$, 279
    wheels, 326
algebraic formality theorem, 364

Bernoulli numbers, 61, 328
bialgebra, 369
bidifferential operator, 26

coalgebra, 251
    cocommutative coalgebra, 251

coderivation, 253
cofree coalgebra of V, 252
cofree cocommutative coalgebra of V, 251–252
coproduct, 251
counit, 251
complex manifold, 226
    holomorphic vector bundle, 227
        Atiyah class, 230
    structure sheaf, 227
configuration spaces, 272
covariant derivative, 149

deformation quantisation, 35
    $\star$-product, 50
    deformation quantisation of TG, 136
    deformation quantisation of a parallelisable manifold, 132
    deformation quantisation problem, 50
    equivariant deformation quantisation, 218
        equivariant $\star$-product, 218
    holomorphic $\star$-product, 230
    Lecomte-de Wilde quantisation, 156, 170
    Moyal $\star$-product, 25–26
    Moyal algebra, 35

Neroslavski–Vlassov
    theorem, 125
Nest-Tsygan classes, 233
PBW quantisation, 60
quantum momentum
    map, 222

Fedosov quantisation, 178
  Abelian connection, 189
  algebra $\widehat{W}_\hbar^*(M)$, 184
  algebra $\widehat{W}_\hbar(M)$, 183
  algebra $\widehat{W}_\hbar(M)$, 186
  elliptic elements in
      Fedosov's algebra,
      213
    algebraic index, 214
  Fedosov $\star$-product, 200
  Fedosov characteristic
      class, 203
  Fedosov connection, 186,
      189
  Fedosov index formula,
      216
  Fedosov isomorphism, 197
  Fedosov iteration process,
      193
  Fedosov theorem, 189
  Fedosov trace, 214–215
  operator δ, 190–191
  operator δ^*, 191
  quantum parametrix, 213
  Weyl algebra bundle
    bundle $\widehat{W}_\hbar^\mathbb{C}(M,\omega)$, 183
    bundle $\widehat{W}_\hbar^\mathbb{C}(M,\omega)$, 185
  Weyl algebras bundle, 181
formal deformation of an
    algebra, 109
  classification theorem,
      115
  equivalence of $\star$-products,
      115
  equivalence of
      deformations, 114
  obstruction classes, 112,
      113, 115
  secondary obstructions,
      124

Gerstenhaber algebra, 363

Hamilton equations, 2
Hamiltonian, 2
Heisenberg equation, 8
  Heisenberg equation in
      Fedosov algebra, 206
Hochschild cohomology, 79
  ∪-product, 101
  brace operation, 369
  differentiable Hochschild
      complex, 93
  Gerstenhaber bracket,
      102
  graded Hochschild
      cohomology, 117, 131
  graded Lie algebra
      structure, 107
  Hochschild coboundary,
      79
  Hochschild–Kostant–
      Rosenberg map, 96,
      268
  local Hochschild complex,
      93
  smooth Hochschild–
      Kostant–Rosenberg

theorem, 95, 108
Hochschild homology, 81
  algebraic Hochschild–
    Kostant–Rosenberg
    theorem, 89, 91
Homological algebra, 65
  Čech complex, 228
  bar-construction, 84
  chain complex, cochain
    complex, 65
  chain homotopy, 66
  chain map, 66
  Chevalley–Eilenberg
    complex, 71
  cohomology of a Lie
    algebra, 72
  de Rham cohomology, 67
    the ∧-product, 74
  Dolbeault cohomology, 67
  Dolbeault complex, 229
  equivariant cohomology,
    223, 225
    Cartan equivariant
      complex, 223
  groups $Ext_R^k(M, N)$, 83
  groups $Tor_k^R(M, N)$, 83
  homology groups of a
    chain complex, 66
  homology perturbation
    lemma, 199
  Lichnerowicz–Poisson
    cohomology, 69
  projective module, 82
  projective resolution, 82
  sheaf cohomology, 228
  sheaf on a manifold, 227
homotopy associative

H-space, 242
Hopf algebra, 374

Index theory, 208
  algebraic index theorem,
    216
  Atiyah difference
    construction, 211
  Atiyah–Singer formula,
    210, 212
  differential operator on
    vector bundles, 208
  elliptic operator, 209
  Fourier transform, 15
  Fredholm operator, 209
  index of a Fredholm
    operator, 209
  parametrix, 210
  principal symbol of a
    differential operator,
    209
  pseudo-differential
    operators, 14–15, 210
  Schwarz functions, 14
  Sobolev space, 209

Jacobi identity, 5, 38, 42, 52
  graded Jacobi identity, 47
  graded Jacobi identity of
    Gerstenhaber
    bracket, 104
jet spaces, 300
  r-jet bundles
    $J^r(M)$, $J^r(E)$, 301
  r-jet of function, $j_p^r(f)$,
    301
  r-jet spaces,
    $J_p^r(M)$, $J_p^r(E)$, 301

infinite jet bundle
$J^\infty(M)$, 301
Maurer–Cartan
connection, 303
the bundle of formal
affine coordinate
systems M^{aff}, 303,
307
the formal coordinate
charts bundle M^{coord},
302

Kazhdan and Etingof
theorem, 375
Kontsevich quantisation, 269
Kontsevich ⋆-product,
269
Kontsevich
L_∞-morphism, 283,
304
Kontsevich formality
theorem, 268
Kontsevich global
formality theorem,
306
Kontsevich quantisation
of $\mathbb{R}^d$, 269

Lagrangian, 3
Lie algebra, 52
 – of formal functions on
$\mathbb{R}^{2n}$, $\mathfrak{P}_{2n}$, 143
 – of formal symplectic
vector fields on
$\mathbb{R}^{2n}$, $\mathfrak{S}_{2n}$, 143
cohomology of $\mathfrak{P}_{2n}$, 148
DGL algebra, 107
formal DGL algebra,
267
homotopy equivalence
of DGL algebras, 266
quasi-isomorphism of
DGL algebras, 260
Gelfand–Fuks
cohomology, 144
Lie algebroid, 232
Lie bialgebra, 372
local Vey class, 146
subalgebra $\mathfrak{s}_{2n} \subseteq \mathfrak{S}_{2n}$,
144
Vey class S^3_Γ, 149, 153,
155
linear connection, 149
Christoffel symbols, 152,
159
complex connection, 230
curvature, 159
Bianchi identity, 159
equivariant symplectic
connection, 219
Fedosov connection, 186
holomorphic connection,
230
local gauge potential, 159
symmetric connection,
150, 159
symplectic connection,
150
torsion-free connection,
150
loop space, 239
composition of loops, 240
homotopy associativity,
241
iterated loop space, 240,
333

Maurer–Cartan equation, 256
   equivalence of the
      Maurer–Cartan
      elements, 264
   Maurer–Cartan element
      in $\mathfrak{g}^{\cdot}$, 256
   the set $\widehat{\mathfrak{MC}}(\mathfrak{g}^{\cdot})$, 266
minimal action principle, 3
momentum map, 221

operad theory, 337, 340
   $\mathbb{S}$-object in a category, 339, 346
   $\mathbb{S}$-vector space, 339
   $\circ$-product, 340, 346
   $\mathcal{O}$-algebra, 342
   n-cube spaces, 333
   algebra over operad, 337, 342
   bar-cobar resolution of an operad, 349
   coalgebra over a cooperad, 348
   coideal, 351
   cooperad, 348
   dual operad of a cooperad, $\mathscr{C}^{\dagger}$, 348
   formal $\mathcal{O}_{\infty}$-algebra, 357
   free algebra over an operad, $\mathbb{F}_{\mathcal{O}}(V)$, 344
   free coalgebra over a cooperad $\mathscr{C}$, 348
   free cooperad, 348
   free operad, $\mathbb{T}(\mathcal{V})$, 341
   ideal in an operad, 349
   intrinsically-formal $\mathcal{O}$-algebra, 357
   little n-cubes operad, $\mathscr{E}_n$, 337
   morphism of $\mathcal{O}$-algebras, 342
   morphism of operads, 340
   operad $\mathcal{B}_{\infty}$, 370
   operad $\mathcal{G}_{\infty}$, 363
   operad $\widetilde{\mathcal{B}}$, 371
   operad in monoidal category, 340, 346
   operad of associative algebras, <u>ASS</u>, 345, 351, 353, 355
   operad of commutative algebras, <u>COM</u>, 345, 351, 353, 355
   operad of endomorphisms, $\mathscr{E}_V$, 341
   operad of Gerstenhaber algebras, $\mathcal{G}$, 363
   operad of Lie algebras, <u>LIE</u>, 346, 351, 353, 355
   quadratic cooperad, 351
   quadratic dual operad, 352
   quadratic operad, 350
   $\mathcal{O}_{\infty}$-algebra, 353
   Koszul operad, 355
   operad $\mathcal{O}_{\infty}$, 354
   quadratic dual cooperad, 352, 353
resolution of an operad, 348
Schur functor, 340
topological operad, 337
unital operad, 340

path integral, 294
Peetre's theorem, 94
Poisson σ-model, 296
Poisson structure, 44
    Casimir function, 311
    deformation of exact Poisson structure, 165
    deformation of Poisson structure, 141, 163
        obstructions theory, 142
    formal Poisson structure, 268
    Kirillov–Kostant–Souriau structure, 53
    Poisson bivector, 49
    Poisson bracket in $\mathbb{R}^{2n}$, 4
    Poisson bracket on symplectic manifold, 41
    Poisson structure on manifold, 44
    Weinstein's theorem, 49
polydifferential operators, 97
    symbol of polydifferential operator, 98
    the DGL algebra $\mathscr{D}_{poly}^{\cdot}(M)[1][[\hbar]]$, 267
polyvector fields, 45
    Schouten–Nijenhuis bracket, 47, 108
    the DGL algebra $\mathscr{T}_{poly}^{\cdot}(M)[1][[\hbar]]$, 267
Pontryagin classes, 160
    class $p_1(M)$, 157, 160

quantisation, 10
    $\star$-product in $\mathbb{R}^n$, 36
    canonical commutation relations, 11
    canonical quantisation, 15
    classical quantisation problem, 11
    Moyal product, 22, 24
        associativity of Moyal product, 26
    uncertainty principle, 13
    Weyl quantisation, 16
        Weyl commutation relations, 17
        Weyl representation, 18
    Wintner's theorem, 12

Schrödinger equation, 8
spaces $\bar{C}_{n,m}$, $\bar{C}_n$, 273
symplectic structure, 39
    canonical symplectic form on vector space, 18
    Darboux theorem, 42
    exact symplectic structure, 163
    Hamilton vector field, 40, 147
    holomorphic symplectic structure, 230
    symplectic form on a manifold, 39
    symplectic structure on manifolds, 39

Thom space, 211

universal enveloping algebra, 55
    center of universal enveloping algebra, 315

Duflo's isomorphism, 317
Poincaré–Birkhoff–Witt

theorem, 55, 120, 319
symmetrisation map, 59